AF664154

Recent Advances in Seed Spices

Recent Advances in Seed Spices

Y.Ravindrababu
R.K. Jaiman
K.D. Patel
N.R. Patel
S.B.S. Tikka

2024

Daya Publishing House®
A Division of
Astral International Pvt. Ltd.
New Delhi – 110 002

First Published, 2011

Reprinted, 2024

ISBN: 978-81-7035-689-9 (HB)

Disclaimer:

Every possible effort has been made to ensure that the information contained in this book is accurate at the time of going to press, and the publisher and author cannot accept responsibility for any errors or omissions, however caused. No responsibility for loss or damage occasioned to any person acting, or refraining from action, as a result of the material in this publication can be accepted by the editor, the publisher or the author. The Publisher is not associated with any product or vendor mentioned in the book. The contents of this work are intended to further general scientific research, understanding and discussion only. Readers should consult with a specialist where appropriate.

Every effort has been made to trace the owners of copyright material used in this book, if any. The author and the publisher will be grateful for any omission brought to their notice for acknowledgment in the future editions of the book.

Published by : **Daya Publishing House®**
A Division of
Astral International Pvt. Ltd.
– ISO 9001:2015 Certified Company –
4736/23, Ansari Road, Darya Ganj,
New Delhi-110 002
Ph. 011-43549197, 23278134
E-mail: info@astralint.com
Website: www.astralint.com

Printed at : Replika Press Pvt. Ltd.

Foreword

A spice is a dried seed, fruit, root, bark or vegetative substance used in nutritionally insignificant quantities as a food additive for flavor, color or as a preservative that kills harmful bacteria or prevents their growth. In India a wide variety of spices are grown and many of them are native to the subcontinent and also known as "Home of Spices". Out of 109 spice crops, twenty crops have been identified as seed spices out of which cumin, fennel, coriander and fenugreek are major seed spices. Both Gujarat and Rajasthan accounts for about 80 per cent area and production in India.

The major concern in seed spices are low production, effect of global warming, maintenance of quality and indiscriminate use of chemical fertilizers and pesticides. The residue of undesirable chemicals is major hurdle for export.

The promotion of good agricultural practices has become prevalent in an increasing number of horticultural products more so in grapes and pomegranates. Now, efforts are on to extend good agricultural practices to a larger number of horticultural crops and spices. To tackle the issue of contamination in seed spices a traceability framework on the lines of Grape-Net (developed by the Agricultural and Processed Food Product Export Development Authority (APEDA) may be developed in seed spices. Promoting organic farming, awareness about good agricultural practices, on farm primary processing and adding new and upgrading terminal markets would definitely help export promotion.

This book entitled "Recent Advances in Seed Spices" is an attempt to bring some of these issues to the forefront and identify opportunity to enhance productivity, quality and export promotion. I am sure this book would be helpful to the students, researchers, traders and exporters.

R.C. Maheshwari

Vice Chancellor,
S.D. Agricultural University,
Sardarkrushinagar – 385 506, Gujarat

Preface

The present book *"Recent Advances in Seed Spices"* is an attempt to compile information on status of seed spices, utilization of genetic resources, production technologies for enhancement of productivity & quality, protection technologies for the management of various diseases and insects, technologies for farm mechanization in seed spices, post harvest technologies for minimization of post harvest losses, interventions of biotechnological tools for spices improvement in terms of quality, transfer of technologies from research farm to farmers field, and use of information technology for spices trade.

Hope that this book will be very useful to the students, spices research workers, biotechnologists, traders and exporters.

Y. Ravindrababu

R.K. Jaiman

K.D. Patel

N.R. Patel

S.B.S. Tikka

Contents

Recent Advances in Seed Spices (2011) *Pages* **1–13**
Editors: **Y. Ravindrababu, R.K. Jaiman, K.D. Patel, N.R. Patel and S.B.S. Tikka**
Published by: **DAYA PUBLISHING HOUSE, NEW DELHI**

1

Seed Spices Research in Gujarat: Present Status, Constraints and Future Strategies

S.B.S. Tikka[1], Y. Ravindrababu[2] and D.B. Prajapati[2]
[1]Director of Research and Dean P.G. Studies, S.D. Agriculture University, S.K. Nagar, Jagudan – 382 710, Mehsana, Gujarat
[2]Centre for Research on Seed Spices, SDAU, Jagudan – 382 710, Mehsana, Gujarat

From the immemorial, India has been famous for his quality spices, which now earn foreign exchange through the trade of 310830 tonnes valued Rs.3450.50 crores, which go up by 15 per cent in rupee value and six per cent in quantity compared to last year. Seed spices and its value added products contribute about 28.93 per cent in quantity and 58.44 per cent in value of total spices export. (2009, Spices India). Thus, seed spices play an important role in national economy. India is the largest producer, consumer and exporter of quality spices as well as spice products and known as "Land of spices". The seed spices are annual herbs, whose dried fruits or seeds are used as spices. The seed spices are mostly used for culinary, confectionery, perfumery industries and also considered indispensable in seasoning and of habit foods, cosmetics and medicines. Due to change in food habit, the demand of seed spices increased at the rate of 10-12 per cent per year. More than 80 per cent of total seed spices produced in Gujarat and Rajasthan, hence emerged as "Seed Spices Bowl". The list of major seed spices are enumerated in details in Table 1.1.

Distribution

The seed spices are grown in different parts of the world covering mainly Mediterranean region, south Europe and Asia. Almost all the seed spices crops are cultivated in India and known as the largest producing country in the world. The prominent states where seed spices produced are Gujarat and Rajasthan. While other states where commonly grown are Haryana, Punjab, M.P., Bihar, U.P., West Bengal, Orissa, Tamil Nadu and Karnataka.

Table 1.1: List of the Seed Spices

Sl.No.	English Name	Vernacular Name	Botanical Name	Chromosome Number	Family
1.	Cumin*	Jiru, Jira	*Cuminum cyminum* L.	2n = 14	Apiaceae
2.	Fennel*	Moti saunf	*Foeniculum vulgare* Miller	2n = 22	Apiaceae
3.	Coriander*	Dhania	*Coriandrum sativum* L.	2n = 22	Apiaceae
4.	Fenugreek*	Methi	*Trigonella fonum-graecum* L.	2n = 16	Fabaceae
5.	Dill seed	Suwa	*Anethum graveolens* L. *Anethum sowa* Kurz.	2n = 22	Apiaceae
6.	Ajwain	Bishop's weed	*Trachyspermum ammi* (L) Sprague	2n =18	Apiaceae

* Mandatory Crops under AICRP.

Area and Production

(A) India

India is the world's largest producer of seed spices account for about 12.43 and 7.0 per cent of the total area and production of spices in the country. About 8.61 lakh hectare is grown in seed spices in the country, of which 3.25 lakh hectares is covered in Gujarat. The total production of seed spices is 5.62 lakh MT, whereas that of Gujarat is 2.24 lakh MT (2005-06).The seed spices are well distributed over different agro-climatic regions in India. But the major belt spread from semi arid to arid covering larger area in Gujarat and Rajasthan. The share of Gujarat in area and production is 81 and 93 per cent in fennel and 67 and 74 per cent of cumin, respectively. Similarly in minor seed spices, the share of Gujarat in area and production is 51.7 and 67.5 per cent in dill and 15.6 and 29.2 per cent of Ajwain, respectively. The contribution to fenugreek production is 16.6 per cent (Table 1.2). The productivity of all seed spices crops in Gujarat go up highest in the country.

Table 1.2: Area, Production and Productivity of Seed Spices in India (2005-06)

Crop	Area '000 ha		Production '000 MT		Productivity kg/ha	
	India	Gujarat	India	Gujarat	India	Gujarat
Cumin	403.0	267.9 (66.5)	199.9	147.6 (73.9)	496	551
Coriander	356.2	8.67 (2.4)	254.9	8.42 (3.3)	716	971
Fennel	45.0	36.3 (80.6)	61.3	57.23 (93.4)	1499	1577
Fenugreek	32.6	4.9 (15.0)	35.7	5.9 (16.6)	1095	1204
Ajwain	14.5	2.3 (15.6)	4.4	1.3 (29.2)	303	565
Dill	09.4	4.8 (51.7)	5.7	3.8 (67.5)	606	792
Total	860.7	324.87	561.9	224.25		

(...) Share of Gujarat.

Table 1.3: Distribution of Seed Spices in Gujarat (2007-08)

Crop	*Area '00 ha*	*Production, '00mt*	*Av. Productivity Kg/ha*	*Highest Productivity Kg/ha.*	*Major Districts*
Cumin	2624.62 (72.21)*	1985.01 (53.88)*	760	1250 (Bhavnagar)	Surendranagar (22.4[a], 22.25[b],400[c]), Banaskantha (12.19, 14.51, 900)[†], Patan (10.15, 8.72, 650), Jamnagar (8.69, 7.35, 640) Mehsana (6.86, 6.35, 700) Ahmedabad (6.62, 4.09, 470), Rajkot (5.53, 11.67, 700)
Fennel	744.68 (20.49)	1268.28 (34.43)	1700	2000 (Kheda, S.K.)	Banaskantha (45.0, 46.22, 1750) Mehsana (28.20, 24.84, 1500), S.K. (9.51, 11.17, 2000), Surendranagar (5.29, 5.74, 1850), Patan (4.87, 4.58, 1600) Kheda (3.14,3.67, 2000)
Coriander	122.11 (3.67)	215.48 (5.85)	1760	2300 (Kutch)	Kutch (37.88, 49.38, 2300), Junagarh (28.58, 25.17, 1550), Jamnagar (11.06, 6.14, 980), Porbander (6.92,4.92, 1250) Rajkot (5.68, 4.82, 1500)
Fenugreek	45.03 (3.36)	61.17 (4.38)	1358	2200 (Amreli)	Banaskantha.(18.88, 66.34, 1250), Anand (16.43, 4.59, 1000), Mehsana (9.55, 4.80, 1800), Amreli (8.82, 2.46, 2200) Porbander (6.55,3.11, 1700) Junagarh (6.0, 2.51, 1500), Dahod (4.77, 1.33, 1000)
Dill Seed#	48.32 (1.58)	38.19 (1.62)	790	1400 (Kutch)	Ahmedabad (51.59, 32.62, 500), Patan (24.83, 37.70, 1200), Surendranagar (7.55, 6.70, 701), Kutch (6.72, 11.91, 1400)
Ajwain	98.22 (2.70)	53.87 (1.46)	550	1000 (Ahmedabad, Tapi)	Jamnagar (84.61, 80.21, 520), Mehsana (4.07, 5.57, 750), Bhavnagar (4.07, 5.57, 750), Amreli (2.36, 2.38, 550) Tapi (1.63, 1.63, 1000) Ahmedabad (0.55, 1.0, 1000)
Total	**3634.66**	**3683.81**			

DOA, GOG (2009).

* Data in parenthesis indicates per cent contribution of total seed spices

† a, b= per cent contribution in area and production compared to state respectively c = productivity kg/ha

\# = Distribution in the year of 2005-06.

2. Sowing is done late in North Gujarat (late Nov. or early Dec.) resulting in to yield loss, because of heavy attack of aphids and high temperatures during grain filling stage, where as in Saurashtra it is growing early (2nd fortnight of Oct.) facing high temperatures during germination resulting in to poor germination.

(B) Suitability in Gujarat

Gujarat is the top most in India so far as the production of minor seed spices is concerned. The statistics on area and production of minor seed spices are presented in Table 1.3. Among the seed spices cumin and fennel are the most popular cash crops and the economy of the North Gujarat depends to a considerable extent on these crops due to short duration nature and have greater potential to grow even under restricted water availability gets better income compared to other traditional crops (Mehta and Edison, 1993). During 2007–08, Gujarat produced 3.68 lakh MT of seeds spices from an area of 3.63 lakh hectares. Cumin contributed maximum followed by fennel, coriander, fenugreek, ajwain and dill. The major area falls in North Gujarat (Mehsana, Patan, Banaskantha, Sabarkantha and Gandhinagar districts), Saurashtra region (Rajkot, Surendranagar, Jamnagar and Junagadh districts) and Kutch (Table 1.3).

Cumin

Area under cumin was 2.62 lakh hectares which produced 1.99 lakh MT with the productivity of 760 kg/ha. Surendranagar, Patan, Jamnagar, Mehsana, Ahmedabad, and Rajkot are the major districts of cumin cultivation. The highest productivity of 1250 kg/ha was recorded in Bhavnagar district followed by Banaskantha (900 kg/ha)

Fennel

1.27 lakh MT of fennel was produced from an area of 0.75 lakh hectare with an average productivity of 1700 kg/ha. The highest productivity of 2000 kg/ha recorded in Kheda and Sabarkantha districts followed by Surendranagar (1850 kg/ha). Banaskantha, Mehsana, Sabarkantha, Patan, Surendranagar and Kheda are the major districts for the fennel cultivation.

The area of coriander, fenugreek, dill and ajwain is 12.21, 4.5, 4.8 and 9.8 thousands ha. with production 21.54, 61.17, 3.8 and 5.4 thousand tonnes, respectively. Kutch, Junagadh, Jamnagar, Porbander, Rajkot for coriander, Banaskantha, Anand, Porbander, Junagarh and Dahod for fenugreek. Ahmedabad and Patan for dill and Jamnagar, Bhavnagar, Mehsana, Amreli and Tapi are the major districts for ajwain cultivation.

The productivity of Gujarat in case of Cumin (760 kg/ha) and fennel (1700 kg/ha) is the highest in the World whereas that of coriander, fenugreek, dill and ajwain is highest in the country (Table 1.3).

Production Constraints

India is the largest producer of seed spices in the world but the productivity in the country is very low. So far, major general constraints are enumerated as details.

1. Cultivation of spice crops in marginal lands
2. Insufficient quantity of good seed material of high yielding, disease resistant varieties.
3. Poor knowledge of farmers about the standard agro techniques
4. Lack of farmer's training and technology dissemination programmes on spices.

Cumin

Specific Constraints

1. Inherent poor germ inability and initial slow growth habit

3. Improper time and method of controlling blight as well as powdery mildew.
4. Delay in second irrigation required for better germination.
5. Moisture stress during flowering stage causing heavy loss to yield and quality.
6. Epidemic form of the blight disease causes complete failure of crops.
7. None availability of aphid resistance variety

Fennel

Specific Constraints

1. Use of immature or over matured seedlings for transplanted crops.
2. Lack of earthing up of soil causing heavy lodging of the crop.
3. Improper time and method of controlling blight and seed midge.
4. Continuous heavy irrigation and fertilization during December and January resulting into sugary secretion there by poor seed setting.
5. Dense population of drilled crop resulting into poor growth and yield.
6. No facility for value addition and postharvest handling of spices.
7. All varieties of *kharif* are of long duration and tall varieties.
8. All varieties are susceptible to major pests and diseases
9. None availability of genetically superior variety for quality parameters

Coriander

Specific Constraints

1. Unavailability of high yielding varieties.
2. All varieties are susceptible to Powdery mildew disease of the crop

Fenugreek

Specific Constraints

1. Low productivity of the popular local variety
2. Powdery mildew is only serious disease of the crop
3. Termite is the major pest

Dillseed

Specific Constraints

1. Susceptible to Powdery mildew, aphids of the crop
2. Long duration crop (6 months)
3. Low productivity of the local variety

Ajwain

Specific Constraints

1. Susceptible to Powdery mildew and aphids to the crop

2. Long duration crop
3. Unavailability of distinctly high yielding variety

Spectrum of Activities

1. Collection, conservation and evaluation of germplasm (indigenous and exotic types) of different seed spice crops and its utilization in crop improvement programme. More weightage to be giver. to develop varieties suitable for different value added products.
2. Breeding for resistance to biotic and abiotic stresses, multiple disease resistance, high yield potential and superior grain quality.
3. Development of suitable crop production, protection and postharvest technologies conductive to high yield and better quality of produce.
4. Evolving dual purpose variety in coriander, fenugreek for green leaves and seed and development of fennel varieties suitable for early and mid-late season.
5. Screening of germplasm against pest and diseases, their utilization.
6. Planting material production
7. A well organized extension and training programme supported with a sound seed production programme for quick adoption of improved varieties as well as package of practices.

Highlights of Research Attainments

Landmarks of Centre

The research activities on spices crops were not well managed prior to 1950 in Gujarat. However, Agricultural Institute, Anand initiated preliminary research work on fennel and cumin as results the first, two varieties of S-7-9 and S-404 fennel and cumin, respectively were developed in 1952. The potential attempts towards their crop improvement were started in 1961, simultaneously in Gujarat, Delhi and Rajasthan.

In Gujarat a scheme for research in spices and condiments was sanctioned with headquarter at Manund (Patan district). It was shifted to Vijapur in 1965 and research work on cumin, fennel, coriander, fenugreek was initiated. In 1974-75 AICRP on spices was sanctioned which started functioning actively in 1978-79.

During this period (1961 to 75) cumin varieties MC-43 (1969), coriander variety GCo-1 (1974), fennel variety PF-35 (1973) were developed. The first systematic work on these spices in the state started only when the research project were shifted to Jagudan in 1980. (Agalodiya *et al*, 2007)

Crop Improvement

Genetic Resources

One of the primary mandates of the centre is survey, collection, conservation and utilization of germplasm for further breeding programme. So far, the collection of germplasm was started in 1963. However, the systematic germplasm resource activity with long term objectives started functioning actively in 1980-81 with the establishment of a centre of the All India Coordinated Spices Improvement Project at Jagudan. Table 1.4 gives the present status of the germplasm holding of each of the six seed spices at this Centre.

Table 1.4: Genetic Resources Holding in Various Seed Spices

Sl.No.	Crop	Exotic	Indigenous	Released	Total
1.	Cumin	7	207	4	218
2.	Fennel	4	41	3	48
3.	Fenugreek	–	74	2	76
4.	Coriander	22	84	2	108
5.	Dillseed	–	77	3	80
6.	Ajwain	–	60	1	61
	Total	33	543	15	591

Table 1.5: Varieties Released in Different Seed Spices for Cultivation

Sl.No.	Crop/Released Variety	Year of Release	Average Yield (kg/ha)	Yield Potential# (Kg/ha)	Distinguish Characteristic
1.	Cumin				
	Guj. Cumin–1	1982	541	958	High Yield
	Guj. Cumin–2	1992	622	841	High Yield, Better Grain Quality
	Guj. Cumin–3	1998	620	777	High oil content, Resistant to wilt disease
	Guj. Cumin – 4*	2004	1250	1875	High oil content, High yield, Bold grain, Resistant to wilt
2.	Fennel				
	Guj. Fennel–1	1984	1724	2114	Early maturity
	Guj. Fennel–2	1997	1940	2816	High yield, High oil content
	Guj. Fennel -11	2003	2489	3483	High yield, Heat tolerance, suitable for Rabi cultivation
3.	Coriander				
	Guj. Cori.–1	1974	1082	1558	High yield and stable
	Guj. Cori.–2	1984	1443	2007	High yield, bold grain
4.	Ajwain				
	Guj. Ajwain- 1	1987	1475	2269	High yield potential, medium dwarf
5.	Dillseed				
	Guj. Dillseed–1	2000	1596	2077	Suitable for irrigated condition
	Guj. Dillseed–2	2000	935	1150	Suitable for rain fed condition
	Guj. Dillseed–3	2007	1062	1659 (Irrig)	Suitable for rain fed and irrigated condition, 1296 kg/ha (R.F.)
6.	Fenugreek				
	Guj. Methi–1	1999	1864	1927	Medium bold
	Guj. Methi- 2*	2006	2001	2079	High yield, bold lusterus grain moderately resistant to powdery mildew

The improved varieties have great potential in achieving greater levels of productivity.

* Identified at national level..

Crop Improvement

Fifteen Varieties

Research work carried out at Jagudan Centre under the AICRP and SDAU'S network on the conventional breeding work resulted in the release of fifteen high yielding improved varieties in the seed spices (coriander-2, cumin-4, fennel-3, fenugreek-2, Dill seed -3 and Ajwain-1) (Table 1.5).

- ☆ Gujarat Cumin–3 is first wilt resistance variety developed from an introduction from West Germany for cultivation and recommendation for Gujarat and Rajasthan.
- ☆ Gujarat Cumin–4 is second wilt resistance variety developed from local germplasm and recommendation for state and national level. Very much popular in Gujarat and other state for its cultivation and remarkable impact for enhancing the state productivity.
- ☆ Gujarat Methi–2 is high yielding, moderately resistant to powdery mildew, lustrous and bold seeded having good medicinal properties.
- ☆ Gujarat Fennel-11 having synchronous flowering habit and late heat tolerance, for rabi fennel cultivation in the state. It has given on an average 2489 kg/ha yield which was 12 per cent higher than the GF-2.
- ☆ Gujarat Dill-3: A new dwarf, early dill cultivar having good economic traits in irrigated as well as rainfed conditions in Gujarat. The variety GD-3 produced 1062 kg/ha, which was 18.79 and 16.19 per cent higher than the check variety GD-1 (Irrigated) and GD-2 (Rainfed), respectively. The yield potential of GD-3 under irrigated and rainfed conditions is 1659 and 1296 kg/ha, respectively.

Impact of the Technology/Variety Development

(A) Crop Improvement

Achievement and Impact through New Fenotypes

Cumin

Variety GC-2 released in 1992 with high yielding capacity, better grain quality and GC-3 released in 1998 with wilt resistance have become popular among the farmers. The productivity gains at the farmers field increased from 315 kg/ha during 1992-93 to 422 kg/ha in 2001-02. Also during the same period, the area under cultivation has increased from 1.16 lakh ha to 1.45 lakh ha. Last released variety GC-4 (2003) have high yield potential wilt resistance and better grain quality, compared to GC-2 and GC-3; which make cumin crop still more profitable and tremendous increase in productivity up to 760 kg/ha.(2007-08)

Fennel

Kharif fennel variety GF-2 released in 1997, being instrumental for increasing the productivity from 1316 kg/ha (1992-93) to 1503 kg/ha (1998-99). This variety is also grown under Rabi season. During 2003, variety Guj. Fennel-11 released for Rabi sowing has 12 per cent grain increases over GF-2. and increasing the productivity from 1579 kg/ha (2005-06) to 1700 kg/ha (2007-08). The variety Guj. Fennel -11 is further helped in gaining more productivity on farmer's field.

Fenugreek

Variety Guj. Methi -1 released in 1999 have helped in increasing the productivity from 1034 kg/ha to 1138 kg/ha as well as area under cultivation from 5800 ha to 7280 ha during 1992-93 to 2001-02.

Recently released variety Guj. Methi-2 (2006) has high yield potential, Powdery mildew resistance and better grain quality, compared to Guj. Methi-1; which has made great impact in fenugreek crop stills more profitable and tremendous increase in productivity up to 1358 kg/ha. So far, the variety enables to make crop more remunerative.

(B) Maintenance

Except for fenugreek (which is self–pollinated and hence maintenance of germplasm is easy) all the other seed spices are cross–pollinated. In these crops, each accession is grown in every year in two rows. To check cross-pollination, one meter selection of each row is covered by a muslin cloth cage before the initiation of anthesis. The seed produced in the cage is saved as a representative of the accession.

(C) Evaluation

To provide a sound base material for breeding programme the germplasm of six crops were collected and evaluated for various traits and volatile oil content.

Table 1.6: Promising Accessions Identified for Different Characters (2007-08)

Crop	*Important Characters*			
	Early Maturity	*Bold Seeds*	*High Volatile Oil Content*	*High Yield*
Cumin (218)	JC-2002-31, JC-2000-66, 2002-37, 99-3, 95-125, GC-1, JC-95-75, 2002-24, 2000-62, 96-40	JC-96-5, JC-2002 -17, JC-99-32	EC-279081, JC-99-24, JC-2000-22, JC-2000-21, GC-3, GC-4	JC-2002-31, 2000-66, 2002-37, 99-3, 95-125, GC-1, JC-95-75, 2002-24, 2000-62, 96-40
Fennel (48)	JF-575, JF-544, JF-473 JF-472-2-3, JF-500-1	JF-359-1, JF-500-1, JF-582, JF-592-1, JF-528	GF-11, JF-200, JF-420-3-2, JF-525, JF-550 -1-1, GF-560	EC-206924, JF-537, JF-546, JF-472-5
Fenugreek (76)	Anantpur, JFg-240, JFg-245, JFg-191	JFg-241, JFg-239, JFg-209, GM-2	–	JFg-245, GM-2, JFg-80, Anantpur
Coriander (108)	JCr-360, Lam-87, EC-363980, JCr-381, Lam-6, JCr-387, 402, Austrilian and EC-363972	JCr-344, JCr-327, JCr-342, JCr-329	JCr-371, JCr-378	Lam-4, JCr-392, 401, Dhana-98, EC-343370, JCr-350, 690, Russian
Dillseed (80)	JD-95-73, JD-01-11, JD-sel-3, JD-01-18, JD-02-11, GD-2,GD-3	JD-95-41, JD-95-45, JD-95-72, JD-95-134, JD-01-9	JD-01-14, JD-02-10, JD-02-12, GD-2, GD-3	JD-01-2, JD-01-5, JD-01-19, JD-95-72, JD-02-12, GD-1, GD-2, GD-3
Ajwain (61)	JA-148, JA-174, JA-182, JA-183, JA-214, GA-1	JA-131, JA-115, JA-182, JA-214, JA-229, GA-1	JA-106, JA-215, JA-220, JA-230, GA-1	JA-131, JA-169, JA-182, JA-220, JA-230, GA-1.

Promising Accessions

Cumin

- ☆ Among them nine entries observed tall type, having more than 40 cm plant height.
- ☆ Eight entries had good branches *i.e.* more or equal seven branches per plant.

- ☆ Nine entries isolated as high yielder recorded more than 1225 kgha-1 grain yield (Yield kg/ha)
- ☆ Two selections of cumin have been isolated for high yield and resistance to wilt and two for blight
- ☆ Thirteen entries have none splitting grain and bold seeded.

Coriander

- ☆ Among them nine entries observed dwarf type, having less than 68 cm plant height.
- ☆ Six entries had good branches *i.e.* more than 6 branches per plant.
- ☆ Seven entries isolated as high yielder recorded more than 2330 kgha-1 grain yield.
- ☆ Seven entries for green seededness and another for dual purpose.

Fenugreek

- ☆ Among them 11 entries were found tall types having more than 63 cm plant height.
- ☆ Thirteen entries had recorded more or equal 6 branches per plant.
- ☆ Seven entries were found more than 40 pods per plant. Long pod length *i.e.* >12 cm more seed per pod *i.e.* >20 were found in 11 and 8 entries, respectively.
- ☆ The twelve entries were matured before 110 days.
- ☆ Nine entries found promising for yield *i.e.* more than 3400 kg/ha
- ☆ Ten entries found leafy as well as grain purpose

Dillseed

- ☆ In dill, five for Indian type (compact umbel, better yield and suitable for rain fed and irrigated conditions) and four for European (suitable for export purpose) have been identified.

Ajwain

- ☆ In ajwain five selections have been identified for earliness, high yield, resistance to powdery mildew.

Technology Dissemination through FLD'S

The frontline demonstrations were conducted to enhance the productivity of the Districts in Mehsana, Banaskantha and Patan districts each of 0.25 ha. area on farmer's field during 2007-08 to demonstrate the performance of newly developed variety GC-4 of cumin and high yielding variety of GF-11 with local varieties in rabi season.

The yield increment of cumin, GC-4 ranged from 20.76 to 53.16 per cent. The wilt incidence in GC-4 was 10 per cent as against 63 per cent of GC-2.During field days variety GC-4 was appreciated by large number of farmers who attended the field or visited the plots. The yield increment of variety of GF-11 ranged from 13.46 to 15.18 per cent. During field days variety GF-11 was appreciated by large number of farmers for its earliness, more branches, umbels and synchronized maturity. Overall results indicate that there is wide scope to increase productivity and present gaps between achieved and achievable productions.

Table 1.7: Front-line Demonstrations in Cumin and Fennel (2007-08)

Sl.No.	Name of the Crops	District	Average Yield (kg/ha)		Per cent IOC (GC-2)	Wilt Disease Intensity (Per cent)		No. of Farmers on Field Day
			CC-4	GC-2		CC-4	GC-2	
1.	Cumin	Patan (3)	1100	908	21.14	3.3	34.7	125
		Banaskantha	1012	838	20.76	5.0	32.0	87
		Mehsana	968	632	53.16	10.0	63.0	63
		Average	**1026.7**	**792.7**	**29.51**	**6.1**	**43.2**	**92**
2.	Fennel	Mehsana (4)	1534	1352	13.46	–		62
		Patan	1728	1480	16.76	–		125
		Average	**1631**	**1416**	**15.18**	–		

( ) indicating the no. of demonstration in district.

Quality Parameters Required for Export Produce

Wide scope for seed spice processing industry in Gujarat and may enhance the export of seed spices up to two to three times more to the present level if,

Quality is Upgraded as per Quality Specification/Regulation of the Importing Countries

A. Mode of harvest at physiological stage of seed spice crops to ensure quality

B. Value addition through processing and Postharvest Technology

- ✩ Seed size, color, shape, luster, physical purity
- ✩ Cleanliness for dead insects, animal excreta, other foreign material
- ✩ Threshing, Drying, Cleaning and Grading in cement-concrete threshing yard, Polishing and Packaging through machine.
- ✩ Immediate Transportation or Marketing and Safe and sanitary storage
- ✩ Green house drying up to 10-11 per cent moisture level hence check the microbial contamination and able to maintain the color, aroma and essential oil.

C. To Encourage the value added products in the India

- ✩ Encourage to entrepreneur by government through suitable policies for agro based industries
- ✩ Oil and oleoresins are free from contamination and can be stored for long period without deteriorating quality. They are easily transportable.

E. Residue effect

- ✩ To make produce free from pesticide residue avoids pesticides and fungicides before harvesting the crop. *e.g.* BHC, DDT etc.
- ✩ Low aflatoxin level and prevent microbial load through solar drying

Organic Spices

The growing concern and awareness of health and hazards through chemical residual toxicity led the development of international trade for organic crops and the organic spices trends may developed. India is now one of the countries exporting organic spice. Europe is the world's leading market for organically produced spices. Its current share in global market is around 2-35 which is expected to increase to 10 per cent in the coming years. Major consumers are USA (30,000 tonnes), Europe (2100 tonnes) and Japan (6000 tonnes) wherein the existing market, potential for organic spices is 10 per cent of the conventional spices. Organic cultivation being more expensive, the producer should get premium price for the produce.

Research on Organic Farming and Value Addition

Fenugreek

The farmers of North Gujarat Agro-Climatic zone interested for organic farming of fenugreek (GM-1) are advised to apply FYM @ 10 t/ha and treat the seeds with Azospyrillum culture @ 5.0 kg/ha for getting higher yield (1950 kg/ha) and net income return (Rs.16110/ha) than recommended dose of fertilizers (1588 kg/ha).

Fennel

To ensure the higher volatile oil yield (22.19 kg/ha) and seed yield (1565 kg/ha) of rabi fennel it should be harvested at physiological maturity *i.e.* more than fifty per cent umbels converted in to yellow colour.

Ajwain

Harvest the ajwain crop at physiological maturity stage (more than 50 per cent umbels turning yellowish brown and seeds grey brown) recorded maximum seed yield (1272 kg/ha) and volatile oil yield (90.37 kg/ha)

Dillseed

Harvest the dillseed crop at physiological maturity stage (more than 50 per cent umbels turning yellowish brown and seed appearing grey brown) produced 28 and 55 per cent higher seed and volatile oil yield respectively, over delayed harvesting stage.

Future Strategies

Crop Improvement

- ☆ Enhancement of genetic variability by importing new germplasm from centre of origin *i.e.* Mediterranean countries.
- ☆ Collection, evolution and characterization of seed spice.
- ☆ Breeding for resistance to biotic and abiotic stresses multiple disease resistance, high yield potential and superior grain quality.
- ☆ More focus will be given to develop varieties suitable for different value added products.
- ☆ Evolving dual purpose variety in coriander, fenugreek for green leaves and seed and development of fennel varieties suitable for early and mid-late season.

Crop Production

- ☆ Development of suitable productivity enhancing technologies through agronomic research *i.e.* integrated use of organic manure, inorganic fertilizer and biofertilizer sustaining the productivity of soil will be worked in cumin and fennel based crop sequence.

- ☆ Use of biofertilizer, bio control agents and soil amendment to manage soil pathogen and organic production will be worked out.
- ☆ Development of water management technology aiming at judicious economic use of water for higher yield and better grain quality.

Crop Protection

- ☆ Main efforts of crop protection group would be to assist in developing different spices varieties resistant to pests and diseases.
- ☆ Technology development to manage diseases through integrated approaches will be continue
- ☆ Formulation of forecasting models for blight and powdery mildew diseases of seed spices.
- ☆ Formulation of forecasting models for seed midge and aphids insect/pests of seed spices.

Postharvest Technology

- ☆ To develop postharvest technologies such as threshing, cleaning, grading, drying as well as safe storage of seed spices at farmers and traders level for production of exportable quality seed.

Transfer of Technology

- ☆ To impart training to seed growers for production and protection technologies, on farm processing for value addition to increase domestic and export value.
- ☆ To produce large quantity of seedlings of newly released varieties for quick replacement of old varieties and increase production.
- ☆ To conduct on farm demonstration of new varieties and technologies.

References

Agalodiya, A.V., Prajapati, D.B. and Jaiman, R.S., 2007. Seed spices research in Gujarat: Present status and thrust area. In: *Oral Paper Pesented in National Seminar on "Production, Development, Quality and Export of Seed Spices: Issues and Strategies*, February, 2-3, held at NRCSS, Tabiji, Ajmer" organised by Society of Medicinal and Aromatic Plants and NRCSS, p. 79.

Deputy Director, 2009. Spices exports go up by 15 per cent in value. *Spices India*, A Journal Published by Spices Board, Cochin, 22(1): 7.

D.O.A., 2009. *District Wise Area and Production of Seed Spices in Gujarat*. Directorate of Agriculture, Government of Gujarat, Gandhinagar.

Mehta, K.G. and Edition, S., 1986. Improved varieties of seed spices suitable for Gujarat. *Indian Cocanut and Spices Journal*, 27(1 and 2): 3.

Recent Advances in Seed Spices (2011) *Pages* **14–20**
Editors: **Y. Ravindrababu, R.K. Jaiman, K.D. Patel, N.R. Patel and S.B.S. Tikka**
Published by: **DAYA PUBLISHING HOUSE, NEW DELHI**

2

Cumin and Fennel Production in Gujarat Status and Strategy

A.V. Agalodiya, N.R. Patel and R.K. Jaiman
Centre for Research on Seed Spices, SDAU,
Jagudan – 382 710, Mehsana, Gujarat

ABSTRACT

Cumin (*Cuminum cyminum*) and fennel (*Foeniculim vulgare*) are most important seed spices crops of Gujarat state together contributing 56.90 and 87.43 per cent towards total area and production of seed spice in the state (2005-06). Cumin is grown on an average in 2.04 lakh hectares and produces 0.92 lakh MT. By and large cumin cultivation is concentrated in Patan, Mehsana and Banaskantha districts of North Gujarat region and Surendranagar, Rajkot, Junagadh, Jamnagar and Amreli districts of Saurashtra region. The productivity level ranged between 400-545 kg/ha (Average of 2001-02 to 2005-06). The average productivity obtained in Frontline Demonstrations and yield potential of the variety is 827 and 1234 kg/ha respectively, which is quite high. There has been significant increase in area and production during last 15 years. The average area and production increased from 77500 ha and 33000 MT (1986-87 to 9091) to 2.04 lakh ha and 92400 MT (2001-02 to 2005-06) respectively. The fennel cultivation in the state is mainly concentrated in Mehsana, Sabarkantha, Banaskantha, Patan and Gandhinagar districts of North Gujarat region and Kheda district of Middle Gujarat. The productivity level (Average of 2001-02 to 2005-06) ranged from 1255 to 1461 kg/ha with an average productivity 1344 kg/ha. Similar to cumin the area (10050 to 29700 ha) and production (13600 to 41300 MT) in the state increased significantly. The average productivity (1917 kg/ha) obtained in frontline demonstrations and yield potential of recently released variety (3483 kg/ha) indicate high productivity gap between actual and realizable yield. Poor soil fertility, lack of adequate irrigation water, poor seed replacement rate (10 to 15 per cent), lack of proper irrigation, weed, disease and pest management and climatic factors play vital role in influencing productivity of cumin and fennel. After completion of Narmada and Sujalam-Sufalam irrigation projects adequate irrigation water will be available in major parts of North Gujarat and Saurashtra region where both these crops are cultivated. These will lead to increase in area and productivity of both cumin and fennel crops. In the study, an attempt has been made to analyze the production scenario and suggest development strategy for acceleration of cumin and fennel production in the state.

Introduction

Gujarat is leading cumin and fennel growing state in the country contributing 56.9 and 60.8 per cent toward area and production of cumin and 85.7 and 90.1 of fennel, respectively. During 2005-06 Gujarat produced 1.48 lakh MT of cumin from an area of 2.68 lakh ha with the productivity of 551 kg/ ha which is the highest in the world. North Gujarat region (Patan, Mehsana and Banaskantha) and Saurashtra region (Surendranagar, Rajkot, Junagadh and Jamnagar) is the cumin belt. Where as fennel is grown in 36251 ha and produces 57236 MT with an average productivity of 1579 kg/ha, which is again highest in the world. Banaskantha, Mehsana, Sabarkantha and Kheda are the major districts for fennel cultivation.

Though the productivity of cumin and fennel in Gujarat is the highest in the world, there is a wide gap between realized yield and yield potential of the recently released varieties of both cumin and fennel. The major general constraints for low productivity of cumin and fennel are:

1. Non availability of sufficient quantity of good quality planting material of high yielding, disease resistant varieties.
2. Lack of proper agronomic management practices like sowing method and time, manures and fertilizers, irrigation, weed control and plant protection.
3. Insufficient and improper extension system.
4. Cultivation of these crops in marginal lands

Specific Constraints

Cumin

1. Inherently poor germinability and initial slow growth habit
2. Sowing is done late in North Gujarat (late Nov. or early Dec.) resulting in to yield loss, because of heavy attack of aphids and high temperatures during grain filling stage, where as in Saurashtra it is growing early (2nd fortnight of Oct.) facing high temperatures during germination resulting in to poor germination.
3. Improper time and method of controlling blight as well as powdery mildew.
4. Delay in second irrigation required for better germination.
5. Moisture stress during flowering stage causing heavy loss to yield and quality.

Fennel

1. Use of immature or over matured seedlings for transplanted crops.
2. Lack of earthing up of soil causing heavy lodging of the crop.
3. Improper time and method of controlling blight and seed midge.

4. Continuous heavy irrigation and fertilization during December and January resulting into sugary secretion there by poor seed setting.
5. Thick sowing of drilled crop resulting into poor growth and yield.

Cumin (*Cuminum cyminum*)

Production Scenario

Cumin is grown in the state in acerage of 2.04 lakh ha with an annual production of 0.92 lakh MT. the average production of the crop stand at a level of 4.52 q/ha. Major cumin growing areas are concentrated in North Gujarat and Saurashtra region. Patan, Mehsana and Banaskantha districts of North Gujarat, whereas Surendranagar, Rajkot, Junagadh and Jamnagar of Saurashtra are the important cumin growing districts. The productivity of the crop show some variations across major growing districts.

Data (Table 2.1) show that average productivity of cumin in the state increased from 427 kg/ha during 9th five year plan to 452 during 10th five year plan. During these periods increased trend in productivity was observed in all the districts except Sabarkantha and Surendranagar. During 10th Five Year Plan the productivity of the state was between 400-545 kg/ha. In North Gujarat the productivity of only Mehsana district (471 kg/ha) was higher than state average (457 kg/ha) while in the remaining districts *i.e.* Patan, Banaskantha and Sabarkantha was lower. The cumin crop in Patan is mostly grown in conserved moisture whereas in Banaskantha light textured soils with poor fertility is the major limiting factor for low productivity. In Sabarkantha district poor soil and limited irrigation water resulted into low yields. Among the important cumin districts of Saurashtra, productivity of Junagadh (545 kg/ha) and Rajkot (498 kg/ha) is much more higher than the state productivity. Lower yields in Surendranagar and Jamnagar districts are due to water stress and poor crop management.

Table 2.1: Area, Production and Productivity of Cumin in Gujarat

District	Area '00ha		Production '00MT		Productivity (kg/ha)	
	I	II	I	II	I	II
Mehsana	241.8	184.2	100.6	86.8	419	471
Patan	175.9	385.9	80.7	170.8	459	443
Banaskantha	255.6	334.9	92.6	133.9	355	400
Sabarkantha	14.8	41.8	8.2	17.0	550	406
Suredranagar	186.2	285.9	98.8	120.6	507	422
Rajkot	115.4	225.8	56.4	112.6	472	498
Junagadh	13.2	84.4	6.0	46.0	426	545
Jamnagar	–	121.1	–	57.4	–	474
Ahmedabad	175.8	207.4	69.0	84.5	400	407
Kutchh	34.2	64.9	15.0	30.0	–	461
Other	73.4	59.4	27.8	42.5	431	–
State	1124.4	2044.4	480.6	924.3	427	452

I: Mean 1996-97 to 2000-01; II: Mean 2001-02 to 2005-06.

The frontline demonstrations conducted in Mehsana, Sabarkantha, Banaskantha, Patan, Kutchh and Ahmedabad districts during 2004-05 and 2006-07 indicated better scope for increasing productivity and reducing gaps between achieved and achievable productions.

Development Strategy

(I) Productivity Enhancement

For increasing cumin production in the state, first priority should be laid on improving productivity in inefficient areas. This requires large scale seed multiplication of newly released wilt resistant variety GC-4 to dissemination on larger area, efficient uses of manures and fertilizers as well as irrigation water to realize the yield potential of the improved varieties. Adequate protection to crop against weed, insect-pest and diseases would further result in enhancing the productivity of the varieties. In districts of Patan, Surendranagar, Ahmedabad and Sabarkantha where irrigation water is one of the major limiting factor for low production there is good possibility of increasing productivity up to 20-30 per cent as soon as the Narmada project irrigation will be available in this areas, whereas in Junagadh, Rajkot and Jamnagar districts because of more irrigation water available due to check dams the productivity may be increased up to 15 per cent. An estimate indicates that a modest increase of 10 and 20 per cent by the end of 11th Five Year Plan from North Gujarat region additional of production of 4110 and 8160 tonnes would be obtained respectively. Similarly Saurashtra region would result in additional production of 3360 and 6730 tonnes respectively. District wise production possible through productivity enhancement is presented in Table 2.2.

Table 2.2: Additional Production of Cumin Possible through Area Expansion

Districts	*Mean Current Area '00ha*	*Mean Current Yield kg/ha*	*Projected Average Area '00ha*		*Expected Additional Yield '00t*	
			I	*II*	*I*	*II*
Mehsana	184.2	471	202.6	221.0	8.6	17.3
Patan	385.9	443	424.5	463.1	17.3	34.3
Banaskantha	334.9	400	368.4	401.9	13.5	26.6
Sabarkantha	41.8	406	46.0	50.2	1.7	3.4
Surendranagar	285.9	422	314.5	343.1	12.2	24.2
Rajkot	225.8	498	248.4	271.0	11.1	22.4
Junagadh	84.4	545	92.8	101.2	4.6	9.2
Jamnagar	121.1	474	133.1	145.3	5.7	11.5
Ahmedabad	207.4	407	228.1	248.9	8.3	16.8
Kutchh	64.9	461	71.4	77.9	2.9	5.9
Total expected additional production ('00tonnes)					**85.9**	**171.6**

I: With 10 per cent increase during eleventh Five Year Plan (2007 to 2012).

II: With 20 per cent increase during eleventh Five Year Plan (2007 to 2012).

(II) Area Expansion

Cumin inspite of its high production risk factor and low productivity significant increase in the area observed in all the districts between 2000-01 and 2001-02 to 2005-06. The per cent increase in area ranged from 18.00 per cent (Ahmedabad district) to 525.5 per cent (Jamnagar district). Low cost of production and high return are major factors for increase in area. Further the area will be increased significantly in the Narmada command area (Surendranagar, Patan, Mehsana, Banaskantha and Ahmedabad districts). In Sabarkantha the productivity of wheat is less than state average there is good possibility of replacing wheat by cumin. A modest district wise estimate for increasing cumin production through area expansion by the end of 11th Five Year Plan *i.e.* 2007-12 is presented in Table 2.3.

Table 2.3: Additional Production of cumin possible through productivity enhancement

Districts	*Mean Current Area '00ha*	*Mean Current Yield kg/ha*	*Projected Average Yield * kg/ha*		*Expected Additional Yield '00t*	
			I	*II*	*I*	*II*
Mehsana	184.2	471	518	565	8.62	17.3
Patan	385.9	443	487	531	17.13	34.1
Banaskantha	334.9	400	440	480	13.5	26.8
Sabarkantha	41.8	406	447	488	1.7	3.4
Surendranagar	285.9	422	464	506	12.1	24.1
Rajkot	225.8	498	548	597	11.1	22.2
Junagadh	84.4	545	599	655	4.6	9.3
Jamnagar	121.1	474	521	569	5.7	10.5
Ahmedabad	207.4	407	448	489	8.4	16.9
Kutchh	64.9	461	507	553	2.9	5.9
Total expected additional production ('00tonnes)					**85.8**	**170.5**

I: With 10 per cent increase during eleventh Five Year Plan (2007 to 2012).

II: With 20 per cent increase during eleventh Five Year Plan (2007 to 2012).

Fennel (*Foeniculum vulgare*)

On an average fennel is grown in 30,730 ha and production is 41,300 tonnes with an productivity of 1344 kg/ha. The area is restricted to North Gujarat (Mehsana, Patan, Banaskantha, Sabarkantha and Gandhinagar districts) and Kheda district. Mehsana, Banaskantha and Sabarkantha contributes 83.76 per cent in area and 84.86 per cent in production. The productivity of Mehsana, Patan and Banaskantha is more than the state average whereas that of Sabarkantha and Kheda is lower than state average. The reasons for low productivity in Sabarkantha are (1) more area under *rabi* cultivation and (2) inadequate irrigation water whereas in Kheda district poor crop management is the major constraint for the low productivity (Table 2.4).

Development Strategies

(A) Productivity Enhancement

The frontline demonstration conducted in Mehsana, Sabarkantha, Banaskantha, Gandhinagar and Kheda districts indicated that the productivity can be increase by 2.67 per cent (Patan district) to 58.88 per cent (Sabarkantha district) if the proper production technology is adopted and if the newly improved variety is used. An estimate indicates that the modest yield increase of 10 and 20 per cent by the end of 11th five year additional production of 4120 and 8320 tonnes will be obtained, respectively. Mehsana will produce 2420 and 4140 tonnes, Banaskantha 760 and 1830 tonnes and Sabarkantha 750 as well as 1530 tonnes, respectively (Table 2.5).

Table 2.4: Area, Production and Productivity of Fennel in Gujarat

District	*Area '00ha*		*Production '00MT*		*Productivity (kg/ha)*	
	I	*II*	*I*	*II*	*I*	*II*
Mehsana	59.6	118.4	105.4	163.8	1768	1443
Patan	12.7	11.8	21.5	16.9	1784	1461
Banaskantha	4.8	77.5	7.0	109.3	1458	1372
Sabarkantha	60.0	61.5	73.4	77.4	1223	1255
Kheda	31.6	21.8	28.8	24.1	911	1137
Gandhinagar	–	12.8	–	16.4	–	1335
Other	6.6	3.5	9.4	5.1	1224	1457
State	162.6	307.3	224.0	413.0	1378	1344

I–Mean 1996-97 to 2000-01.

II–Mean 2001-02 to 2005-06.

Table 2.5: Additional Production of Fennel Possible Through Productivity Enhancement

Districts	*Mean Current Area '00ha*	*Mean Current Yield kg/ha*	*Projected Average Yield * kg/ha*		*Expected Additional Yield '00t*	
			I	*II*	*I*	*II*
Mehsana	118.4	1443	1588	1738	187.0 24.2**	205.2 41.4**
Patan	11.8	1461	1607	1753	18.9 2.0**	20.7 3.8**
Banaskantha	77.5	1372	1509	1646	116.9 7.0**	127.6 18.3**
Sabarkantha	61.5	1255	1381	1507	84.9 7.5**	92.7 15.3**
Gandhinagar	12.8	1322	1458	1591	18.7 * 2.3**	20.4 4.0**
Kheda	21.8	1137	1251	1365	27.3 3.2**	329.8 5.7**
Total expected additional production ('00tonnes)					**453.7 46.2****	**796.4 88.5****

I: With 10 per cent increase during eleventh Five Year Plan (2007 to 2012).

II: With 20 per cent increase during eleventh Five Year Plan (2007 to 2012).

(B) Area Expansion

Inspite of high remuneration, from fennel crop irrigation water is the major limiting factor for area increase in fennel. However, the major area falls in command area of Narmada and Sujalam-Sufalam irrigation project, there is a better possibility of area expansion in Sabarkantha, Patan and Banaskantha districts. Considering an estimate increase of 10 and 20 per cent by the end of 11th Five Year Plan; additional production of 4130 and 8250 tonnes would be obtained respectively (Table 2.6).

Table 2.6: Additional Production of Fennel Possible Through Area Expansion

Districts	*Mean Current Area '00ha*	*Mean Current Yield kg/ha*	*Projected Average Yield * kg/ha*		*Expected Additional Yield '00t*	
			I	*II*	*I*	*II*
Mehsana	118.4	130.2	142.0	1443	187.9 24.1**	204.9 41.1**
Patan	11.8	30.0	0.2	1461	19.0 2.1**	20.8 3.9**
Banaskantha	77.5	85.3	92.8	1372	117.03 8.6**	127.3 18.0**
Sabarkantha	61.5	67.7	73.8	1255	84.9 7.5**	92.6 15.2**
Gandhinagar	12.8	14.1	15.4	1325	18.7 2.3**	20.4 4.0**
Kheda	21.8	24.0	26.2	1137	27.3 3.2**	29.8 5.7**
Total expected additional production ('00tonnes)					**454.8 47.8****	**495.8 87.9****

I: With 10 per cent increase during eleventh Five Year Plan (2007 to 2012).

II: With 20 per cent increase during eleventh Five Year Plan (2007 to 2012).

References

Anonymous, 2006. *District-Wise Area and Production of Seed Spices in Gujarat*. Director of Agriculture, Krushibhavan, Gandhinagar, Gujarat.

Kumavat, S.M., Chauhan, J.S. and Sahu, M.P., 2007. *Cumin Production in Rajasthan: Status and Strategy, Production, Development, Quality and Export of Seed Spices*. S.K. Malhotra and B.B. Vashishtha, NRCSS, Ajmer, pp. 218–223.

Patel, K.P., Patel, N.R., Patel, M.M. and Agalodiya, A.V., 2005. Gujarat Cumin 4: A new wilt resistant cumin variety having high yield potential and better seed quality. In: *Paper Presented in National Symposium on Current Trends in Onion, Garlic, Chillies and Seed Spices Production, Marketing and Utilization* held at NRC for Onion and Garlic, Rajgurunagar (Pune) during 25-27 November.

Sahu, M.P. and Rathore, P.S., 2003. Seed spices production in India: Status and strategy. In: *Presentation Note on Seed Spices*, Organized by Directorate of Research on Seed Spices, Calicut.

Recent Advances in Seed Spices (2011) *Pages* **21–30**
Editors: **Y. Ravindrababu, R.K. Jaiman, K.D. Patel, N.R. Patel and S.B.S. Tikka**
Published by: **DAYA PUBLISHING HOUSE, NEW DELHI**

3

Seed Production Standards in Seed Spices

Y. Ravindrababu, G.M. Patel, K.P. Patel and B.S. Prajapati
Centre for Research on Seed Spices
S.D. Agricultural University, Jagudan – 385 710, Gujarat

Seed is vital input in agriculture. It is the beginning and the end in agriculture. Agriculture might have changed to high-tech form but the seed remains the same. To feed ever increasing population in India, there is a need to sustain and consistent increase in yield and spices crops are no exception. Therefore, for higher productivity quality seed should be made available to the farmer. The general seed certification standards applicable to other field crops followed for seed spices which constitutes minimum isolation etc. However, the eligibility requirement for certification is that only notified varities are allowed for certification. In seed spices only two varieties *i.e.,* Gujarat Cumin-4 and Gujarat Fennel-11 are notified vide notification in Gujarat state. Hence seed certification standards for these two crops discussed in following paragaraphs.

Classes and Sources of Seed

(*a*) Nucleus Seed

Nucleus seed is the original seed produced for the first time by a plant breeder. The genetic purity of this stage is cent percent pure and it is used as a source for breeder seed multiplication. It is produced in extremely small quantities.

(*b*) Breeder Seed

Breeder seed is the first step in generation system of seed multiplication. The production of breeder seed is directly controlled by originating or sponsoring plant breeder of the breeding programme or seed technologist. The breeder seed production programme is monitored by a team of scientists. The breeder seed production programme does not fall under the perview of seed act.

The tag used for breeder seed is golden yellow tag.The genetic purity should be sufficient to maintain the genetic purity of foundation seed.

(*c*) Foundation Seed

Foundation seed is the progeny of breeder seed. Foundation seed production programme is supervised and approved by the seed certification agency. The foundation seed should have minimum genetic purity and conform to seed certification standards. The colour of the tag is white.

(*d*) Certified Seed

It is progeny of foundation or certified seed itself that does not exceed two generations. It is last step in generation system of seed multiplication. The production is supervised by the seed certification agency.The colour of the tag is azure blue.

Seed Source Verification

The seed producer must submit the tag and purchase records to the certification agency for verification.

Phase of Seed Certification

Only two varieties in seed spices crops *viz.*, Gujarat Cumin-4 and Gujarat fennel-11 are notified therefore certified programme is being taken for these varieties. For seed certification programme seed producer must pass through the following phase.

1. Registration of application and verification of application
2. Verification of seed source (original bill), stage of seed(tag and bag of seed)
3. Selection of seed production plot according to seed certification standard
4. Field visit of the seed certification agency for rouging and removal of offtype plants at flowering and time of harvesting
5. Seed sampling and analysis including genetic purity test and seed health test
6. Approval of Seed certification and issue of tags

Seed Certification Standards

(A) Seed Production Plot Condition/Field Standards

(1) Land Requirement

Field plot where seed production programme is to be taken should be free from any other crop plant/other variety of the same crop and in previous year same crop should not be taken in this field.

(2) Isolation Distances

Seed spices crops *viz.*, Cumin, Fennel, Corriander, Ajwain and Dill seed comes into family apiaceae and are cross pollinated in nature. The cross pollination occurs mostly through insects. The coriander and cumin are considered as a often cross pollinated whereas fennel, ajwain and dill are cross pollinated crops. The fenugreek is self pollinated in nature. Based on cross and self pollination crops isolation distance for foundation and certified stage seed are as follows.

Sl.No	Name of Crop	Isolation Distance (Meter)	
		Foundation	Certified
1.	Cumin	800	400
2.	Fennel	1000	800

(3) Field Inspection

Field inspection is essential for maintaining of Genetical purity of seed. Field inspection in these crops is done at the following crop stage.

(i) Before Flowering Stage

At this stage inspection is done to check minimum isolation distances, recommended planting ratio, volantier plant and for removal of disease plants.

(ii) Flowering Stage

This inspection done to check the days to flowering, plant of other variety/crop and disease infested plant and rouging of that plant.

(iii) Seed Setting Stage

This inspection is done for remove of off type and disease infected plant

(iv) Before Harvesting

At harvesting stage inspections is done to remove off type plants in terms of plant morphological characters, undesirable plants, diseased plants, for confirmation of varietal purity.

The specific requirement for seed crop of seed spices showing maximum permitted limits for off types, objectionable weeds and designated diseases are as Table 3.1.

Table 3.1: Specific Seed Requirement for Seed Spices Crops

Sl.No.	Crop	Maximum Limit (Per cent)					
		Foundation			Certified seed		
		Off Type	Objectional Weed	Disease Infested	Off Type	Objectional Weed	Disease Infested
1.	Cumin	0.10	0.01	0.10	0.20	0.02	0.50
2.	Fennel	0.10	-	0.10	0.20	-	0.50

(B) Seed Standards in Seed Spices Crops

Soon after completion of seed processing the composite sample is taken for analysis of the seed standards such as percentage of pure seed, inert matter, other crop seed, total weed seed, germination and moisture. The prescribed permittable seed standards limit of seed spices crops is given in Table 3.2.

Table 3.2: Minimum Seed Standards for Seed Spice Crops

Sl.No.	Crop	Seed Purity Standards (Per cent)											
		Genetical Purity (Minimum)		Inert Matter (Maximum)		Other Crop Seed in Per cent (Maximum)		Weed Seed in Per cent (Maximum)		Germination Per cent (Minimum)		Moisture Per cent (Maximum)	
		F	C	F	C	F	C	F	C	F	C	F	C
1.	Cumin	95	95	5	5	0.05	0.10	0.10	0.20	65	70	8	8
2.	Fennel	95	95	5	5	0.05	0.10	0.10	0.20	65	70	8	8

Note: (a) Insect damage seeds should not be more than 0.5 per cent in each sample (b)The moisture per cent for vapour proof container should be 2 per cent less than the prescribed limita (c) F and C stands for Foundation and Certified seed categories

Agricultural Practices for Seed Production Programme

Cumin	
Date of sowing	First fort night of November
Seed rate	12 to 16 kg per hacter
Spacing	30 cm or Broad casting
Recommended fertilizer dose	15 + 15 + 0 kg N P K at the time of sowing (Basal dose) 15 kg N at 30 days after sowing
Fennel (*kharif*)	
Date of sowing	Nursery middle of June Transplanting at second fort night of August
Seed rate	1 to 1.5/100 square meter
Spacing	90 x 60 cm
Recommended fertilizer dose	40 + 60 + 0 kg N P K at the time of sowing (Basal dose) 30 kg N at 30 days after sowing 30 kg N at 60 days after sowing
Fennel (*rabi*)	
Date of sowing	Last week of October to first week of November
Seed rate	4 to 5 kg per hacter
Spacing	45 cm or 30 x 22.5
Recommended fertilizer dose	45 + 30 + 0 kg N P K at the time of sowing (Basal dose) 22.5 kg N at 30 days after sowing 22.5 kg N at 60 days after sowing

Irrigation Schedule

Cumin

First irrigation: Immediately after sowing

Second: 8-10 days after sowing

Third: 30 days after sowing

Fourth: 45 days after sowing

Fifth: 60 days after sowing

Fennel

The field is irrigated immediately after seeding if there are no rains. Subsequent irrigation are given at regular intervals of 15-18 days depends upon soil type. Generally the total number of irrigation given in North Gujarat six to eight in *kharif* and four to five in *rabi* crops.

Plant Protection

Cumin

1. Blight

Blight normally appears at the time of flowering. Dark brown necrotic spot at tips of young leaves and stems. The spots enlarge, collapse and become black as the disease advances. The affected seeds are shriveled and dark coloured.

Control Measure

(*a*) Seed treatment with thiram or mancozeb or amisen@ 3 gm/kg of seed

(*b*) Crop rotation

(*c*) Spray Mancozeb on 35 – 40 DAS@ 35mg/10 lit. of water

(*d*) Disease resistant Variety

2. Wilt

The diseases appear in patches and gradually spreads to adjacent areas. Roots are primary point of infection and plant show varying degree of yellowing. Leaf colour changes from green to yellow. In severe attack whole plants become yellow and finaly death of entire plants.

Control Measure

(*a*) Seed treatment with carbendazim @ 3 gm/kg of seed

(*b*) Deep ploughing in summer

(*c*) Crop rotation with cluster bean or sorghum

(*d*) Apply FYM 10 ton/hac or castor/mustard cake/poultry mannure @ 2.5 ton/hac

(*e*) Grow resistant variety GC –3 and GC -4

3. Powdery Mildew

Initially white to grey coloured specks appeared on the lower leaves and stem. Later whole plant become almost white with a powder consisting of abundun spores. Seed development affected seeds became small and shrivelled

Control Measure

(*a*) Application of 300 mash sulphur @ 25 kg/ha at 45 DAS

(*b*) Spray calixine 30 EC @ 7 ml/litre

Fennel

1. Alternaria Blight

Initial the floral buds are affected which dry after yellowing and browning at finally inflorence and its stalk dry out.

Control Measure

(*a*) Seed treatment with thiram or mancozeb or amisen@ 3 gm/kg of seed
(*b*) Crop rotation
(*c*) Spray Mancozeb on 35 – 40 DAS@ 35mg/10 lit. of water
(*d*) Disease resistant Variety

2. Ramularia Blight

Minute angular brown necrotic lesions on the lower surface of the older leaves which later show whitesh growth, infection spreads to stem and then cover whole plant with a ash coloured spot and letter spots collapse and drying of leaves.

Control Measure

(*a*) Seed treatment with thiram or mancozeb or amisen@ 3 gm/kg of seed
(*b*) Crop rotation
(*c*) Spray Mancozeb on 35 – 40 DAS@ 35mg/10 lit. of water
(*d*) Disease resistant Variety

3. Root Rot

The main symptoms appear as yellowing and wilting of apical or terminal portion and then complete drying of plants.

Control Measure

(*a*) Deep the seedling in to COC solution(4gm/lit water) before transplanting
(*b*) Deep ploughing in summer
(*c*) Soil solarization

4. Sugary Disease

The disease is spread by insect vector aphid. Control measure:

(*a*) It is Physiological disorder controlled by avoiding excess use of nitrogenous fertilizer and irrigation water.
(*b*) Spray Dimethoate/Methyl-o-demeton/Phosphamidon@10ml/10 lit of water when incidence of thrips

5. Powdery Mildew

The crop is attacked at flowering stage in cloudy weather during Feb.-March. Initially symptoms appears on the leaves and then on stem, powdery mass of fungus can be visualized. Later whole plant with white coating powders.

Control Measure

(*a*) Application of 300 mash sulphur @ 25 kg/ha at 45 DAS

(*b*) Spray calixine 30 EC @ 7 ml/litre

6. Phyllody

Control Measure

(*a*) Avoid sowing after sesamum

(*b*) Uprooting and burning of the infected plant

Cumin

1. Crop: Cumin
2. Botanical name: *Cuiminum cyminum* L.
3. Family: Apiaceal
4. Chromosome No.: 2n=14
5. Origin: Mediterranean region
6. Notified variety: Gujarat Cumin -4
7. Year of release: 2003
8. Yield attributes
   (*a*) No. of branches per plant: 7.9
   (*b*) No. of umbels per plant: 40.1
   (*c*) No. of umbellates per umbel:: 5.4
   (*d*) No. of seeds per umbellates: 6.2
   (*e*) 1000 seed weight: 4. 9
9. Quality characters
   (*a*) Essential oil per cent: 4.2
   (*b*) Seed colour: Ash
   (*c*) Seed hardness: Non–spliting
10. Description of variety
    (*a*) Days to 50 per cent flowering: 60
    (*b*) Days to maturity:110
    (*c*) Plant height (cm): 31
    (*d*) Distinguishing characters: Medium tall, bushy habit, basal branching from first node, dark purple petals and dark green foliage

Fennel

1. Crop: Fennel
2. Botanical name: *Foeniculum vulgare* Miller.
3. Family: Apeaceal

4. Chromosome No.: 2n=22
5. Origin: Mediterranean region
6. Notified variety: Gujarat Fennel-11
7. Year of release: 2003
8. Yield attributes
   (*a*) No. of branches per plant: 8
   (*b*) No. of umbels per plant: 18
   (*c*) No. of umbellates per umbel:: 32
   (*d*) No. of seeds per umbellates: 30
   (*e*) 1000 seed weight: 7.4
9. Quality characters
   (*a*) Essential oil per cent: 1.8
   (*b*) Seed size: Medium bold
10. Distinguishing characters: Medium tall, compact seeds, more umbellate central umbellate is short.

- ☆ Phases of seed certification
- ☆ Field inspection
- ☆ Unit of certification
- ☆ Processing
- ☆ Seed treatment
- ☆ Sample and sampling of seed
- ☆ Packing
- ☆ Seed testing
- ☆ Seed Production technology of Cumin
- ☆ Family
- ☆ Botanical name
- ☆ Chromosome number
- ☆ Land Requirement
- ☆ Varieties
- ☆ Sowing time
- ☆ Seed rate
- ☆ Fertilizers
- ☆ Intercultural operations
- ☆ Diseases and their control
- ☆ Insect pests and their control
- ☆ Field inspection

- ☆ Field standards
  - (*a*) Isolation
  - (*b*) Specific requirements
    Off types
- ☆ Seed Standards
- ☆ Harvesting and Storage
- ☆ Yield

Seed Testing Procedure

Seed testing procedures are developed to assess the quality of seed before its sown. The quality of seed ultimately determines its planting value and is important to the farmers and seed producers as well as seed organizations for labeling and marketing as well.

The seed testing rules focus mainly on following parameters.

- ☆ Seed sampling
- ☆ Physical purity analysis
- ☆ Germination test
- ☆ Viability test
- ☆ Seed health testing
- ☆ Determination of moisture content
- ☆ Seed vigour test
- ☆ Genetic purity test

Seed Sampling

The first step in seed testing is sampling. The object of sampling is to obtain a sample of a suitable size for tests. A grain spear can be used to collect a representative sample of the grain. The objective is to obtain sample from a seed lot by taking a small portion at random from different places and combining them. The sampling procedures for a particular situation should be followed as prescribed by the seed testing laboratoryof the locality.

Seed Lot

It is a uniformly blended specified quantity of seed physically identifiable in respect of which seed lot certificate may be issued either in bag or in bulk. The maximum quantity based on seed size of seed spices is 20,000 kg per lot.

Each Sample Should Carry Following Information

- ☆ Name and address
- ☆ Kind of seed and variety
- ☆ Year of harvest
- ☆ Type of test required (*e.g.* germination, purity), and them forward the sample to be tested on the notified seed laboratory of locality

Weight of Submitted Sample

The minimum weight for submitted samples for various tests as follows.

Moisture Test

For moisture determination submitted samples of 100 grams for those species that have to be ground and 50 grams for all other species should be taken.

Recent Advances in Seed Spices (2011) *Pages* **31–41**
Editors: **Y. Ravindrababu, R.K. Jaiman, K.D. Patel, N.R. Patel and S.B.S. Tikka**
Published by: **DAYA PUBLISHING HOUSE, NEW DELHI**

4

Plant Genetic Resources in Improvement of Spices in India: An Overview

V.A. Parthasarathy*, K.N. Shiva and B. Krishnamoorthy
Indian Institute of Spices Research, Calicut – 673 012, Kerala

Introduction

India has been a traditional producer, consumer and exporter of spices in the world. About 60 spices are grown out of 109 spices listed by ISO (International Organization for Standardization). Blessed with varied climatic conditions, almost all the states grow one or more spices. In India, spices are grown in an area of 24.10 lakh ha with an annual production of 39.44 lakh tonnes (2006-07). With variety of spices in the production list, almost one-third of the world demand is being met by Indian exports. During the year 2002-03, the spices export quantity has touched an all-time high of 264107 tonnes. India has reached another milestone in export of spices and spice products during 2007-08 by achieving a target of 4,44,250 tonnes valued at Rs. 4435.50 crores (US$ 1.0 billion). Trade contributes 37 per cent share in volume and 23 per cent in Value. More than 90 per cent of the spices produced in the country are consumed domestically and the rest is exported in the form of raw and value- added products.

India has 12 biogeographic zones rich in all aspects of biodiversity: ecosystem, species and genetic diversities. India is rich in plant wealth with over 45,000 species, which comprise of 12 per cent of the global wealth of plants. Nearly 33 per cent of these are endemic, mainly located in 26 centres (Khosoo, 1995).

---

* Director, E-mail: parthasarathy@spices.res.in.

Genetic Resources of Spices

Spices have played a vital role in the history of human culture and civilization. Spices condiments and aromatic plants were the first articles traded by the ancient people. These prompted the foreigners, to the land of spices – India and finally Vasco-da- Gama landed on the Malabar Coast (Kappad beach) on May 20th 1498–an event that changed Indian history in the centuries that followed. Diversity of spices grown in India are given in Table 4.1.

Table 4.1: Diversity of Spices Grown in India

Sl.No.	Botanical Name	Part Used	English Common Name	Family
1.	*Acorus calamus* L.	Rhizome	Sweet-flag	Acorceae
2.	*Allium cepa* L.	Bulb	Onion	Alliaceae
3.	*Allium sativum* L.	Bulb	Garlic	Alliaceae
4.	*Alpinia galanga* Willd.	Rhizome	Greater alangal	Zingiberaceae
5.	*Amomum subulatum* Roxb.	Fruit/seed	Large cardamom, Sikkim cardamom, Nepal cardamom	Zingiberaceae
6.	*Anethum graveolens* L.	Fruit and leaf	Dill seed	Apiaceae
7.	*Apium graveolens*	Root	Celery seed	Apiaceae
8.	*Armorica rusticana* Gaet.	Root	Horse-radish	Apiaceae
9.	*Brassica juncea* (L.) Czernajev and Cosson	Seed	Indian mustard	Brassicaceae
10.	*Brassica nigra* (L.) Koch	Seed	Black mustard	Brassicaceae
11.	*Bunium persicum* (Bioss.) B. Fedtsh	Seed, tuber	Black caraway	Apiaceae
12.	*Capparis spinosa* L.	Unopened flower bud	Caper	Capparidaceae
13.	*Capsicum frutescens* L. *Capsicum annuum* L.	Fruit	Chillies or Capsicum	Solanaceae
14.	*Carum bulbocastanum* L.	Fruit	Black caraway	Apiaceae
15.	*Carum carvi* L.	Fruit	Caraway seed	Apiaceae
16.	*Cinnamomum cassia* B. and Presl	Bark, leaf	Cassia, China or Cassia	Lauraceae
17.	*Cinnamomum tamala* (Buch-Ham.) Nees and Ebermaier	Bark and leaf	Cassia, Indian	Lauraceae
18.	*Cinnamomum verum* B. and Presl.	Bark	Cinnamon	Lauraceae
19.	*Coriandrum sativum* L.	Leaf and seed	Coriander	Apiaceae
20.	*Crocus sativus* L.	Stigma	Saffron	Iridaceae
21.	*Cuminum cyminum* L.	Fruit	Cumin seed	Apiaceae
22.	*Curcuma longa* L.	Rhizome	Turmeric	Zingiberaceae
23.	*Elettaria cardamomum* (L.) Maton	Fruit and seed	Cardamom	Zingiberaceae
24.	*Foeniculum vulgare* P. Miller	Fruit	Fennel	Apiaceae
25.	*Garcinia gummi-gutta* (L.) N. Robson	Pericarp-lobes (fruit rind)	Camboge (Malabar tamarind)	Clusiaceae
26.	*Garcinia indica* (Thouars) Choisy	Pericarp-lobes (fruit-rind)	Kokum	Clusiaceae

Contd...

Table 4.1–Contd...

Sl.No.	Botanical Name	Part Used	English Common Name	Family
27.	*Hyssopus communis* L.	Leaf	Hyssop	Lamiaceae
28.	*Juniperus communis* L.	Berry	Juniper berry	Cuppressaceae
29.	*Origanum majorana* L.	Leaf and flowering top	Sweet marjoram	Lamiaceae
30.	*Mentha arvensis* L.	Leaf	Japanese mint	Lamiaceae
31.	*Mentha spicata* L.	Leaf	Spear mint Garden mint	Lamiaceae
32.	*Mentha piperita* L.	Leaf	Peppermint	Lamiaceae
33.	*Murraya koenigii* L.	Leaf	Curry leaf	Rutaceae
34.	*Myristica fragrans* Houtt.	Aril Kernel	Mace Nutmeg	Myristicaceae
35.	*Nigella sativa* L.	Fruit	Black cumin	Ranunculaceae
36.	*Ocimum basilicum* L.	Leaf	Sweet basil	Lamiaceae
37.	*Papaver somniferum* L.	Seed	Parsley	Apiaceae
38.	*Petroselinum crispum* (Miller) Nyman ex A.W. Hill	Seed	Parsley	Apiaceae
39.	*Pimenta dioica* (L.) Merr.	Fruit and leaf	Pimento or allspice	Myrtaceae
40.	*Pimpinella anisum* L	Fruit	Aniseed	Apiaceae
41.	*Piper nigrum* L.	Fruit	Black pepper, white pepper	Piperaceae
42.	*Punica granatum* L.	Seed (dried with flesh)	Pomegranate	Punicaceae
43.	*Rosmarinus officinalis* L.	Leaf	Rosemary	Lamiaceae
44.	*Salvia officinalis* L.	Leaf	Sage	Lamiaceae
45.	*Saturega hortensis* L.	Stem, leaf Flowering top	Savory	Lamiaceae
46.	*Sinapis alba* (L.) Biossier	Seed	White mustard, yellow mustard	Brassicaceae
47.	*Tamarindus indica* L.	Fruit	Tamarind	Cesalpiniaceae
48.	*Thymus vulgaris* L.	Leaf and flowering top	Thyme	Lamiaceae
49.	*Trachyspermum ammi* (L.) Sprague ex Turril	Fruit	Bishops weed or Ajowain	Apiaceae
50.	*Trigonella foenum-graecum* L.	Seed	Fenugreek	Fabaceae
51.	*Vanilla fragrans* (Salis.) Ames (=*Vanilla planifolia* Andrews)	Pod	Vanilla	Orchidaceae
52.	*Zingiber officinale* Rosc.	Rhizome	Ginger	Zingiberaceae

The tropical evergreen forests of Western Ghats are home for at least two major spices – black pepper and cardamom. It is also a secondary center of origin of another spice cinnamon. A wide range of diversity is available in the spices in the form of varieties/cultivars/improved varieties, species and related taxa in the fragile ecosystem of Andaman and Nicobar group of Isalnds. The important genera include *Alpinia, Amomum, Costus, Curcuma, Garcinia, Globba, Hedychium, Kaempferia, Myristica, Piper, Vanilla* and *Zingiber*, of which maximum diversity is reported in the genera *Piper, Zingiber, Garcinia* and *Myristica* (Shiva *et al.*, 2003).

The depletion of forest area, rampant destruction of the forest trees and undergrowth and changes in the agro-ecological conditions has led to a sharp decline in the populations of wild pepper and cardamom and their related taxa. If not collected and conserved, these valuable genetic resources may be lost for ever. The Indian Institute of Spices Research (IISR), Calicut, All India Coordinated Research Project on Spices (AICRPS) and National Research Centre on Seed Spices (NRCSS), Ajmer (Rajasthan) are giving prime importance in collection, evaluation, utilization and conservation of the biological diversity of spices from all over the country and abroad.

India is blessed with varied agro-climatic conditions to grow almost all spices, starting from tropical-black pepper to temperate-saffron. For majority of the spices, wide diversity exists in the form of species, varieties/cultivars/land races and ecosystems in the country (Table 4.2) except for tree spices and seed spices, for which the genetic base is limited, which can be broadened by introducing promising exotic varieties/types (Table 4.3) from centre of origin/diversity.

Table 4.2: Origin and Centre of Diversity in Spices

Spices	*Origin/Centre of Diversity*	*Form of Diversity and Distribution in India*
Black pepper	Origin-Tropical evergreen forests of Western Ghats; Diversity – Northeastern and South-western regions of India	110 species; >100 cultivars; Kerala and Karnataka
Cardamom (S)	Origin- Evergreen forests of W.G.; Diversity-Malaysian-Indonesian region (8 spp.), Western Ghats and Silent Valley forests	*E. cardamomum;* Limited species; and ecosystem diversity
Ginger	Origin-South Asia; Diversity-Indo-Malayan Region (24 spp.)	8 spp.(W.Ghats); 50 land races; NE and Ker.
Turmeric	Origin-Indo-Malayan region	40 spp.(15-endemic); 60 cvs./land races/ morphotypes; NE and South India
Nutmeg	Origin–Moluccas Islands of Indonesia; Diversity-Indo-Malayan region and W. Ghats Forests	12 spp. (mostly endemic); 2 cultivars; W. Ghats and A and N Islands
Clove	Origin- Moluccas Islands of Indonesia	Many spp. of *Syzygium* (W. Ghats); Variants-dwarf, bushy, narrow leaves and king cloves (T.N. and Kerala); 2 exotic collections (Sri Lanka and Zanzibar)
Cinnamon	Origin-Sri Lanka and South Western Ghats of S.I. (secondary centre of origin)	26 spp. (13 spp.- S.I.-W.Ghats); few varieties
Chinese cassia	Diversity-China,Vietnam,Thailand and upto the Malaysian region	Elite trees introduced in Valparai (T.N.)
Indian Cassia	India–NE region	Elite trees
Garcinia	Diversity-Malaysia and Africa, India-NE Region (Assam-Burma) and W. Ghats	36 spp.(17 spp. + 3 var. – endemic – W. Ghats, A&N Isds. and NE); 7-local var. of 'Kodumpuli'
Tamarind	Origin-Tropical Asia (India) but also occurs in Central Africa	Monotypic genus; few elite genotypes/ varieties

Contd...

Table 4.2–Contd...

Spices	*Origin/Centre of Diversity*	*Form of Diversity and Distribution in India*
Curry Leaf	Origin-India; Diversity/distribution-Northeastern parts and Foothills of Himalayas in north to Kerala in south Except arid regions	3 spp. (India); 2 types (Midrib-pink and green; Leaf shape-broad and narrow)
Vanilla	Origin-Mexico (Tropical America)	5 species
Major Seed spices (Coriander, cumin, fennel and fenugreek)	Central Asian and Mediterranean regions collections	No wild species; Traditional varieties +exotic
Paprika	South and Central America	Exotic collections and Kt-Pl-19
Paprika like chillies	Dharwad (Karnataka); Warangal (A.P.)	Land races, Arka Abhir, KDC-1

Table 4.3: Promising Exotic Varieties/Types of Seed Spices

Sl.No	*Spice*	*Essential Oil (Per cent) and Country*	*Major Components/ Types with Major Producing Countries*
1.	Coriander	1.4 – 1.7 (Norwegian)	Linalool, alpha–pinene, gamma -terpinene, geranyl acetate
2.	Cumin	4.8 (Exotic variety: UC – 198 – Egypt)	Cuminalehyde – 15 to 40 per cent
3.	Fennel – [Indian fennel *F. v.* (Miller) Thellung var. Panmorium]	2.5 – 6.5	Common fennel -*Foeniculum vulgare* var. amara Bitter fennel – *F. vulgare* var. vulgare (Central Europe, India, Japan, Argentina and North America)
			Sweet fennel (French/Roman fennel) – *F. vulgare* var. dulce [South Europe (France, Italy)]
4.	Fenugreek	0.02	Diosgenin: 0.20 to 0.35 per cent (Algeria, Morocco and Ethiopia)

Collection of Germplasm in Spices

Extensive survey conducted over a period of three decades by the centres under AICRPS and IISR in the spice growing tracts of India led to a substantial collection of mandatory spices. Apart from indigenous collections, exotic germplasm from the Mediterranean and Western countries were also introduced to broaden the genetic base of the seed spices. At present, the gene repository of IISR consists of 5943 accessions comprising wild accessions, land races and local cultivars of spices (Table 4.4). AICRPS centers consist of 691 accessions of black pepper, 305 cardamom, 660 ginger, 1280 turmeric, 195 tree spices and 4102 seed spices. Besides, NRCSS is holding 1099 accessions of seed spices (Table 4.5).

Table 4.4: Germplasm of Spices Conserved at IISR, Calicut

Spice	*Number of Accessions*
Black pepper	2347
Cardamom	436
Ginger	684
Turmeric	1040
Nutmeg	484
Cinnamon	408
Clove	233
Garcinia	86
Allspice	2
Vanilla	93
Paprika	130
Total	**5943**

Table 4.5: Genetic Resources of Spices Maintained Inder AICRPS Centers

Crop/Center	*Indigenous*		*Exotic*	*Total*
	Cultivated	*Wild and Related sp.*		
Black pepper				
Panniyur	214	72	3	289
Sirsi	96	20	1	117
Chintapalle	58	–	–	58
Yercaud	125	10	–	135
Pundibari	16	1	–	17
Dapoli	70	5	–	75
Total	579	108	4	691
Cardamom				
Pampadumpara	152	1	–	153
Mudigere	152	–	–	152
Total	**304**	**1**	**–**	**305**
Ginger				
Pottangi	172	2	3	177
Solan	286	–	–	286
Dholi	47	–	–	47
Kumarganj	58	–	–	58
Pundibari	42	–	–	42
Raigarh	44	–	–	44
Total	**655**	**2**	**3**	**660**

Contd...

Table 4.5–Contd...

Crop/Center	*Indigenous*		*Exotic*	*Total*
	Cultivated	*Wild and Related sp.*		
Turmeric				
Pottangi	197	–	–	197
Jagtial	273	–	–	273
Dholi	87	2.	–	89
Raigarh	42	–	–	42
Kumarganj	126	–	–	126
Pundibari	136	14	–	150
Solan	145	–	–	145
Coimbatore	258	–		258
Total	**1264**	**16**	**–**	**1280**
Clove				
Pechiparai	22	–	–	22
Dapoli	2	–	–	2
Yercaud	13	–	–	13
Total	**37**	**–**	**–**	**37**
Nutmeg				
Pechiparai	22	–	–	22
Dapoli	97	–	–	97
Total	**119**	**–**	**–**	**119**
Cinnamon				
Pechiparai	12	–	–	12
Dapoli	11	–	–	11
Yercaud	16	–	–	16
Total	**39**	**–**	**–**	**39**
Coriander				
Jobner	753	–	102	855
Jagudan	74	–	21	95
Coimbatore	274	–	–	274
Guntur	248	–	–	249
Hissar	251	–	–	251
Dholi	97	–	–	97
Kumarganj	75	–	–	75
NRC Seed Spices	283	–	18	301
Total	**2055**	**–**	**141**	**2196**

Contd...

Table 4.5–Contd...

Crop/Center	*Indigenous*		*Exotic*	*Total*
	Cultivated	*Wild and Related sp.*		
Cumin				
Jobner	370	–	6	376
Jagudan	240	–	7	247
NRC Seed Spices	132	–	8	140
Total	**742**	**–**	**21**	**763**
Fennel				
Jobner	261	–	20	281
Jagudan	135	–	4	139
Hissar	122	–	–	122
Dholi	31	–	–	31
Kumarganj	44	–	–	44
NRC Seed Spices	245	–	19	264
Total	**838**	**–**	**43**	**881**
Fenugreek				
Jobner	353	–	12	365
Jagudan	64	–	–	64
Guntur	125	–	–	125
Hissar	240	–	–	240
Dholi	106	–	–	106
Kumarganj	77	–	–	77
NRC Seed Spices	205	–	2	207
Total	**1170**		**14**	**1174**
NRC Seed Spices				
Ajowan	79	–	–	79
Dill	55	–	4	59
Nigella	16	–	–	16
Anise	9	–	3	12
Celery	9	–	2	11
Caraway	7	–	3	10
Total	**175**	**–**	**12**	**187**
Grand Total	**7977**	**127**	**238**	**8332**

Source: AICRPS (2007); NRCSS (2007).

Characterization and Evaluation of Spices Germplasm

All these germplasm collections were subjected to cataloguing and evaluation under Initial evaluation trials (IET), Comparative yield trials (CYT) and Coordinated varietal trials (CVT)/Multi-location trials (MLT). The characterization of germplasm of spices has expressed high amount of variability with regard to black pepper, ginger, turmeric, cardamom, cinnamon and reasonable amount of variability in other spice crops. Only in vanilla and clove, the variability is very much limited. Descriptors to characterize the germplasm have been developed for most of these crops and are available with IPGRI (1994, 1995) for black pepper and cardamom, Indian Institute of Spices Research, Calicut for ginger, turmeric and tree spices, ICRI, Myladumpara for vanilla, NRC on Seed Spices, Ajmer (Rajasthan) for seed spices and also with AICRPS-centres/SAU's. The lines having unique characters have been registered with NBPGR.

Utilization of Spices Germplasm

The promising entries have resulted in development/release of about 25 varieties in major and tree spices under IISR and about 90 improved varieties including three hybrids in 9 spice crops for different agro-climatic conditions of the country under AICRPS (Tables 4.6 and 4.7). At NRCSS, about 10 high yielding varieties (one each in fennel, coriander, nigella, anise and celery and two each in fenugreek, Ajowan and dill) have been developed. In these, some of the varieties are having source of tolerance/resistance to biotic and abiotic stresses and a few with high quality parameters and components. Besides, some of the lines are under pipeline for release and several lines have been identified for yield, quality and pest/disease and drought resistance or tolerance from the germplasm of spices.

Table 4.6: Improved Varieties Released Under IISR for Cultivation

Spices	*Total No.*	*Varieties*
Black pepper	9	Subhakara, Sreekara, Panchami, Pournami, PLD-2, IISR- Thevam, IISR-Malabar Excel, IISR-Girimunda, IISR-Shakthi
Cardamom	3	IISR-Kodagu Suvasini, IISR-Avinash, IISR–Vijetha
Ginger	3	IISR – Varada, IISR – Mahima, IISR – Rejatha
Turmeric	7	Suvarna, Suguna, Sudharsana, Prabha, Prathiba, IISR-Alleppey Supreme, IISR-Kedaram
Cinnamon	2	IISR – Navashree, IISR–Nithyashree
Nutmeg	1	IISR–Viswashree
Total	**25**	

Conservation of Spices Germplasm

Since most of the spices except seed spices are vegetatively propagated, the germplasm is being maintained in clonal repositories (nursery/field genebanks) of IISR, Calicut. IISR has been using information collected through surveys and GIS has identified the ecological niches and β and γ diversity centers in major spices and this information is being passed on to the Department of forests, in the concerned states for protecting the sites from disturbances. Resource mapping was also done in black pepper and cardamom for *in situ* conservation of their habitat. The seed spices germplasm are being maintained in seed/gene banks (seed to seed). All these spices germplasm are being conserved by *ex situ* method at various centres of SAU's under AICRPS (Table 4.5). Apart from this, Indian Institute of Spices Research, Calicut is having largest germplasm collections in black pepper, ginger, turmeric and tree spices at its National Conservatories (as *ex situ* and *in vitro* methods) (Table 4.4) and some portions by NBPGR and its Regional Stations (in field gene bank, seed bank and *in vitro*), Indian Cardamom Research Institute (Spices Board), Myladumpara and Gangtok (in field and *in vitro*).

Table 4.7: Improved Varieties Developed/Recommended for Release Through AICRPS

Spices	Total No.	Varieties
Black pepper	7	Panniyur-1, Panniyur-2, Panniyur-3, Panniyur-4, Panniyur-5, Panniyur-6, Panniyur-7
Cardamom	10	Mudigere 1, Mudigere 2, PV-1, ICRI-1, ICRI-2, ICRI-3, ICRI-4, PV-2, ICRI -5, ICRI – 6
Ginger	4	Suprabha, Suruchi, Surabhi, Himgiri
Turmeric	9	CO-1, BSR-1, BSR-2, Roma, Suroma, Ranga, Rasmi, TCP-2, Rajendra Sonia
Coriander	22	CO. 1, CO. 2, CO. 3, CS-287, Guj. Coriander-1, Guj. Coriander-2, Rajendra Swathi, RCr-41, Sadhana, Swathi, Sindhu, RCr-20, RCr-435, RCr-436, RCr-684, Hissar Anand, UD-446, Hissar Sugandh, Hissar Surabhi, Sudha, RCr-480, DH-228 (leafy type)
Cumin	9	Guj. Cumin-1, Guj. Cumin-2, Guj. Cumin-3, RZ-19, MC-43, S-404, RZ-223, Guj. Cumin-4, RZ-341
Fennel	10	PF-35, Guj. Fennel-1, Guj. Fennel-2, CO.1, RF-101, Rajendra Sourabha, HF-33, GF-11, RF – 143, RF-178
Fenugreek	15	CO. 1, CO. 2, Rajendra Kanti, RMt-1, Lam Sel. -1, Hissar Sonali, Guj. Methi-1, RMt-303, Rajendra Khushba, Hissar Suvarna, Hissar Madhavi, Hissar Mukta, RMt-305, Guj. Methi-2, RMt-351.
Cinnamon	2	YCD-1, PPI (Ci) -1
Nutmeg	1	Konkan Sugandha

Source: Anandaraj *et al.* (2006); Parthasarathy *et al.* (2005); Ravindran *et al.* (2006);

Bioprospecting of Spices

The medicinal properties of major and minor spices are well-documented. Among the related species of turmeric, *Curcuma aromatica* is important in the preparation of medicine and cosmetic articles. *C. amada* (Mango ginger) is used as a vegetable for making pickles, while *C. zeodoaria* (Indian arrow root) is a major source of starch in many parts of India. Besides, *C. angustifolia, C. caulina, C. montana, C. leucorrhiza, C. decipiens, C. raktakanta, C. pseudomontana, C. erubescens, C. xanthorrhiza, C. malabarica* and *C. harita* are reported to be useful for arrow root preparation (Velayudhan *et al.*, 1999b). Some of the wild *Myristica* species are used in the dye industry (eg. *M. magnifica* and *M. malabarica*) and also in tribal medicine.

The dried rind of Kokum fruit is the traditional kokum or 'binda' of commerce used for garnishing curries. Value-added products such as squash (amrut kokum), syrup, kokum powder, kokum butter, concentrates, pigments etc. are obtained from kokum. It is also reported to be a good source for making wine. Camboge fruit has excellent therapeutic value and the dried rind is a popular spice used in cookery. The true vanilla obtained from the fruits (beans) possess a gently, rich spicy aroma, which are widely used as a flavouring for ice-cream, soft drinks, chocolate, baked goods, puddings, liquors and also in perfumery.

Conclusion

Even though conventional approaches remain as a basic strategy for germplasm collection, modern methods like GIS would be helpful to locate species richness. The future thrust areas would be: Intensive survey and documentation of spices diversity, which are facing extinction, rare and threatened and endemic species in view of future threats (Man made – deforestation, overexploitation, etc. and Natural – Flood, disease epidemics, drought, etc.); Characterization of germplasm through conventional methods and molecular techniques to eliminate duplicates and to guard against biopiracy and to identify useful genes or genes of our interest; Widening the genetic base especially of seed spices and tree spices; Evolving efficient and cost effective conservation strategies and promoting participatory (farming community and institution) mode of conservation of germplasm. Keeping in view of the IPR era, bioprospecting of the biological resources is getting importance in recent times.

References

AICRPS, 2007. *Annual Report: 2006–2007 of All India Coordinated Research Project on Spices.* Indian Institute of Spices Research, Calicut, Kerala.

Anandaraj, M., Johny, A. Kallupurackal and Shiva, K.N., 2006. New spices varieties identified. *Spice India*, 19(8): 25–28.

IPGRI, 1994. *Descriptors for Cardamom (Elettaria cardamomum Maton).* International Plant Genetic Resources Institute, Rome, 51 p.

IPGRI, 1995. *Descriptors for Black Pepper (Piper nigrum L.).* International Plant Genetic Resources Institute, Rome, 39 p.

Khoshoo, T.N., 1995. Census of India's biodiversity: Tasks ahead. *Curr. Sci.*, 69: 14–17.

NRCSS, 2007. *Annual Report: 2006–2007 of National Research Centre on Seed Spices*, Ajmer, Rajasthan.

Parthasarathy, V.A., Shiva, K.N. and Johny, A.K., 2005. Research and development in major seed spices under AICRPS. In: *A Course Manual for Winter School on Advances in Seed Spices Production*, (Ed.) S.K. Malhotra, 2–22 September, NRC Seed Spices, Ajmer (Raj.), pp. 255–271.

Ravindran, P.N., Nirmal Babu, K., Shiva, K.N. and Johny A. Kallupurackal (Eds.), 2006. *Advances in Spices Research: History and Achievements of Spices Research in India Since Independence.* Agrobios, Jodhpur (Raj.), xxi+994 p.

Shiva, K.N., Suryanarayana, M.A. and Medhi, R.P., 2003. Genetic resources of spices and their conservation in Bay islands. *Indian J. Plant Genet. Resour.*, 16(2): 91–95.

Velayudhan, K.C., Muralidharan, V.K., Amalraj, V.A., Gautam, P.L., Mandal, S. and Kumar, Dinesh, 1999a. *Curcuma Genetic Resource.* Scientific Monograph No. 4. NBPGR, New Delhi.

Velayudhan, K.C., Asha, K.I., Mithal, S.K. and Gautam, P.L., 1999b. Genetic resources of turmeric and its relatives in India. In: *Biodiversity, Conservation and Utilization of Spices, Medicinal and Aromatic Plants*, pp. 101–109.

Recent Advances in Seed Spices (2011) *Pages* **42–66**
Editors: **Y. Ravindrababu, R.K. Jaiman, K.D. Patel, N.R. Patel and S.B.S. Tikka**
Published by: **DAYA PUBLISHING HOUSE, NEW DELHI**

5

Status of Characterization and Evaluation in Coriander

E.V. Divakara Sastry*, Dhirendra Singh and Shyam Singh Rajput
Department of Plant Breeding and Genetics, S.K.N. College of Agriculture, Rajasthan Agricultural University, Jobner – 303 329, Rajasthan

India is the land of seed spices. The diversity in the geographical conditions of our country provide ideal situation for the cultivation of several types of spices ranging from Saffron which is cultivated in Himalayas to black pepper down south, cumin and fennel from the extreme west in our country to large cardamom exclusively grown in North East. Further these spices range from small plants (saffron, cumin) to large trees (nutmeg and cloves) and vary from annuals to perennials (Sastry, 2007). Thus each of the spice crops requires special attention for their cultivation and the maintenance of their genetic resources. Further these crops also differ in their reproductive mechanisms, ranging from strictly asexual reproduction (ginger, turmeric) to sexual reproduction, in which both self (fenugreek) to cross pollination (vanilla, coriander) and all the degrees in between occur. Thus spices group is a varied group of plants requiring special attention in maintaining their resources. If we look at the acreage and their distribution, we find that their distribution is wider with pockets of concentrated cultivation. Such type of cultivation is both boon as well as bane for the germplasm resource curator as well as plant breeder. It is boon because single variety culture can be easily practiced, but bane because, more location specific problems are to be countered particularly with regard to diseases and even local adaptations. To a germplasm curator it is bane because, any monoculture practice is bound to replace the mosaic of the forms that are found at a given place particularly so in a spices. Further there is always a threat of pollen contamination in maintaining these resources. As such most of the spices have cross pollination as a mode of reproduction and insects are important in this regard.

* Corresponding Author: E-mail: evdsastry@yahoo.com.

Let us look at what constitutes the genetic resources. Broadly speaking genetic resources include every single variety, cultivar, cultigens, even wild species that are related to a given crop. The germplasm resources of any crop include wide variety of material as listed below (Thomas and Mathur, 1991):

1. Those derived from centers of diversity
   (*a*) Primitive cultivars,
   (*b*) Natural hybrids between cultigens and wild relatives,
   (*c*) Wild relatives (wild and weedy races), and
   (*d*) Related species and genera.
2. Those derived from areas of cultivation
   (*a*) Commercial types,
   (*b*) Obsolete varieties,
   (*c*) Primitive varieties, and
   (*d*) Special purpose types.
3. Those derived from breeding programmes
   (*a*) Pure line from farmer's stock,
   (*b*) Elite varieties or hybrids,
   (*c*) Breeding lines,
   (*d*) Mutants,
   (*e*) Polyploids, and
   (*f*) Intergeneric and interspecific hybrids.

A list of common seed spices is given in Table 5.1. Let us discuss the status of genetic resources in terms collection, evaluation and conservation. Of all the seed spices listed in Table 5.1, Coriander, cumin, fennel and fenugreek are commonly grown hence are termed as major seed spices, rest are called minor seed spices (Sharma, Bhati and Bhutt, 2001). This article will concentrate on coriander.

Collection

If we look at the centers of origin of the seed spices, India is not a home for these spices. The major as well as minor seed spices have mostly originated in Mediterranean region (Vavilov, 1951, 1926) (Table 5.2). India received these crops as introductions primarily through Arab invaders. Hence, the variation found in India is always expected to be limited in comparison to primary centers of origin. Let us examine coriander as a case. Vavilov (1992) reported coriander to have originated in central Asia, the Near East and Abyssinia. On the other hand four distinct regions namely (*i*) India, (*ii*) North America, (*iii*) Central Asia and (*iv*) Abyssinia were recognized as cradles of distinct types where different types of coriander have developed (Ivanova and Stoletova, 1990). In a detailed study of different collections obtained from different countries in coriander, Diederichsen (1977) noted differences in the specific morphological characters among different collections obtained from different countries. He has proposed probable route of diversity in coriander as given in Figure 5.1. Nevertheless, India has several regions which can be classified as diversity rich regions for coriander (Gangopadhyay, Yadav and Karihaloo, 2005) (Table 5.3).

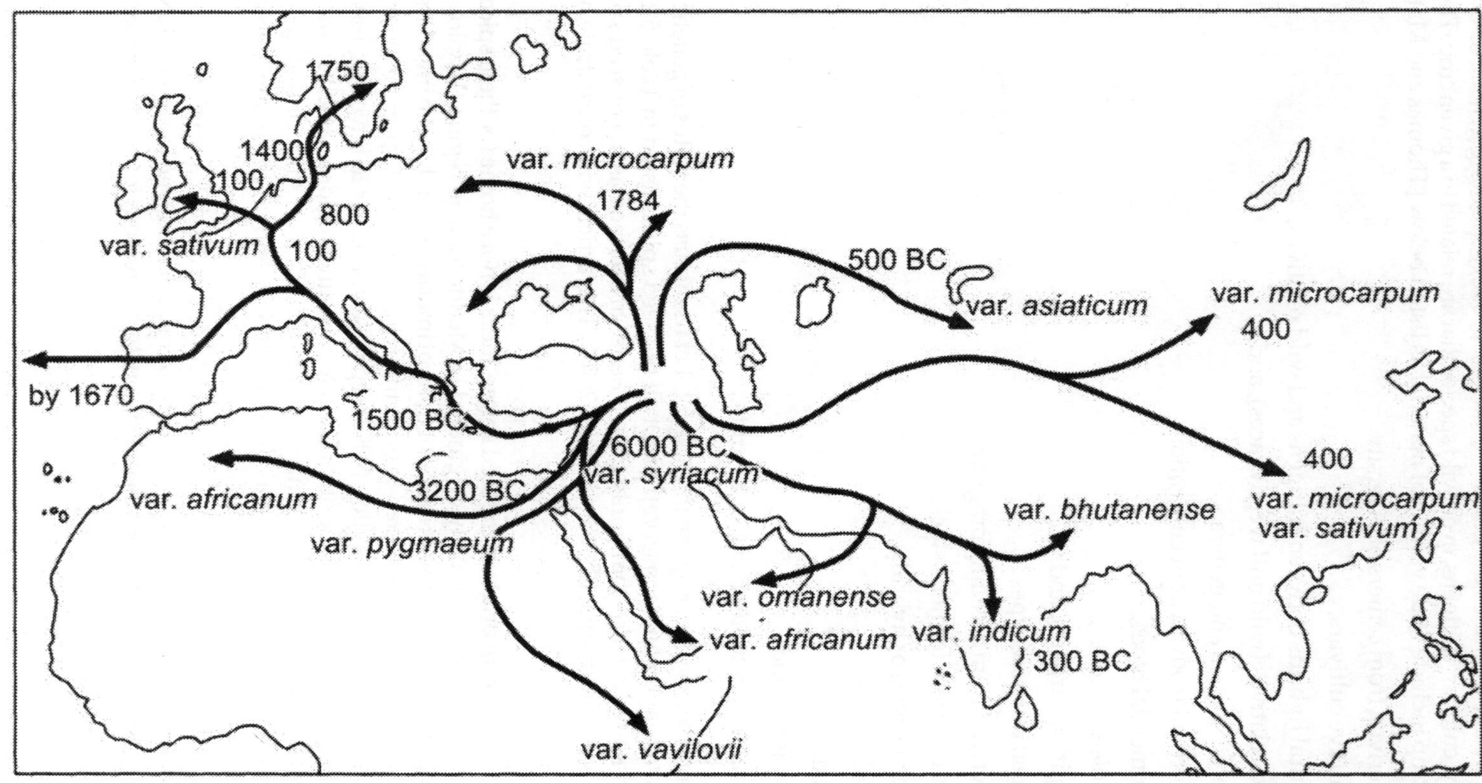

Figure 5.1: Possible Path of Travel of Coriander from Mediterranean Region to Other Parts of World (from Diederichsen and Hammer, 2003).
Figures indicate historical accounts according to Laufer (1919), Schultze-Motel (1968–1987, 1993, 1994), Sinskaja (1969), Zohary and Hopf (1993).

Table 5.1: List of Seed Spices

Sl.No.	*Common Name*	*Botanical Name*	*Family*
1.	Ajawain, Bishops weed	*Trachyspermum ammi* (L.) Sprague	Apiaceae
2.	Anise	*Pimpinella anisum* L.	Apiaceae
3.	Bitter fennel	*Foneiculum vulgare* Miller	Apiaceae
4.	Black mustard	*Brassica nigra* (L.) Koch	Brassicaceae
5.	Black caraway	*Bunicum persicum* (Boiss) B. Fedtsch	Apiaceae
6.	Black caraway	*Carum bulbocastanum* L.	Apiaceae
7.	Black cumin (kalongi)	*Nigella sativa* L.	Ranunculaceae
8.	Celery, garden celery	*Apium graveolens* L. Var. dulce (Miller) Pers.	Apiaceae
9.	Caraway	*Carum carvi* L.	Apiaceae
10.	Coriander	*Coriandrum sativum* L.	Apiaceae
11.	Cumin	*Cuminum sativum* L.	Apiaceae
12.	Dill	*Anethum graveolens* L.	Apiaceae
13.	Indian Dill (sowa)	*Anethum sowa* Kurz	Apiaceae
14.	Fenugreek/Kasuri Methi	*Trigonella foenum-graecum* L. *Trigonella corniculata* L.	Fabaceae
15.	Indian mustard	*Brassica juncea* (L.) Czernj et cosson	Brassicaceae
16.	Poppy seed	*Papaver somniferum*	Papaperaceae
17.	Parsley	*Petroselinum crispum* (Miller) Nyman ex. A.W. Hill	Apiaceae
18.	Sesame, gingley	*Sesamum indicum* L.	Pedaliaceae
19.	Sweet fennel (European fennel)	*Foeniculum vulgare* Miller ssp. *Capillaceum* var. *dulce*	Apiaceae
20.	White mustard	*Sinapsis alba* L.	Brassicaceae

Table 5.2: Seed Spice Crops and their Centre of Origin

Seed Spices	*Centre of Origin*	*Distribution*
	Major group	
Coriander	Mediterranean region	India, Morocco, Russia Bulgaria, Mexico, Argentina, China, Romania, Japan, Italy
Cumin	Mediterranean region	India, Turkey, Iran, Egypt, Pakistan, Syria, Italy
Fennel	South Europe and Mediterranean region	India, Egypt, China, Romania, Russia
Fenugreek	South East Europe and West Asia,	India, Morocco, Bulgaria
	Minor group	
Ajowain	Mediterranean region and South West Asia	India, Iran, Egypt, Afghanistan
Dill	Europe, Africa and Asia	India, Germany, Hungary, Netherlands, Pakistan and USA
Celery	Mediterranean region	India, Southern France, China and Egypt
Aniseed	Eastern Mediterranean region	Bulgaria, Cyprus, Germany, Russia and India
Nigella	Eastern Mediterranean	India, Pakistan, Sri Lanka, Bangladesh, Nepal, Iraq, Egypt

Table 5.3: Diversity Rich Areas of Coriander in India (from Gangopadhyay, Yadav and Karihaloo, 2005)

State	*District/Zones*
Rajasthan	Baran, Bundi, Jaipur, Alwar, Sikar, Kota, Jhalawar, Chittorgarh
M.P. and Chattisgarh	Guna, Rajgarh, Mandsoar
Andhra Pradesh	Kurnool, Cuddapah, Ananthapur
Tamil Nadu	Trichy
Gujarat	Dantiwada, Junagadh

The second factor which influences the collection methodology is reproductive system. Fortunately, in all the seed spices, the reproduction is primarily sexual and it is through cross pollination in coriander.

The third factor which influences the collection is the distribution of crop itself. Seed spices are traditionally grown in marginal lands in isolated pockets. None of the crop is widely sown unlike commercial (like cotton, or pulses) or food crops (like cereals or millets). Even in areas where the cultivation is supposed to be intensive, its acreage is very limited. In such a scenario, the variability is supposed to be limited. Moreover, since, the crops have been cultivated on marginal lands with little or no inputs, the populations tend to develop characters that help them survive rather than provide high yields. This may be one of the reasons, that the yield jumps are limited in spices, unlike several of the pulses or oilseeds. Nevertheless, efforts to collect available genetic resources have been taken up by several institutes, particularly with the establishment of All India Coordinated Research Project on Spices. Table 5.4 lists the available collections at several of these locations. Literature search also indicates that germplasm collections of the seed spices are also maintained at CIMAP, Lucknow, Defense Research Laboratories* as well as other institutes not connected by All India Coordinated Research Project on Spices.

Good collections of coriander are maintained in China**, Australia, Germany, Netherlands, USA as well as the countries of Mediterranean region namely Morocco, Egypt, Iran as well as horn of Africa (Ethiopia).

Coming to the collections present in the country in various centers, there is no picture with regard to duplicates. Duplicates are expected because of intra exchange of material between institutes and then renumbering or because of duplicate sampling. Unfortunately, the documentation of the collections at various centres is still incomplete for various known or unknown reasons.

Characterization and Evaluation

Characterization and evaluation of the germplasm are essential features which help in assessing the genetic worth of the crop genetic resources as well as help in selecting desirable types. However, for proper identification of the lines, vivid descriptives are essential, hence, for any given crop preparation of descriptor lists are essential.

---

* Defense Research & Development Organization. http://www.drdo.org/labs/darl/achieve.html.

** Chinese Crop Germplasm Information System. http://icgr.caas.net.cn/.

Table 5.4: Germplasm Holdings of Major Seed Spices at All India Coordinated Research Project on Spices Centers (Anandraj *et al.*, 2007)

Crop/Center	*Indigenous*		*Exotic*	*Total*
	Cultivated	*Wild and Related sp.*		
Coriander				
Jobner	693	–	102	795
Jagudan	74	–	21	95
Coimbatore	224	–	–	224
Guntur	244	–	5	249
Hissar	190	–	–	190
Dholi	85	–	–	85
Raigarh	20	–	–	20
Kumarganj	72	–	–	72
Total	**1602**	**–**	**128**	**1730**

Diederichsen (2004) states that a rational and standardized terminology for species or infraspecific groups of plants is highly desirable for effective communication of genetic diversity hence it is imperative that, genebanks must be current with recent developments in taxonomy and nomenclature. The importance of appropriate taxonomical and nomenclatural principles for the successful management and use of plant genetic resources was already elaborated and emphasized by Vavilov (1931). Taxonomic concepts for cultivated plants were further developed (Mansfeld, 1953; 1954; Danert, 1962; Harlan and de Wet, 1971). The important role of taxonomy for genetic resources has recently been emphasized by Hazekamp and Guarino (1999), and is also evident from the architecture of modern computerized databases, such as the Genetic Resources Information Network Database (GRIN) of the United States Department of Agriculture (Shands, 1995).

Taxa, considered at one point of time as species, are now often considered as infraspecific taxa, or not as formal taxonomic groups at all. The historical dimension of genebank work causes these institutions to be conservative in accepting nomenclatural changes. However, new insights into the genetic makeup and composition of germplasm sources, in particular from molecular genetic investigations, require continuous evaluation of existing characterization methods for genetic diversity. Information associated with genebank accessions, from taxonomic revisions or other historic documentations of diversity will be important for detection of changes in biodiversity. An example in this direction is a botanical variety of coriander collected in Ethiopia, taxonomically described and documented in herbaria by Vavilov and his colleagues, which could recently not be found in cultivation. This demonstrates the use of reliable methods to transfer genebank collections and associated information to future generations.

Characterization

Any germplasm collection is evaluated for two types of characteristics namely characterization data and evaluation data. In most of the other crops, clear and elaborate internationally standardized descriptors are available for describing the entries in a collection. This is still lacking in seed spices. Where such documentation is available, is limited only to certain morphological traits and even the description is vague and ambiguous. One thing is sticking, the resistance against diseases and pests is very limited or absent.

The fruit size and weight has for a long time been recognized as an important character dividing the species into two subspecies (small-fruited vs. large-fruited), which have different geographical origin and agronomic properties (Alefeld, 1866; Harrod, 1960). The small fruited types are recognized as *C. sativum* L.var. *microcarpum* and the large fruited one are described as *C. sativum* L.var. *vulgare* (Diederichsen and Hemmer, 2003). Diederichsen (1997) however recognized four types of coriander based on seed shape and size as depicted in Figure 5.2. De Candolle (1830) was the first to recognize the small-fruited coriander as being considerably different. In addition, the shape of the mature fruits can be used to distinguish a third group within the species, which originated mostly from the Indian subcontinent. This infraspecific group is morphologically distinguished by ovate fruits and chemotypically characterized by a low proportion or complete lack of camphor in the volatile oil in the fruits. The geographically distinct origin of this third group and the great economic importance of the volatile oil composition support the distinction of this infraspecific group as a third subspecies. Ivanova and Stoletova (1990) proposed a more detailed infraspecific grouping and distinguished four subspecies and nine botanical varieties based on morphological characters as detailed below.

Coriandrum sativum

Subsp. *Sativum*

var. *sativum*

var. *africanum* Stolet.

Subsp. *asiaticum* Stolet

var. *asiaticum* Stolet

var. *anatolicum* Stolet.

var. *afghanicum* Stolet.

Subsp. *vavilovii* Stolet

var. *vavilovii* Stolet

Subsp. *indicum* Stolet.

var. *indicum* Stolet

var. *pygmaeum* Stolet.

The Indian coriander lines belong to *C. sativum* var *indicum.* These are identified by the fruit shape which is ovate (Figure 5.2). The first attempt to create a descriptor list of the seed spices in India was made by All India Coordinated spices and cashewnut improvement project (Anonymous 1983). The descriptor list (Table 5.5) was very minimal and basically is based on evaluation data. Another good effort has been made by NRCSS, Ajmer to prepare a descriptor list for the spices crops (Malhotra and Vashistha, 2006), this is also minimal in its description and needs standardization. Let us examine these two descriptor lists. Of the 19 characters described Anonymous (1983), 4 can be said to be characterization data. These are growth habit, grain colour at maturity, pigmentation on the stem and grain breakability or split habit. Similarly, out of the 28 characters listed by Malhotra and Vashistha (2006), ten can be said to be characterization data, unfortunately most of them are not stable. A study conducted at S K N College of Agriculture, Jobner by Choudhary (1987) has recognized five clear cut growth stages. These are listed below.

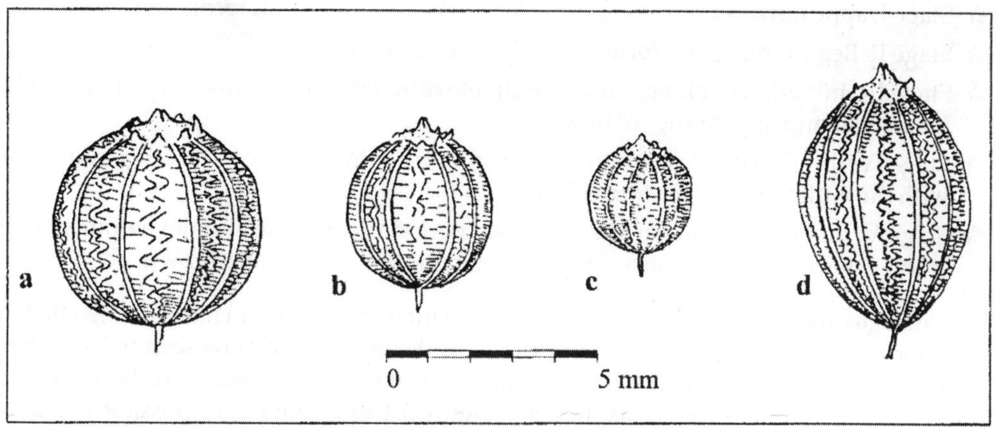

Figure 5.2: Shape of Fruits in Coriander
(*a*) Near Eastern type, (*b*) Ethiopian type, (*c*) Caucasian type and (d) Indian type
(From Diederichsen, 1997)

Table 5.5: Descriptors of Coriander

1. Days to 50 per cent flowering
2. Days to 50 per cent maturity
3. Plant height
4. Primary branches per plant
5- Secondary branches per plant
6. Umbels per plant
7. Umbellets per umbel
8. Male and hermaphorodite florets per umbel
9. Grains per umbellet
10. 1000 grain weight
11. Yield per hectare
12. Early growth vigour (1 poor to 5 best)
13. Biological yield per hectare
14. Growth habit (bushy, erect and semi erect)
15. Grain colour at maturity (colour of dry seeds on the plant)
16. Grain shape (Oblong, oval, spherical)
17. Pigmentation on the stem (at the time of flower initiation)
18. Volatile oil content (Percentage of seed)
19. Grain breakability (0, least breakable to 5 most breakable),

Disease and insect incidence

1. Powdery mildew (0, resistent to 9, most susceptible)
2. Blight (0, resistant to 9 most susceptible)
3. Wilt (per cent age of plant killed)
4. Aphids (0 resistant to 9 most susceptible)

- ✰ Stage I: appearance of first compound lead to beginning of bolt formation.
- ✰ Stage II: Beginning of bolt formation till elongation starts.
- ✰ Stage III: Initiation of elongation to complete differentiation of main shoot into nodes and internodes and appearance of flowers.
- ✰ Stage IV: this flowering stage with the beginning of initiation of flowering and terminates with the initiation of grain formation.
- ✰ Stage V: this stage occurs as the flowering period is complete and development of grains starts. It terminates as soon as grains in main umbels are turning brown.

The five stages are depicted in the Figure 5.3. Evaluation of germplasm has indicated that there is wide variation with regard to the duration of each of the stages as can be seen in Table 5.6. Leaf structure and pattern changes as coriander plant grow. Further, Diederichsen has shown that the leaf shape and size vary across continents as shown in Figure 5.4 and it should be possible to distinguish the origin of an accession simply by observing the shape of the leaves.

Table 5.6: Range and Among a Set of 50 Genotypes for Different Growth Stages in Coriander

Characters	*Range*	*Mean±SEm*
Ground stage duration	28.00–30.67	29.24±0.89
Juvenile stage duration	27.33–40.00	32.80±3.15
Bolting duration	11.67–27.67	19.57±0.75
Flowering duration	36.67–49.33	40.89±3.46
Grain formation duration	16.33–23.00	19.29±1.47
Maturity duration	2.67–8.33	4.00±0.39

If we look at the different leaves of an average coriander plant at stage II (Figure 5.5), the leaves can be divided into three types- the seedling leaves, the basal leaves and the meristem. Of these, maximum variation is found in the leaves 3rd to 5th. The 6th and 7th leaves are still expanding. Thus it seems ideal to use 3rd to 5th leaves to describe the variation in the genotypes.

Evaluation of germplasm collection available at S K N College of Agriculture, Jobner has shown variation both for size, shape and the number of leaflets. Based on these studies, two types of classifications were made. The first based on number of leaflets. Four distinct classes can be made in this regard as shown in Figure 5.6. The second is based on the serration of the laminar edge. Three distinct types were identified and these are depicted in Figure 5.7.

The next important character useful for characterization is the shape of plant. Generally three plant types are recognized, these are prostrate, semi erect and erect types (Figure 5.8). The prostrate types are generally leafy types and make good as leaf yielders (Figure 5.9). These prostrate types maintain their leafiness even when the plant enters stage III and later. Thus these types give bushy appearance later. Hence, it makes sense to classify the germplasm according to this trait. The composition of germplasm entries at S K N College of Agriculture, Jobner based on this character are given in Figure 5.10.

Another important character can be the arrangement of umbellets in umbel. Two types are recognized- compact (Figure 5.11) and loose (Figure 5.12). Some variation is observable in between these two extremes, yet with little practice, the entries are easily identified. In some entries, variation in

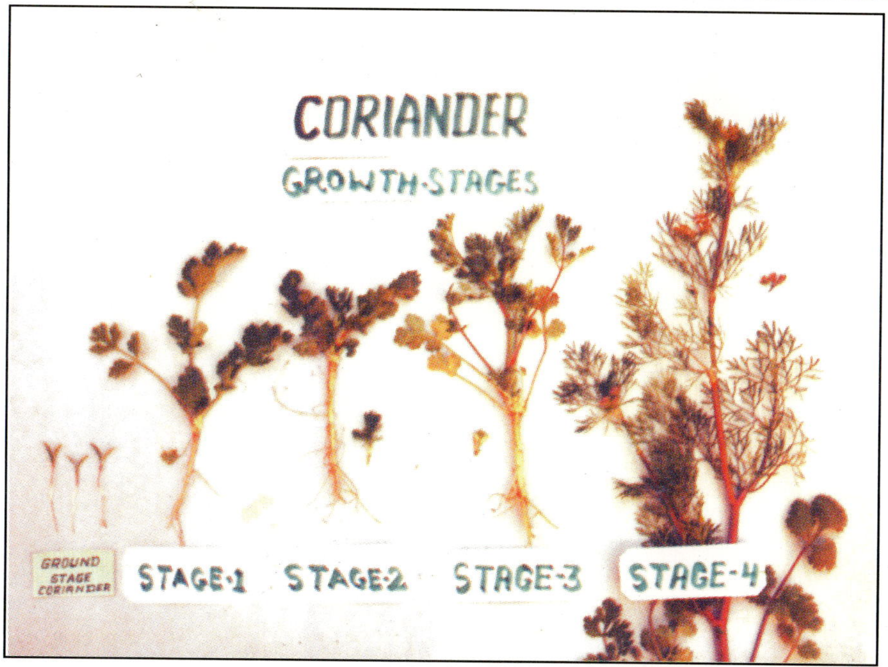

Figure 5.3: Growth Stages in Coriander
(From Choudhary, 1987)

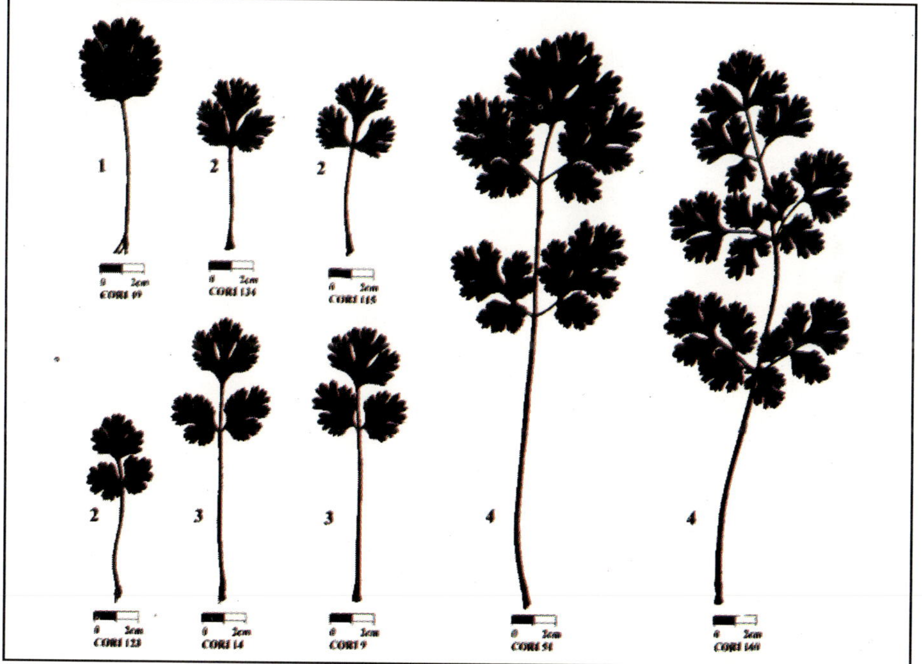

Figure 5.4: Leaf Shape Variation in Different Accessions
From top left, first row, (from Diederichsen, 1997). Place of origin: Cori 49- Spain, Cori 134-Syria, Cori-115-Sudan, Cori 123- Bhutan, Cori-14- Germany, Cori-9- China, Cori 51- Gerogia, Cori- 160- Kasachstan)

Figure 5.5: Different Leaves at Stage I in Coriander

the umbels of different branches can be seen. Hence it is ideal to concentrate on the main umbel only for classification. At our centre we also have a collection, which has no umbellets, only one umbel. This is named as rosette type (Figure 5.13). This is an extreme type. The composition of germplasm lines based on umbel type is given in Figure 5.14.

Figure 5.6: Classification of Leaf Types Based on Number of Leaflets Left to right- single, three and five leaflets. The difference between the third and the fourth is the length of the peduncle between the terminal and middle leaflets.

Different authors have also suggested using the anthocyanin in the petals as well as the stem. We however feel that the expression of anthocyanin is more environment dependent. Entries differ in the expression of the colour. For example even those entries which do not show pigmentation in normal conditions may show pigmentation in stress. This has also been reported by Diederichsen and Hammer (2003). Although it is possible to classify the entries based on presence or absence of Anthocyanin in the petals, it is not a stable character, as true white types are rare and some shade of anthocyanin is always present. With a little observation, the genotypes can be classified based upon these observations.

Similarly there is variation in fruit shape and size. The small seeded types which are described as *C. sativum* var. *microcarpum* are found in limited scale in Kota district of India. Evaluation of germplasm at S K N College of Agriculture, Jobner however, has indicated the differences with regard to seed size. Broadly five types are recognized – small round, small ovate, medium round, medium ovate and large. These are depicted in Figure 5.12.

Figure 5.7: Differentiation Based on Laminar Surface
Left to right- light, medium and deep serration.

Essential oils of coriander are very important and are also the main character used to assess the quality. Higher volatile oil is desirable. The volatile oil is contained in shizogenous channels (vittae) in the seed coat (testa), while the fixed oil is part of the endosperm. The volatile oil content and chemical composition of the volatile oil are highly genotypically determined and some forms are also present without any volatile oil in the fruits (Diederichsen 1997). The fixed oil content is largely influenced by environment and therefore not useful for the characterization or evaluation. It is an established fact that the Indian have less of oil in comparison to European types. The oil content in itself is dependent on environment.

In conclusion, it is stressed that the genotypic classification for characterization should be based on plant type, leaf type, umbel type. These should be included in the descriptors.

Evaluation

Evaluation for morphological characters have been conducted by several authors and the results vary according to the environment and the material used in evaluation.

Figure 5.8: Plant Types in Coriander
Left to right- erect, semi erect and prostrate.

Evaluation of over 300 accessions of coriander including 26 exotic ones at Jobner revealed significant variability for days to flowering, branches/plant, plant height, umbels/plant, 1000 seed weight, seed yield/plant and volatile oil content (Sastry and Sharma, 2001) (Table 5.7). A wide range of variation was observed for all the morpho-physiological characters. Heritability and genetic advance was moderate to high for days to flowering, plant height, umbellets/plant and seed yield/plant. For other characters like branches/plant and umbels/plant, heritability and genetic advance both were low.

Table 5.7: Range, Coefficient of Variation (CV), Heritability (h^2) and Genetic Advance (GA) in Coriander

Characters	*Range*	*CV*	h^2	*GA*
Days to Flowering	3.8-7.3	16.29	90.13	30.0
Branches/Plant	39.9-110.8	16.17	23.78	7.9
Plant Height (cm)	12.2-40.1	29.91	58.18	35.9
Umbels/Plant	19.6-134.0	44.22	8.23	7.5
Umbellets/Plant	19.6-134.0	52.77	19.59	21.3
1000-seedwt. (gm)	5.2-18.5	22.70	94.58	5.2
Seed yield/Plant (g) Volatile oil content (per cent)	0.88-5.4 0.17-0.26	54.45	19.01	21.3

Figure 5.9: Prostrate Type Coriander in Field Condition

Multivariate analysis of characters of 200 accessions showed that maximum genetic diversity among the accessions was contributed by crop duration followed by-plant height, branches/plant and seed weight (Gupta, 1982, Sharma, 1984). Sharma (1982) observed significant genetic divergence in germplasm even in seed characters, for example, seed weight, seed volume and volatile oil content were considered along with seed and straw yield.

Studies on association of different characters with yield, revealed that yield/plant was positively associated with umbels/plant, umbellets/plant and plant height (Sevda, 1980). Similarly. Gupta (1982) and Sharma (1984) observed that branches/plant, days to flowering, umbels/plant, seed/umbel was positively correlated with seed yield and negatively correlated with 1000-seed weight. Path coefficient analysis of the association by Sharma (1984) showed that umbellets/umbel, umbels/plant: branches/plant and 1000-seed weight had high direct effect on seed yield and hence must be considered while making selections.

Further, Sastry *et al.* (1989) noted that expression of yield as well as related characters was significantly affected by the environment. Obviously, the germplasm accessions need to be evaluated over different environments and/or locations before these are finally documented.

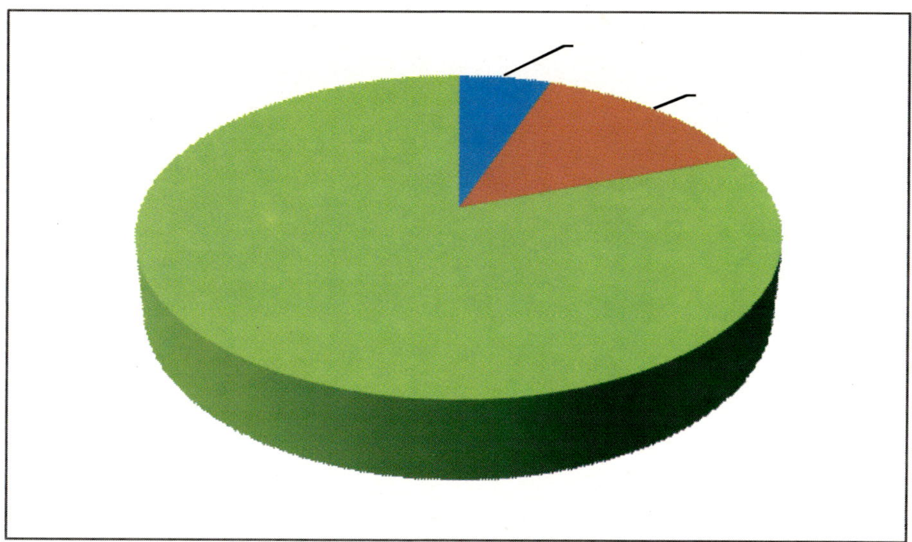

Figure 5.10: Composition of Coriander Germplasm Based on Plant Types at S K N College of Agriculture, Jobner

Figure 5.12: An Example of Loose Type Umbel in Coriander

Evaluation for Biotic and Abiotic Resistance

All the four major spices are affected by a myriad of pest and diseases. The evaluation for biotic and abiotic resistance in seed spices has not yielded any spectacular results. Fungal diseases are an important group affecting spices.

Abiotic Stresses

The seed spices are traditionally grown on low input management conditions. As has already been pointed out earlier, this may be reason that these crops are more adapted to survival than high yield. While all these spices are adapted to moderate temperatures, they are very susceptible to cold and frost. Evaluation of coriander and fennel germplasm has indicated that none of the entries are tolerant to frost (Figures 5.13). Coriander crop is particularly vulnerable to frost during flowering and early seed formation (Pundir and Dahama, 1996). Screening of coriander germplasm under limited moisture conditions has indicated variability with regard to tolerance to drought in coriander (Table 5.8). This indicates possibility of selection of sources tolerant to different abiotic stresses. Infact some of sources have been identified promising for various abiotic stresses (Table 5.9).

Figure 5.13: Rosette Type Umbel in Coriander

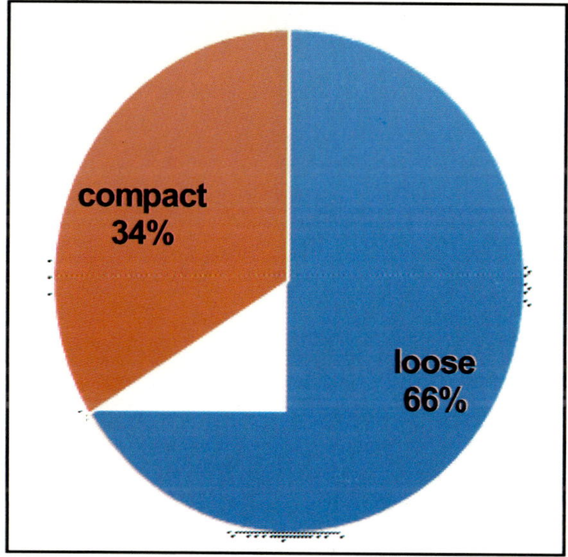

Figure 5.14: Classification of Germplasm Lines Based on Umbel Type

Biotic Stresses

Seed spices are affected by a number of fungal diseases and pests. Most important among them are blights, wilts and mildews. Screening of germplasm has indicated that though variability is found for the resistance against diseases, yet this is partial and the stability of the reaction is questionable. Most of the screenings are based on field reaction.

Figure 5.15: Different Types of Fruit Types in Coriander
Left to right: small round, small ovate, medium round, medium ovate and large.

Figure 5.16: Frost Damage in Coriander

Table 5.8: Stress Tolerance Index (STI) and Stress Tolerance (TOL) of Different Genotypes for Seed Yield (From Anonymous, 2007-08)

Genotypes	*Yield Under Normal Condition (Y_p)*	*Yield Under Limited Moisture Condition (Y_d)*	*Stress Tolerance Index (STI)*		*Stress Tolerance (TOL)*		*Overall Rank*
			STI	*Rank*	*TOL*	*Rank*	
UD-728	37.00	33.33	0.90	5	3.67	5	6
UD-796	46.67	35.00	0.75	13	11.67	14	15
UD-797	51.67	30.00	0.58	16	21.67	17	18
J.Cor-340	56.67	31.33	0.55	17	25.34	19	19
J.Cor-375	50.00	41.00	0.82	9	9.00	10	10
DH-206	48.33	38.00	0.79	10	10.33	13	13
K-Selection	31.67	28.33	0.89	6	3.34	4	7
ND Cor-2	36.33	32.33	1.34	1	40.00	6	5
ICS-1	38.33	35.00	0.91	4	3.00	3	4
LCC-170	29.00	22.33	0.77	12	6.67	8	12
LCC-212	41.00	25.00	0.61	15	16.00	16	17
RD-154	35.00	30.00	0.86	7	5.00	7	8
Hissar Anand	35.00	32.00	0.91	4	3.00	3	3
RCr-20	44.00	36.67	0.83	8	7.33	9	9
RCr-435	48.33	35.33	0.73	14	13.00	15	16
RCr-436	40.00	15.00	0.38	18	25.00	18	19
RCr-446	38.00	28.33	0.75	13	9.67	12	14
RCr-480	44.33	41.67	0.94	3	2.66	2	2
RCr-684	43.33	34.00	0.78	11	9.33	11	11
Local	31.67	30.67	0.97	1	1.00	1	1

STI: $[(Y_p) \times (Y_D)/(Y_p)^2]$; TOL: $(Y_p - Y_D)$

Where,

Y_P: Yields of genotype under normal condition and Y_P: Yields of genotype under limited moisture condition.

Table 5.9: Promising Accessions for Important Characteristics of some Seed Spice Crops (From Gangopadhyay, Yadav and Karihaloo, 2005)

	Important Character	*Promising Accessions*
Coriander	Wilt resistance	ACr-01-250, ACr-01-256, GCr-1
	Stem gall resistance	ACr-01-250, ACr-01-256, RCr-684
	Drought tolerance	RCr-20
	Dual Purpose (Leaf and Seed)	ACr-01-256

Revised Descriptor List for Coriander

Based upon the enumerations presented above it is time we relook at the available descriptor lists and propose amendments. Three descriptor lists are available with us for consideration and adaptation.

1. Descriptor list according to Anonymous 1983
2. Descriptor list according to Malhotra and Vashistha (2006)
3. Descriptor list according to Diderichsen and Hammer (2003).

Of these, as has already been mentioned else where, the first and second are either based on evaluation traits or more arbitrary in classification. The third one is extensive, yet lacks some of the traits that may help in characterization.

We propose the following changes:

Character	*Scale*
Characterization Data	
Plant type (to be recorded after one month of sowing)	1. Prostrate 2. Semi erect 3. Erect 4. Others
Blade shape of the third leaf	1. Entire 2. Medium serration 3. Deeply serrated or lobed
Number of leaflets (3rd leaf)	1. Single leaflet 2. Three leaflets 3. Five leaflets 4. Multiple leaflets
Early growth vigour	Scale of 1 (poor) to 5 (vigorous)
Anthocyanin pigmentation on stem (at the time of flowering)	Present Absent
Inflorescence	Compact Loose
Pigmentation of the petals	Absent Present
Size of fruits	Small Medium Large
Shape of fruits	Round Ovate

Character	Scale
Evaluation Data	
Days to 50 per cent flowering	Number
Days to 50 per cent maturity	Number
Number of leaves (at the start of flowering)	Number
Number of umbels	Number
Umbellets (0nly on main umbel)	Number
Flower whorls in an umbel (main umbel)	Number
Plant height	Cms
Primary branches per plant	Number
Secondary branches per plant	Number
Grains per umbel (main umbel only)	number
Test weight (1000 seeds)	Weight in grams
Volatile oil content in air dried seeds	Percentage
Fixed oil content in air dried seeds	Percentage
Splitting tendency of the mericarps	Scale of 1 (hard) to 5 (easy)
Seed yield per plant	Weight in gms
Biotic and abiotic stresses	
Powdery mildew	Scale of 0 (resistant) to 9 (highly susceptible)
Blight	Scale of 0 (resistant) to 9 (highly susceptible)
Wilt	Based on percentage plants killed in a plot
	Scale of 0 (resistant) to 9 (highly susceptible)
Aphids	Scale of 0 (resistant) to 9 (highly susceptible)
Frost tolerance	Based on percent plants affected in a plot on the third day of occurrence of frost
	Scale of 0 (resistant) to 9 (highly susceptible)

References

Alefeld, Fr., 1866. *Landwirthschaftliche Flora.* Wiegandt and Hempel, Berlin.

Anandraj, M., Parthasarathy, V.A., Shiva, K.N. and Johny, A.K., 2007. Research and development in major seed spices under AICRP spices. In: *Production, Development, Quality and Export of Seed Spices,* (Eds.) S.K. Malhotra and B.B. Vashistha. NRCSS, Ajmer, pp. 370–395.

Anonymous, 1983. *Proceedings of the Sixth Worshop, All India Co-ordinated Spices and Cashewnut Improvement Project.* Central Plantation Crops Research Institute, Kasargod.

Anonymous, 2007–08. *Annual Report, All India Co-ordinated Spices Improvement Project.* Agric. Expi. Station, S.K.N. College of Agriculture, Rajasthan Agricultural University, Jobner.

Choudhary, Chunni Lal, 1987. Association of developmental characters with grain yield in coriander (*Coriandrum sativum* L.). *M.Sc. Thesis,* Sukhadia University, Udaipur.

Danert, S., 1962. Über Gliederungsprobleme bei Kulturpflanzen. *Kulturpflanze,* 10: 350–358.

De Candolle, A.P., 1830. Prodromus systematis naturalis regni vegetabilis. *Treutel et Wu¨rtz*, Paris, 4: 249–250.

Diderichsen, A., 2004. Case studies for the infraspecific classifications in managing germplasm collections of cultivated plants. In: *Proc. XXVI IHC–IVth Int. Symp. Taxanomy of Cultivated Plants*, (Eds.) C.G. Davidson and P. Trehane. Acta Hort 634, ISHS 2004, pp. 127–139.

Diderichsen, A. and Hammer, Karl, 2003. The infraspecific taxa of coriander (*Coriandrum sativum L.*). *Genet. Resor. Crop Evol.*, 50: 53–63.

Diederichsen, A., 1977. Charakterisierung und Evaluierung von Koriander (*Coriandrum sativum* L.) und taxonomische Implikationen. Disseratation zur Erlangung des Doktorgrades des Fachbereichs Agrarwiessenschaften (Landwirtschaftliche Fakultät Göttingen. University of Gottingen, Germany.

Diederichsen, A., 1996. Coriander (*Coriandrum sativum* L.). promoting ıhe conservation and se of underutilized and neglected crops. 3. Institute of Plant Genetics and Crop Plant Research, Gatersleben/International Plant Genetic Resources Institute, Rome.

Diederichsen, A. and Hammer, Karl, 2003. The infraspecific taxa of coriander (*Coriandrum sativum* L.). *Genetic Resources and Crop Evolution*, 50: 33–63.

Gangopadhyay, K.K., Yadav, S.K. and Karihaloo, J.L., 2005. Biodiversity and its utilization in onion, garlic, chillies and seed spices in improvement programme. In: *SYMSACII: National Symposium on Current Trends in Onion, Garlic, Chillies and Seed Spices: Production, Marketing and Utilization*, 25–27 November, Souvenir, pp. 33–47.

Gupta, A., 1982. Genetic divergence of the germplasm of coriander (*Coriandrum sativum* L.) on the basis of morphological characters and grain yield. *M.Sc. (Ag.) Thesis*, University of Udaipur, Udaipur.

Gupta, Kaushalya, Thakral, K.K., Arora, S.K. and Wagle, D.S., 1991. Studies on growth, structural carbohydrate and phytate in coriander (*Coriandrum sativum*) during seed development. *Journal of the Science of Food and Agriculture*. 54(1):43–46.

Harlan, J.R. and de Wet, J.M.J., 1971. Toward a rational classification of cultivated plants. *Taxon*, 20: 509–517.

Harrod, D.C., 1960. A note on coriander of commerce. *J. Phar. Pharmacol.*, 18: 245–247.

Hazekamp, T. and Guarino, L., 1999. Taxonomy and the conservation and use of plant genetic resources. In: *Taxonomy of Cultivated Plants*, (Eds.) S. Andrews, A.C. Leslie and C. Alexander, 3rd Intl. Symposium, Royal Botanical Gardens, Kew, p. 11–17.

Ivanova, K.V. and Stoletova, E.A., 1990. The history of culture and intraspecific taxonomy of *Coriandrum sativum* L. Cf Diederichsen, Axel. 1996.

Laufer, B., 1919. *Sino-Iranica*. Field Museum of Natural History Publication 201, Anthropological Series. Blackstone Expedition, Chicago, 15(3).

Malhotra, S.K. and Vashishtha, B.B., 2006. A manual on minimal descriptors of seed spices crops (for characterization an evaluation). *NNRCSS*, Ajmer.

Mansfeld, R., 1962. U¨ber "alte" und "neue" Systematik der Pflaninzen. *Kulturpflanze, Beiheft*, 3: 26–46

Mansfeld, R., 1954. Zur allgemeinen Systematik der Kulturpflanzen II. *Kulturpflanze*, 2:130–142.

Palanikumar, M., 2008. Evaluation of coriander (*Coriander sativum* L.) genotypes for biomass yield, quality and molecular characterization. *Ph.D. Thesis,* Tamil Nadu Agricultural University, Coimbatore.

Pedro, A., López, Mark, P. Widrlechner, Philipp W. Simon, Satish Rai, Terri D. Boylston, Terry A. Isbell, Theodore B. Bailey, Candice A. Gardner and Lester A. Wilson, 2008. Assessing phenotypic, biochemical and molecular diversity in coriander (*Coriandrum sativum* L.) germplasm. *Genetic Resources and Crop Evolution*, 55(2): 247–275.

Pundir, P. and Dahama, A.K., 1996. *Seed Spices Research in Rajasthan*. Rajasthan Agricultural University, Bikaner.

Roberts, E.H., 1975. Problems of long-term storage of seed and pollen for genetic resources conservation. In: *Crop Genetic Resources for Today and Tomorrow*, (Eds.) O.H. Frankel and J.G. Hawkes. Cambridge University Press, Cambridge, England.

Sastry, E.V. Divakara, 2007. Genetic erosion and germplasm conservation in spices. In: *Souvenir and Abstracts National Symposium to Spices and Aromatic Crops Industry*, (Eds.) R. Dinesh, K.N. Shiva, D. Prasath, K. Nirmal Babu and V.A. Parthasarathy. Indian Society for spices and aromatic crops industry. Indian Society for spices, Calicut, November, pp. 105–127.

Sastry, E.V. Divakara and Sharma, R.K., 2001. Status of seed spices research in Rajasthan. In: *Seed Spices: Production, Quality, Export*, (Eds.) S. Agrawal, E.V. Divakara Sastry and R.K. Sharma. Pointer Publishers, Jaipur, pp. 19–50.

Sastry, E.V.D., Singh, D., Sharma, K.C. and Sharma, R.K., 1989. Stability analysis in coriander (*Conundrum sativum* L.). *Indian J. Genet.*, 49(2): 151–153.

Schultze-Motel, J., 1968–1987. Literatur u¨ber archa¨ologische Kulth turpflanzenreste. *Kulturpflanże* 16 (1968): 215–230; 24 (1976): 159–178; 29 (1981): 447–463; 35 (1987): 401–420.

Schultze-Motel, J., 1993, 1994. Literature on archaeological remains of cultivated plants. *Vegetation History and Botany*, 2(1993): 47–59, 3 (1994): 33–61.

Sevda, P.L., 1980. Study on variability and correlation in local collection of coriander (*Conundrum sativum L.*). *M.Sc. (Ag.) Thesis*, University of Udaipur, Udaipur.

Shands, H., 1995. The US National Plant Germplasm System. *Can. J. Plant Sci.*, 75: 9–15.

Sharma, R.K., Bhati, D.S. and Bhatt, B.N., 2001. *Masalon ki Kheti*. Indian council of Agricultural Research, New Delhi.

Sharma, K.C., 1984. Correlation and path-coefficient analysis in coriander (*Coriandrum sativum* L.). *M.Sc. (Ag.) Thesis*, Sukhadia University, Udaipur.

Sharma, M.M., 1982. Genetic divergence in the germplasm of coriander (*Conundrum sativum* L.) II. On the basis of oil content, seed weight, seed volume and yield. *M.Sc. (Ag.) Thesis.*, Sukhadia University, Udaipur.

Sinskaja, E.N., 1969. *Historical Geography of Cultivated Plants (Russ.)*. Koloṣ, Leningrad.

Thomas, T.A. and Mathur, P.N., 1991. Germplasm evaluation and utilization. In: *Plant Genetic Resources Conservation and Management*, (Eds.) R.S. Paroda and R.K. Arora. IBPGR, Regional Office, New Delhi.

Vavilov, N.I., 1926. Centers of origin of cultivated plants. *Trends Pract. Bot. Gener. Sel.*, 16: 3–24.

Vavilov, N.I., 1951. The origin, variation, immunity, and breeding of cultivated plants. *Chronica Botanica*, 13: 1–366.

Vavilov, N.I., 1931. The Linnean species as concept (Russ.). *Tr. Poprikl. Bot. gen. i. sel.*, 26: 109–134.

Vavilov, N.I., 1992. The phytogeographical bias for plant breeding: Studies of the original material use for plant breeding. In: *Origin and Geography of Cultivated Plants*, (Ed.) V.F. Dorofeev. Cambridge University Press, Cambridge, pp. 316–366.

Zohary, D. and Hopf, M., 1993. *Domestication of Plants in the Old World*, 2nd edn. Clarendon Press, Oxford.

Recent Advances in Seed Spices (2011) *Pages* **67–76**
Editors: **Y. Ravindrababu, R.K. Jaiman, K.D. Patel, N.R. Patel and S.B.S. Tikka**
Published by: **DAYA PUBLISHING HOUSE, NEW DELHI**

6

Breeding Strategies in Seed Spices Crops

R.K. Sharma
Spices Breeder, RAU, Bikaner

Introduction

Among the 20 seed spices ten are important. These are grown in different parts of the world covering mainly Mediterranean region, south European and Asian region. The major producing courtiers other than India are, Morocco, Russia and Bulgaria for coriander; Turkey, Iran, Egypt for cumin; Egypt, China, Romania, Russia for fennel; Morocco and Bulgaria for fenugreek; Iran and Egypt for Ajawain; Germany and Hungary for Dill; South France and China for celery; Bulgaria, Cyprus, Germany and Russia for anise; Pakistan, Sri Lanka and Egypt for *Nigella*. Among these countries India is the largest producer of seed spices, which are being cultivated on 7.8 to 8.9 Lac hectares with production of 4.9 to 5.6 Lac tons. Rajasthan and Gujarat are the two largest growing states accounting for 72 per cent and 86 per cent of the total area and production respectively. Other important seed spices growing states are Madhya Pradesh, Orissa, Tamil Nadu, Andhra Pradesh, Karnataka, Bihar, Uttar Pradesh, Punjab and West Bengal contributing lesser area.

Since 1975 when the ICAR included seed spices in the AICRP, significant achievements have been made in the development of varieties, standardization of agronomic practices and disease and insect pest management, the shift or increase in average productivity has remained low. The important reasons are:

- ☆ Cultivation of these crops traditionally on marginal lands with low fertility
- ☆ Lack of sufficient number of improved varieties to suit different agro-climatic situations.
- ☆ Lack of proper adoption of improved varieties, package of practices and control measures for diseases and pests.

- ☆ Inherent weaknesses of the crops *e.g.* poor and slow germination in case of umbelliferous spices resulting in poor stand.
- ☆ Slow initial growth of the crops resulting into severe weed problem.
- ☆ High incidence of disease, *e.g.* wilt, blight, powdery mildew in cumin; wilt and stem gall in coriander, blight and gummosis in fennel and powdery mildew, downey mildew and mycoplasma like organism (MLO) in fenugreek
- ☆ Lack of proper adoption of harvesting and postharvesting technology adversely affecting the quality and
- ☆ Inadequate marketing facilities are other constraints to production and export of spices.

Obviously for an Overall Development in Spices the Need is for

- ☆ High-yielding, disease and pest resistant varieties with better quality attributes.
- ☆ Improved package of agronomic practices, plant protection and postharvest technology conducive to high yield and better quality of produce. For some of the diseases like for cumin blight- a perfect system of disease forecasting has to be developed, if the disease is to be satisfactorily controlled.
- ☆ A well- organised extension programme supported with a sound seed production programme for quick adoption of improved varieties as well as package of practices.
- ☆ A well-organised and regulated marketing system for the spices where facilities for processing, cleaning, grading, packaging and storage are available.
- ☆ Well-planned export strategies as these crops are important mainly due to their export potential.
- ☆ Product diversification keeping in view the demand of importing countries.

Availability of varieties fitting in different agro-climatic situations ensuring acceptable performance is the basic need for successful cultivation of seed spices.

Important strategies for developing high yielding varieties with desired quality and resistance to diseases are:

1. Enhancement of germplasm collection, evaluation, and conservation.
2. Mutation and biotechnological approaches to enhance the variability.
3. Population improvement vis-à-vis selection
4. Heterosis breeding
5. Breeding for disease resistance
6. Improvement of quality
7. Breeding for performance stability and responsiveness to agronomical inputs.
8. Breeding varieties for mixed cropping

For an effective application of breeding strategies a thorough knowledge of the botanical characteristics of plants, their genetics and association with each other and the seed yield is necessary. Likewise, breeding objectives should be thoroughly spelled and listed priority-wise. A brief description of botanical characteristics of these crops and the list of important objectives follows:

Botanical Description

Coriander

Corriandrum sativum L. (2n = 22) belongs to family Umbelliferae (Apiaceae). Purseglove *et al.* (1981) have given a detailed botanical description of the plant. There are two distinct morphological types: one erect and tall with comparatively stronger main shoot and shorter branches, the other bushy with relatively weaker main shoot and longer, spreading branches. The plants attain heights from 30-100cm, depending upon the variety. The crop comes to bloom in 45 –60 days after sowing and matures in 65-120 days, depending upon variety and cropping situation. Each branch as well as the main shoot terminates in a compound umbel (determinate growth) bearing 3-10 umbellets, each umbellet containing 10-50 pentamerous flowers. The flowers are small, protandrous and difficult to manipulate for controlled pollination. Like other umbelliferous plants, coriander is also a cross-pollinated crop. The degree of cross-pollination has been reported to range from 50 per cent (Ramanujam *et al.*, 1964) to 60 per cent (Dimri *et al.*, 1977). Anuradha Hore (1979) considered poor seed set as a major constraint to yield. Pillai and Nambiar (1982) considered coriander to be andro-monocious. Singh and Ramanujam (1973) reported significant varietal differences in distribution of male and perfect flowers in the umbels. Hermaphrodite flowers opened earlier than males. Selection of a higher proportion of hermaphrodite flowers was considered an effective criterion for higher seed set.

Cumin

Cuminum cyminum L. (2n=14) belongs to the family umbelliferae. It is also considered to be native of Mediterranean region. It is largely cultivated in India, Pakistan, Iraq, Turkey, China, Egypt and Israel etc. Vedmuthu *et al.* (1994) gives a brief description of its botanical characteristics. Cumin plants are delicate much branched, slow growing, herbaceous, attaining an average height of 20-25 cm. The flowers are small, pink rarely white, born in small compound umbels. The plants come into bearing in 60-90 days from sowing. The flowers are so small that artificial emasculation and cross-pollination is extremely difficult. As the artificial hybridization could not be done, the genetic studies are limited only to the study of segregating generations of chance hybrids between pink and white flowers and between corymbose (UC-198) and regular branching type. The study of such crosses revealed that flower color is a single gene controlled character, pink being dominant over white. Similarly, corymbose branching dominant over regular but number of genes could not be ascertained.

Fennel

Foeniculum vulgare Mill. (Syn. *F. officinale*, 2n=22) belongs to family umbelliferae. Two distinct sub species are recognized in *Foeniculum vulgare*. One is *piperitum* (Holmboe) which is a wild type and other one is *copillaceum* (Gillib) with three varieties namely, var. *vulgare* (Miller) Thellung-bitter fennel, Var. *dulce* (Miller) Thellung-sweet fennel, Var. *ozoricum* (Miller) Thellung- culinary florence fennel.

Indian fennel is regarded as distinct variety of *F. vulgare* (Miller) Thellung var. *Panmorium*. It is native of South Europe-Mediterranean region. It is mainly cultivated in Rumania, USSR, Germany, France, Italy, Japan, Argentina, and USA besides India. Fennel plants are stout moderately branched, slow growing in the beginning, attaining an average height 1.0-1.6 M. The flowers are small, distinctly protandrous, with light yellow color. Ramanujam *et al.* (1964) considered fennel to be highly cross pollinated. The plants come into bearing in 80-110 days depending upon variety and date of sowing. In fennel removal of anthers before the stigma matures can be easily done. This facilitated artificially controlled cross-pollination. The technique is being extensively utilized in making controlled crosses at SKN college of Agriculture, Jobner centre of AICRP on spices.

Like other seed spices fennel has also remained almost neglected from genetic studies point of view. The only character, inheritance of which has been reported is stem color. Ramanujam *et al.* (1964) reported that a single gene controls the stem color with green color being dominant over bluish green color. The studies on germplasm collections carried out at different centres of AICRP on spices, and others like those of Bhargava *et al.* (1971), Ramanujam and Joshi (1961), Daniel *et al.* (1962) revealed that in fennel variability was moderate to high for umbels/plant, seeds/umbel, yield/plot and volatile oil content. Variability range was however, narrow for days to flowering, plant height, branches/plant, umbellates/umbel, test weight. None of the varieties, screened under natural incidence of the disease like *Alternaria* and *Rumularia* blights and gummosis, was found to be resistant to these diseases.

Fenugreek

Fenugreek or Methi (*Trigonella foenum-graecum* L. 2n=16) belongs to sub-family *papilionaceae* of *Leguminosae*. It is believed to have originated in South-East Europe or South-West Asian region. It is largely cultivated in Argentina, Egypt, Morocco, Algeria, Ethiopia, and Lebanon besides India. In India its cultivation is concentrated mainly to Rajasthan State which contributes 80 per cent of the total area as well as production. Other states cultivating fenugreek are Gujarat, Tamil Nadu, U.P., Himachal Pradesh, Madhya Pradesh and Andhra Pradesh.

Fenugreek plants are weak, spreading and moderately branched, attaining a height 30 to 50 cm. The flowers are small borne in raceme, and extremely difficult to manipulate for artificial emasculation and pollination. It is self-pollinated crop. Comes in flowering 30 to 50 days after sowing and matures in 110-140 days. Diosgenin content in fenugreek seed, which has been estimated to be 0.78 to 1.00 per cent by Bhavsar *et al.* (1980).) 0.02 to 0.25 per cent by Gupta and Pareek (1981) and 0.40 to 1.26 per cent by Banyai (1973). Such wide differences among studies were attributed to method of estimations yet it was concluded that Algerian and Moroccan types possesses more diosgenin content than the Indian material.

Dill

Dill denotes two related species namely, *Anethum graveolens* L. (Syn. *Peucedanum graveolens*, 2n=22) or European dill and *Anethum sowa* Kurk. 2n=22, Indian dill, belonging to the family *umbelliferae* which are under cultivation in India. Both species have originated in Eurasia and are being extensively grown in England, Finland, Germany, Rumania and USA besides India. In India *Anethum sowa* is commonly grown in un-irrigated conditions while *Anethum graveolens* is cultivated under irrigated conditions in the states of Rajasthan, Madhya Pradesh, Gujarat and Andhra Pradesh. In pharmaceuticals oil of *A. graveolens* is preferred over that of *A. sowa* due to presence of larger percentage of dillapiole and lower percentage of carvone. European dill normally possesses higher percentage of volatile oil than Indian.

Dill plants are stout, erect and moderately branched, attaining a height up to 1.0-1.5 m. The flowers are small, distinctly protandrous with creamy yellow color, closely resembling fennel. It is also highly cross-pollinated crop and honey bees effect pollination. Studies on emasculation and pollination have not been attempted. The crop has remained neglected from genetic studies point of view. Limited studies conducted at Nagla Research Farm (CIMAP) in Tarai area (Gupta, 1982) reveals that the exotic dill entries introduced from European countries, USA and USSR adapts well under Tarai conditions and revealed useful variability in umbels/plant, seeds/umbel, seed yield/plant.

Number of umbels was positively associated with yield. Systematic work on collection, maintenance and evaluation of indigenous germplasm of dill has been started recently at NRC for seed spices TABIJI.

Ajawain

Ajawain (*Carum Copticum* Hiern Syn. *Trachyspermum ammi* (L) Sprague, 2n=18 of the family umbelliferae is believed to have originated in Mediterranean region. It is cultivated as a spice as well as medicinal crop in Iran, Egypt, and Afghanistan, besides India. In India it is cultivated as rainfed crop in Rajasthan, MP, Gujarat, U.P., Andhra Pradesh, Bihar, Bengal and Tamil Nadu. Chittor and Nimbahera area of Rajasthan and adjoining area of Madhya Pradesh (Indore, Mandsor, Ujjain and Gwalior) is the main ajawain growing pocket of India.

Ajwain plants are bushy and highly branched attaining height upto 30 to 50cm. The flowers are small extremely difficult to manipulate for controlled emasculation and pollination. Like other umbelliferous plants ajawain is also a cross-pollinated crop. Ramanujam *et al.* (1964) reported cross-pollination to the extent of 70 to 77 per cent. Protandrous nature of flowers and stigmatic secretions, which attract insect pollinators, promotes the cross-pollination. Sethi (1979) studying limited collection of germplasm revealed that genetic variability exists for yield and yield related characters. This has also been supported by the evaluation trials conducted at centres of AICRP on spices in Andhra Pradesh as well as Gujarat. Systematic work on collection maintenance and evaluation of germplasm and identification of promising material has only been started recently at NRC for seed spices TABIJI. Work on other minor spices is meager.

Status of Genetic Variability in Seed Spices

Crop	*Low*	*Medium to High*
Coriander (683)	✰ Volatile oil ✰ Reaction to disease	✰ Morpho-physiological characters ✰ Seed Size ✰ Yield
Cumin (265)	✰ Volatile oil% ✰ Seed yield ✰ Reaction to diseases ✰ Morphophysiological characters	—
Fennel (162)	✰ Crop duration ✰ Plant height ✰ 1000 seed wt. ✰ Reaction to diseases	✰ Seed/umbel ✰ Yield ✰ Volatile oil%
Fenugreek (308)	✰ Reaction to diseases ✰ Seed Size ✰ Crop duration	✰ Days to flowering ✰ Plant height ✰ Branches/plant ✰ Pods/Plant ✰ Seed Yield

Breeding Objectives and their Priorities–Cropwise

Cumin

1. Resistance to blight
2. Resistance to powdery mildew
3. Breeding for high yield
4. Volatile oil content
5. Seed size, colour and luster

Coriander

1. High yield
2. Volatile oil content
3. Powdery mildew resistance
4. Stem gall resistance
5. Seed size, shape and luster

Fennel

1. Short duration
2. Smaller Height
3. Size of umbels – large
4. Resistance to blight and gummosis
5. High yield
6. Volatile oil content

Specific-Earliness and short duration for mixed cropping.

Fenugreek

1. High yield
2. Resistance to powdery and downey mildew
3. Diosgenin content
4. Early-ness

Likewise the priorities for other crops may be worked out and listed

Breeding Strategies

1. Enhance Germplasm Collection, Evaluation and Conservation

A review of the studies of genetic resources of seed spices (Sharma, 1994) indicates that variability in the germplasm particularly for yield, quality (Volatile oil content) and reaction to diseases was low in all the seed spices. For other characters viz plants height, duration etc it was medium to high in coriander and fennel; medium in fenugreek. But in cumin it is low even for these characters obviously there is an urgent need to enhance the germplasm base to successfully support the crop improvement

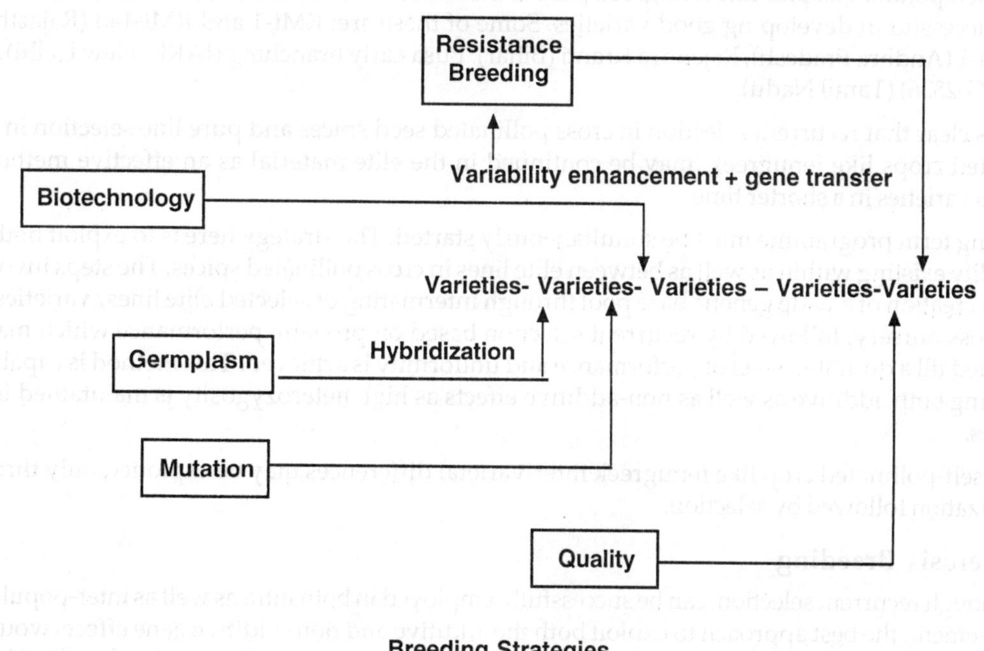

Breeding Strategies

programme. Any delay in this programme may cause an irretrievable loss. Valuable genes so far existing in the land varieties which may be pushed out by improved varieties.

2. Mutation and Biotechnological Approaches to Enhance Variability

Mutations, induces with gamma rays, are useful in creating desirable variability in fenugreek and coriander (Sharma *et al.*, 1996). Noteworthy achievement is a RMt-1 fenugreek mutant, UM. 305, which is resistant to powdery mildew and determinant in growth habit with high yield potential besides 4 others (UM 301 to 304) resistant to powdery mildew and possessing higher yield potential. In coriander useful mutation lines namely 10 Kr-15, 5 Kr.-14, 20 Kr.-5 and 5 Kr.-68 have been derived. However, the results with cumin have not been so encouraging and in fennel and other crops the approaches are yet to be tried. Biotechnological approaches such as somaclonal variations and production of transgenic lines and molecular characterization, at least of the elite germplasm needs to be immediately taken up, to facilitate their use in crop improvement programme.

3. Population Improvement vis-à-vis Selection

Population improvement is the process of pyramiding of the positive genes for desirable characters in a variable population through selection or recurrent selection. The appropriate strategy in umbelliferous (cross-pollinated) spices would be to have short term and long term improvement programme. The short term programme would aim to improve the elite populations through recurrent selection based on the performance of individual plant progenies or even mass-selection. This approach has produced a number of good varieties in coriander, cumin, fennel, ajawain and sowa.

In self pollinated spice like fenugreek pure line selection in the elite variable material has been quite successful in developing good varieties. Some of these are: RMt-1 and RMt-143 (Rajasthan), Lam-sel-1 (Andhra Pradesh), Rajendra Kranti (Bihar). Pusa early branching (IARI – New Delhi), and Co-1 (TG-2336) (Tamil Nadu).

It is clear that recurrent selection in cross pollinated seed spices and pure line selection in self-pollinated crops like fenugreek, may be continued in the elite material as an effective method to develop varieties in a shorter time.

Long term programme must be simultaneously started. The strategy here is to exploit both the variability existing within as well as between elite lines in cross pollinated spices. The steps involved include creation of a wide genetic base pool through intermating of selected elite lines/varieties in a poly-cross nursery, followed by recurrent selection based on progeny performance which may be continued till a desirable level of performance and uniformity is achieved. This method is capable of exploiting both additive as well as non-additive effects as high heterozygosity is maintained in the varieties.

In self-pollinated crop like fenugreek inter-varietal differences may be exploited only through hybridization followed by selection.

4. Heterosis Breeding

Though recurrent selection, can be successfully employed in both intra as well as inter-population improvement, the best approach to exploit both the additive and non-additive gene effects would be the heterosis breeding. Manifestation of heterosis has been reported in coriander (Singh and Ramanujam, 1972), fennel (Ramanujam and Tiwari, 1970) and further studies may reveal its manifestation in other crops. Search for cytoplasmic – genetic types of system of controlling male sterility should be taken up to make the heterosis breeding a reality.

5. Breeding for Disease Resistance

The seed spices are affected by diseases which reduce the yield level which is already low. Some of the diseases like cumin wilt, root rot in fenugreek and gummosis in fennel have so far evaded the effective control measures. Breeding programmes for resistance to these diseases need to be immediately initiated. Identification of pathogen and/or its races causing the diseases, developing techniques to create uniform and artificial disease epiphytotics, use of tissue culture as an aid to accelerate the resistance screening programme are some of the aspects on which research work needs to be initiated to make the resistance breeding effective. Resistance breeding is also directly related to quality of the produce as it obviates the necessity of application of pesticides which often leaves undesirable residues.

6. Improvement of Quality

Quality of the produce needs special emphasis in seed spices as these are exported to foreign markets where the quality standards are very stringent. Volatile oil content, shape, size and luster of grains and their cleanliness constitute the important quality factors. Appropriate weightage has to be given to these quality attributes in the breeding programmes.

5. Breeding for Performance Stability and Responsiveness to Agronomical Inputs

Yield fluctuations over years as well as locations are very wide. Preliminary information obtained from evaluation of varieties/or germplasm suggested that genetic differences among varieties for stability do exist. Factors responsible for differences need to be specially identified and utilized in breeding programme. In general these crops, particularly the cumin, respond very less to agronomic inputs *e.g.* fertilizers as well as irrigation. Attention therefore, needs to be given on identification of responsive types. The high genotype x environment interaction, noted for these crops, also suggests that breeding programmes may be targeted for specific environment and develop suitable varieties for each condition.

It may be concluded that in seed spices, where genetic variability assembled so far has been inadequate, well planned efforts be taken up to enhance the variability. Besides the systematic collection and maintenance of indigenous germplasm, exotic material particularly from centres of origin and secondary centres of variability must be introduced. Biotechnological approaches and mutation breeding should also be employed for enhancing variability. Since the resistance to disease and improvement in quality has special significance in seed spices, a multi-disciplinary collaborative approach be initiated to achieve the objective of developing high yielding, disease resistant varieties with better quality. Breeding approaches must be refined to make the best use of existing variability. Search for cytoplasmic – genetic type of male sterility or temperature sensitive male sterility and development of inbreds be started as an effort targeted towards exploitation of heterosis.

References

Anuratha, Hore, 1979. Improvement of minor (Umbelliferous) spices in India. *Economic Botany*, 33(3): 290–297.

Banyai, 1973. Botanical and quality studies on ecotypes of fenugreek (*Trigonella foenum-graecum* L.) *Agrobotanica*, 15: 187–194.

Bhargava, P.D., Mathur, S.C., Vyas, H.K. and Anwar, M., 1971. Note on screening fennel varieties against aphid infestation. *Indian J. Agril. Sci.*, 42(2): 90–92.

Bhavsar, G.C., Kapadia, N.S. and Patel, N.M., 1980. Studies on *Trigonella foenum-graecum* L. *Indian J. of Pharmaceutical Science*, 42: 39–40.

Dimri, B.P., Khan M.N.A. and Narayan, M.R. 1977. In cultivation and utilization of Medicinal and aromatic plants. RRL. Jammu. Pp. 182.

Gupta, R. and Chadha, K.L., 1995. Medicinal and aromatic plants research in India. In: *Advances in Horticulture, Vol. II,* (Eds.) K.L. Chadha and Rajendra Gupta. Mehrotra Publ. House, New Delhi.

Purseglove, J.W., Brown, E.G., Green, C.L. and Robbins, S.R.J., 1981. *Spices, Vol. II.* Longman, London.

Ramanujam, S., Joshi, B.S. and Saxena, M.B.L., 1964. Extent and randomness of cross pollination in some umbelliferous spices in India. *Indian J. Genet.*, 24(1): 62–67.

Ramanujam, S. and Tiwari, 1970. Heterosis in fennel. *Indian J. Genet.*, 30(3): 732–737.

Sastry, E.V.D., Singh, D., Sharma, K.C. and Sharma, R.K., 1989. Stability analysis in coriander (*Coriandrum sativum* L.). *Indian J. Genet.*, 49(2): 151–153.

Sethi, K.L., 1979. Genetic diversity in relation of geographic distribution in lorage. *Indian J. Agric. Sci.*, 41.

Sharma, R.K., 1994. Genetic resources of seed spices. In: *Advances in Horticulture, Vol. I,* (Eds.) K.L. Chadha and P. Rethinam. Mehrotra Publ. House, New Delhi, pp. 193–207.

Sharma, R.K., Dashora, S.L., Chaudhary, G.R., Agarwal, S., Jain, M.P. and Singh, D., 1996. *Seed Spices in Rajasthan*, Directorate of Research, RAU, Bikaner, Pub. No. DOR/1996/2.

Singh, V.P. and Ramanujam, S., 1972. Gene action involved in cramocarp yield of coriander. *Indian J. Genet.*, 32: 18–26.

Singh, V.P. and Ramanujam, S., 1973. Expression of andromonoecy in coriander, *Coriandrum sativum* L. *Euphytica*, 22: 181–188.

Vedmuthu, P.G.B., Abdul Khader, M.D. and Rajan, F. Salal, 1994. *Advances in Horticulture, Vol. 9*, (Eds.) K.L. Chadha and P. Rethinam. Mehrotra Publ. House, New Delhi.

Recent Advances in Seed Spices (2011) *Pages* 77–84
Editors: Y. Ravindrababu, R.K. Jaiman, K.D. Patel, N.R. Patel and S.B.S. Tikka
Published by: DAYA PUBLISHING HOUSE, NEW DELHI

7

Seed Spices Improvement through Biotechnological Tools

Ramavtar Sharma
Plant Biotechnology Centre, RAU, Bikaner

Out of the total 20 seed spices grown in the country the nine are prominent. These important seed spices are divided into major and minor group of seed spices. The crops covered as major seed spices are coriander, cumin, fennel and fenugreek whereas, ajawain, dill (sowa), nigella (kalongi), celery and aniseed constitute minor group. Most of the above seed spices belong to family Apiaceae except fenugreek and nigella, which belong to family Fabaceae and Ranunculaceae, respectively and have their centre of origin primarily in Mediterranean region and South Europe.

India is the world's largest producer, consumer and exporter of seed spices. Almost all of the seed spice crops are cultivated in India and almost all the states in India grow one or more seed spices. Yet, states of Haryana, Punjab, M.P., Bihar, U.P., West Bengal, Orissa, Tamil Nadu and Karnataka are important. But major growing belt spread from arid to semi-arid regions covering large area in Rajasthan and Gujarat.

Realizing the importance of seed spices in the country's economy these crops were included in the coordinated projects (AICRP) as early as in the year 1975 and later on NRCSS (year 2000) was established to address the problems of these crops exclusively. Despite the efforts made towards the improvement of spices, desired objectives have not been achieved as is evident from marginal or no increase in productivity of these crops. The progress to develop improved varieties has not been achieved due to a number of factors some of them could be listed as under:

- ✫ Lack of adequate genetic variability in the germplasm to support effective breeding programs to improve yield potential, reaction to diseases and quality parameters.

- ☆ Lack of information on the genetics of economically important characters in the seed spices.
- ☆ Inadequate knowledge of population dynamics *viz.* Degree of cross pollination (cumin, coriander and fennel), inbreeding depression and heterosis which is prerequisite for sound breeding methodology to be adopted.
- ☆ Non-availability of resistance sources against important diseases in germplasm collections
- ☆ High degree of genotype-environmental interaction and poor stability of genotypes. Stable genotypes need to be identified.
- ☆ Non-responsiveness of available cultivars to fertilizer application and management practices. There is an urgent need to develop responsive types.
- ☆ Lack of location specific research facilities.
- ☆ Approaches to get increased productivity

In both development of varieties or genetic basé we must immediately shift to non-conventional methods. Strategies could be developed to address various problems through various biotechnological tools like molecular markers, tissue culture and genetic engineering.

Biotechnological Approaches for Seed Spice Improvement

- ☆ Use of molecular markers in germplasm collection, evaluation and conservation.
- ☆ Understanding of genetic structure of population using molecular markers (for varieties developed through population improvement programme)
- ☆ Identification of resistant genes using isolated pathogenic strains and pure genotypes
- ☆ Tagging of genes of economic importance including genes for resistance to biotic and abiotic resistance and quality
- ☆ Enhancement of germplasm through soma-clonal variants, *in vitro* selection schemes for biotic/abiotic resistance.
- ☆ Development of transgenics for biotic and abiotic stress resistance (list of genes attached)
- ☆ Development of various molecular markers for marker assisted selection.
- ☆ Development of biofertilizers for *e.g.* Growth promoting bacteria

Biochemical Studies of Importance

- ☆ In addition to essential oils a number of bio-molecules are present in seed spices, important being, alkaloids and glycosides. While these components impart pleasurable aroma and color to the food, are of great importance for pharmaceutical industries.
- ☆ No simple methods are available to determine quality and quantity of these constituents. Thus there is a need to develop easy assays applicable to fast selection schemes.
- ☆ The information on available diversity in the germplasm for these constituents is meager.
- ☆ Effect of physical conditions during growing season is not well understood especially in relation to genotypes.
- ☆ Development of molecular tags would help in understanding the genetics and improvement of quality and quantity of these constituents.
- ☆ In vitro production technology for these secondary metabolites be developed: since all the genes required for their production are available in all the cells and all the tissues, they may

be made constitutive or up-regulated through mutagenesis or induced by some chemical signaling, what is needed is selection scheme.

☆ Anti-microbial and other pharmaceutical properties of these constituents have been studied but their doses and means of administration up to required doses have not been worked out.

Table 7.1: Transgenic Crops with Herbicide Resistance

Genes	*Herbicides*	*Inhibition*	*Source of Gene*	*Crops*
AroA	Glyphosate	EPSPS	*Salmonella typhimurium*	Soybean, Maize, Rape, Wheat, Cotton
Bar,pat	Glufosinate	Glutamine Synthase	*Streptomyces hygroscopicus, S.viridochromoges*	Tomato,Beet, Wheat, Rice, Potato, Peanut, Maize
Bxn	Bromoxynil	Photosystem II	*Klebsiellaozaene*	Cotton
Gox	Glyphosate	EPSPS	*Achromobacter*	Soybean, Maize, Rape, Wheat, Cotton
Csr 1, Ahas3r	Imidazolinones	Acetolactate synthase(Als)	*E.coli*	Maize, Tobacco, Flax, Rape, Beet, Trefoil
Tfd A	2,4-D	Metabolic	*Alcaligenes eutrohpus*	Cotton, Tobacco

Table 7.2: Transgenic Crops with Abiotic Stress Resistance

Gene	*Origin*	*Host*	*Stress*
Bet A (Choline dehydrogenase)	*E.coli*	Tobacco	Salinity
Bet A	*E.coli*	Potato	Freezing
CodA (Choline dehydrogenase)	*Arthrobactor globiformis*	*Arabidopsis*	Salinity and Drought
P5CS (Proline synthesis)	*Vigna aconitifolia*	Tobacco	Drought
Mltd (Mannitol-1 phosphate dehydrogenase)	*E.coli*	*Arabidopsis*, Tobacco	Salinity
TPS(Trehlose)	*Saccharomyces cerevisiae*	Tobacco	Drought
SacB(Fructan)	*Bacillus subtilis*	Tobacco	Drought
Fad7 (Desaturase)	*Arabidopsis*	Tobacco	Chilling
Des9 (9-Desaturase)	*Anacystis nidulans*	Tobacco	Chilling
HVA1 (LEA)	Barley	Rice	Salinity and Drought
Vhb (Heamoglobin)	*Vitreoscilla stercoraria*	Tobacco	Hypoxia,Anoxia
Afp	Flounder Fish	Tobacco	Freezing
Dismutase	Rice	Tobacco	Oxidative

DREB-Research has been conducted to identify dehydration responsive transcription factors (DREB) that mediate transcription of several genes in responsive to cold and water stress. DREB genes, which are induced by cold and water stresses, encode transcription factors that bind to the *cis*-acting promoter element (DRE) of stress- related genes and turn on their expression. The binding initiates synthesis of gene products implicated in plant acclimation responses to low temperature and water stress. Over expression a fusion of DRE containing promoter from a dehydration induced gene (rd29A) with a DREB gene (DREB1A) in Arabidopsis resulted in marked increase in the transgenic plants tolerance to freezing, water stress, and salinity.

Table 7.3: A List of Few Transgenic Plants Conferring Resistance Against Bacterial Diseases

Gene	*Origin*	*Transgenic Species*	*Resistance to*	*Level of Resistance*
Shiva-1	Giant silk moth	Tobacco	*Ralstonia solanacearum*	Partial
MB39	Giant silk moth	Tobacco	*P. syringae pv. tabaci*	Partial
Attacin E	Giant silk moth	Apple	*Erwinia amylovora*	Partial
Lysozyme	T4 bacteriophage	Potato	*E. carotovora*	Partial
Lysozyme	Human	Tobacco	*P. syringae pv. tabaci*	Partial
Lactoferrin	Human	Tobacco	*Ralstonia solanacearum*	Partial
Tachyplesin	Horseshoe crab	Potato	*E. carotovora*	Partial
Tobotozin resistance protein	*P. syringae pv. tabaci*	Tobacco	*P. syringae*	Total
Phaseolotoxin insensitive OCTase	*P. syringae pv. Phaseolocola*	Bean	*P. syringae pv. Phaseolocola*	Total
Pectate lyase	*Erwinia carotovora*	Potato	*Erwinia carotovora*	Partial
Thionin	Barley	Tobacco	*P. syringae pv. tabaci*	Partial
Xa21	Resistance rice cultivars	Rice	*Xanthomonas oryzae*	Total

Table 7.4: A List of Few Transgenic Plants Conferring Resistance Against Virus Diseases

Transgene	*Crop*	*Mode of Action*	*Virus*
Coat Protein (CP)	Tobacco	Replication inhibition	TMV, AMV, CMV,
Coat Protein (CP)	Potato	Replication inhibition	PVX, PVY, Ring spot
Replicase	Potato	Enzyme competition	PVY
Replicase	Pea	Enzyme competition	PSbMV
Replicase	Rice	Enzyme competition	RYMV
Movement Protein(MP)	Tobacco	Movement interference	TMV
Movement Protein(MP)	Potato	Movement interference	PLRV, PVX
Viral Protease	Potato	Polyprotein processing	PVY
Antisense RNA	Potato	Translation inhibition	PLRV
Ribozyme	Tomato	RNA cleavage	CEVd
Satellite RNA	Tobacco	Competes for capsids	CMV
RIPs Protein	Tobacco	Inhibits rRNA (60s)	TMV, PVX
2,5 oligoadenylate	Tobacco	Degrades ds RNA	CMV
Nucleocapsid (NC)	Tobacco	Replication inhibition	TSWV

Estimation of Genetic Diversity Using Molecular Markers

Characterization of available germplasm provides tools for identification of different entries and evaluation of genetic variation among them. Characterization consists of recording those characteristics that are highly heritable, can be seen by the eye and are expressed in all environments–IBPGR, (1981). Thus, the process of defining plant genetic variation and identification of plants are dependent process for which usually plant's morphological traits are used. The characterization of cultivars requires a large set of phenotypic data that is often difficult to assess, almost unavailable at early stages of plant growth and sometimes variable due to environmental influences. Thus, genetic marker analysis was shifted from morphological markers to the protein markers (1930).

Table 7.5: A List of Few Transgenic Plants Conferring Resistance Against Fungal Diseases

Genes	*Pathogen*	*Crop*
Chitinase (Bean, Tobacco)	*Rhizoctonia solani,*	Brassica, Tobacco
Glucanase (Soybean, Barley)	*Rhizoctonia solani,*	Tobacco
RIPs	*Rhizoctonia solani, Alternaris, Botrytis*	Tobacco
Lysozyme peptide	*Phytophthora infestans*	Potato
HEWL	*Fusarium oxysporium*	Tobacco
Human lysozyme	*Erysiphe cicoracearum*	Tobacco
Rice chitinase	*Botrytis cineraria*	Cucumber
Glucose oxidase	*Phytophthora infestans*	Potato
Osmotin	*Phytophthora infestans*	Tobacco

Protein markers are independent of cultivar physiology and morphology and offer significant advantages over morphological methods of variety and/or species identification that they are rapid, relatively cheap, eliminate the need to grown plants to maturity and are largely unaffected by the growth environment. Two types of bio-chemical method can be used for the varietal identification (Sammour, 1991).

1. The direct multi-locus approach. This involves the electrophoresis examination of protein display as substantial degree of molecular polymorphism. The analysis of seed storage protein is a good example of this approach.
2. The indirect, single locus approach. This involves the utilization of protein which, despite, polymorphism are derived from a single locus. Such proteins will generally be enzymes and are known as isozymes (or allozymes).

These markers are useful in cultivar identification they are limited by the number of informative markers and give no direct assessment of the potential variation present in the genome and the profiling influenced by tissue specificity and developmental stage. This disadvantages can be overcome by using the electrophoresis markers of a conservative proteins *e.g.* seed proteins, (Sammour, 1991). Extraction of storage proteins from grains has been used for many years involves separation of proteins by gel electrophoresis. This process is slow and resolution is inadequate to resolve closely related varieties. Because of all these drawbacks and genetic marker analysis was shifted from use of protein markers to DNA marker in 1980. DNA markers allow direct assessment of variation in genotypes. Varietal profiling methods that directly utilize DNA could potentially address all of the limitations associated with morphological and biochemical data DNA based profiling are not affected by the growth and environment effects and the basis of inheritance can be understood. Profiles can be interpreted in terms of the presence or absence of specific alleles. Cost and speed of data acquisition are acceptable for variety identification.

During 1980s and 1990s a variety of molecular markers have been developed which are being used for various purposes like crop improvement, characterization, identification of cultivar, assessment of genetic variability, assessment of genetic relationship, parental identification, genetic map construction.

Alec Jeffery introduced the term DNA finger printing for the first time in 1985 to describe bar-code-like DNA fragment patterns generated by multilocus probes after electrophoresis separation of genomic DNA fragments. The emerging patterns make up a unique feature of the analyzed individual and are currently considered to be the ultimate tool for biological individualization.

Various types of molecular markers are utilized to evaluate DNA polymorphism and are generally classified as hybridization-based markers and polymerase chain reaction (PCR) based markers. In the former, DNA profiles are visualized by hybridizing the restriction enzyme-digested DNA, to a labeled probe, which is a DNA fragment of known origin or sequence. RFLP (Restriction fragment length polymorphism) is a hybridization based marker. RFLP markers were used for the first time in the construction of genetic maps by Botstein *et al.* (1980). RFLPs, being codominant markers, can detect coupling phase of DNA molecules, as DNA fragments from all homologous chromosomes are detected. They are very reliable markers in linkage analysis and breeding and can easily determine if a linked trait is present in a homozygous or heterozygous state in an individual, an information highly desirable for recessive traits. However, their utility has been hampered due to the large amount of DNA required for restriction digestion and southern blotting. The requirement of radioactive isotope makes the analysis relatively expensive and hazardous. The assay is time-consuming and labour-intensive and only one out of several markers may be polymorphic, which is highly inconvenient especially for crosses between closely related species. Their inability to detect single base changes restricts their use in detecting point mutations occurring within the regions at which they are detecting polymorphism.

A rapid expansion of identification abilities came from a profusion of new methods that utilize the polymerase chain reaction (PCR) process. The development of PCR (Mullis *et al.* 1986; Mullis and Faloona, 1987) removed the necessity for probe hybridization steps thus making it more amenable to automation. PCR-based markers involve in vitro amplification of particular DNA sequences or loci, with the help of specifically or arbitrarily chosen oligonucleotide sequences (primers) and a thermostable DNA polymerase enzyme. The amplified fragments are separated electrophoretically and banding patterns are detected by different methods such as staining and autoradiography. Ever since thermostable DNA polymerase was introduced in 1988 (Saiki, 1988), the use of PCR in research and clinical laboratories has increased tremendously. The primer sequences are chosen to allow base specific binding to the template in reverse orientation. PCR is extremely sensitive and operates at a very high speed. Its application for diverse purposes has opened up a multitude of new possibilities in the field of molecular biology.

Randomly-amplified polymorphic DNA markers (RAPD), DNA amplification fingerprinting (DAF), Arbitrary primed polymerase chain reaction (AP-PCR), Amplified fragment length polymorphism (AFLP), are PCR based markers that utilize primers of sequence that are arbitrary to that of the target genome. The others PCR based markers that use primers synthesized to specific, complementary sequences in the target DNA of varieties, are simple sequence repeat (SSRs), Inter simple sequence repeat markers (ISSR), Sequence-tagged sites (STS), variable number of tandem repeats (VNTRs), Sequence characterized amplified regions (SCAR).

In which micro-satellites or simple sequence repeats (SSR) are short DNA sequences consist of 1-6 bp monomer sequences that is repeated several times. The term micro-satellite was introduced by Litt and Lutty, 1989. Owing to ubiquity, hyper variability, abundance and genome wise distribution, SSR loci represent a new generation of powerful genetic markers for eukaryotes. Use of these marker system, however is hampered by the requirement for the sequence information from flanking regions, from which primers are designed for PCR amplification. Discovery and characterization of a large numbers of SSRs is therefore time consuming and expensive for many texa. SSR is a both hybridization based and PCR based marker system. SSRs are now converted into PCR based markers like, inter simple sequence repeat (ISSR). This technique was reported by Zietkieqicz *et al.* (1994). Primers based on to freedom from necessity of obtaining flanking genome sequence information, ISSR analysis is technically simpler than many other marker systems. This method provides highly reproducible results and generates abundant polymorphism in many systems.

Genomic resources in Apiaceae are relatively underdeveloped. Sporadic studies mainly used anonymous markers like RAPD and AFLP for diversity studies Pedro *et al.* (2008). The recent availability of a large set of PCR-based codominant markers, such as simple sequence repeats (SSR), from carrot however, could help in the genotyping of related taxa (Pablo *et al.*, 2009).

Creation of Genetic Diversity Using Biotechnological Tools

All plant genome projects to date seem to find a remarkably high number of novel genes. Novel in this context means genes that cannot be assigned to any known gene family and without any significant match in another genome. Moreover, genome evolution is a continuous and dynamic process in which genomes expand and contract, duplicate in whole or in part, and gain and lose genes at the same time as transposons amplify and are excised. Transposon amplification, gene loss, and changes in gene expression (the latter perhaps induced by epigenetic changes such as alterations in DNA methylation and histone modification) alter both the structure of the genome and the function of genes. We are only beginning to glimpse how these processes and events correlate with patterns of diversification and how they may provide the substrate for evolution.

However, information is indicative enough of greater flexibility in genome structure and function. Consequently, provides ample evidences in favor of evolutionary methods of crop improvement and biotechnology in order to generate novel genotypes and molecules fulfilling the needs of modern civilization. Evolutionary biotechnology strategies coupled with properly developed selection techniques are expected to provide continued solutions to ever arising crisis.

References

Botstein, D., White, R.L., Skolnief, M. and Davis, R.W., 1980. Construction of a genetic linkage map in man using restriction fragment length polymorphism. *American Journal of Human Genetics*, 32: 314–331.

Sammour R.H., 1991. Using electropharetic techniques in varietal identification, biosystematic analysis, phylogenetic relations and genetic resources management. *Journal of Islamic Academy of Sciences*, 4(3): 221–226.

Mullis, K.B. and Faloona, F.A., 1987. Specific synthesis of DNA in vitro via a polymerase catalysed reaction. *Methods of Enzymology*, 255: 336–350.

Mullis, K.B. and Faloona, F.A., Scharf, S., Saiki, R., Horn, G. and Erlich, H., 1986. Specific enzymatic amplification of DNA *in vitro*: The polymerase chain reaction. *Cold Spring Harbor Symp. Quant. Biol.*, 51: 263–273.

Saiki, R.K., Walsh, P.S., Levenson, C.H. and Erlich, H.A., 1989. Genetic analysis of amplified DNA with immobilized sequence specific oligonucleotide probes. *Prac. Natl. Acad. Sci., USA*, 86: 6230–6234.

Jeffreys, A.C., Brookfield, J.F.Y. and Semeonoff, R., 1985. Positive identification of an immigration test case using human DNA fingerprints. *Nature*, 317: 818.

Litt, M. and Lutty, J.A., 1989. A hypervariable microsatellite revealed by *in vitro* amplification of a dinucleotide repeat within the cardiac muscle actin gene. *Am. J. Hum. Genet.*, 44: 397.

Pablo, F., Cavagnaro, Sang-Min Chung, Aamir Ali, Mehtap Yildiz, Maria S Alessandro, Philipp, W. Simon, 2009. Development and characterization of novel SSR markers in carrot (*Daucus carota* L.) and their application for mapping and diversity analysis in apiaceae. *Plant and Animal Genomes XVII Conference.*

Pedro, A. López, Mark P. Widrlechner , Philipp W. Simon, Satish Rai, Terri D. Boylston, Terry A. Isbell, Theodore B. Bailey, Candice A. Gardner and Lester A. Wilson, 2008. Assessing phenotypic, biochemical, and molecular diversity in coriander (*Coriandrum sativum* L.) germplasm. *Genetic Resources and Crop Evolution*, 55, 2/March.

Recent Advances in Seed Spices (2011) *Pages* **85–93**
Editors: **Y. Ravindrababu, R.K. Jaiman, K.D. Patel, N.R. Patel and S.B.S. Tikka**
Published by: **DAYA PUBLISHING HOUSE, NEW DELHI**

8

Action Plan for Organic Seed Spices Production

Arun K. Sharma*
Central Arid Zone Research Institute, Jodhpur – 342 003, Rajasthan

ABSTRACT

Quality production of seed spices is becoming major criterion for export as well as indigenous market. Quality of these produce can be improved with organic farming which is otherwise also getting demand from international and national market. Increase in organic production can be possible through four dimensional action plan *i.e.* improvement and effective implementation in policy, market, research and at production level. Policy plays major role as it has more effect than other three. Policy in terms of water use for spices only, subsidy, allocation of fund and personnel can be effective. Similarly better marketing facility, certification and developing models of organic farming for various agro-climatic zones can help in promotion of organic farming. Adoption of principal and practices of organic farming at production level will complete the efforts to made available organic food to the consumers.

Seed spices are the crops which requires lesser resources yet gives more profit as compared to conventional food grain or cash crops. This character makes them special crop. However, quality of the produce is the only criterion that decided profit. More the quality more the profit will be possible because better price is possible by export and only best quality can be exported. Fortunately seed spices of India have great demand world over and it will be least affected by ups and downs of world

* E-mail: arun.k_sharma@yahoo.co.in.

economy as they are the part of basic need *i.e.* food. A few decades ago India was having monopoly in some of the seed spices like cumin, but now several other countries like China, Iran, etc. have also started export.This started competition in world market and only quality produce can withstand. Quality has become more important after the advent of chemical farming in which residues of pesticides remains in the produce. Use of chemical fertilizers deteriorate the quality in terms of aroma, essential oil, content, taste etc. Therefore, the quality of seed spices deteriorate in chemical farming by two ways:

1. Residues of pesticides
2. Reduction in inherent quality

Increasing health awareness in consumers is further increasing the demand of safe food. To produce pesticide free, with good inherent quality of seed spices, organic farming can be the best viable option.

Quality Organic: An Imperative Option

Production of spices in organic farming and maintaining quality is becoming a compulsion for standing in international and domestic market because:

1. As mentioned earlier demand of organically produced species is increasing.
2. With the scientific advancement, many of our monopoly crops is being grown by several other countries and quoting lower rates in international market. For example, cumin was a monopoly crop of Indian but now it is being growns by China, Iran, Turkey and Egypt.
3. In this era of global open economy, domestic consumers are free to buy a quality with low priced produced from international market.

Therefore for maintaining our monopoly or rather competitiveness for international as well as domestic market, economic as well as quality production is becoming imperative.

Besides, for meeting the demand, organic farming helps in two way to get low cost quality produce. These are:

1. In organic production system, no external synthetic chemical is used, moreover emphasis is given on recycling of locally available resource. With this approach cost of production can be reduced upto 60 per cent as compared to conventional chemical farming (Lampkin, 1994).
2. There are several example of experiments (Leclare *et al.*, 1991; Sharu and Meerbai, 2001) and farmer's experience which shows that due to balanced nutrient supply through organic sources, the quality of organic produce increases many fold in terms of aroma, essential oil content, texture, taste and shelf life. Author himself experienced a much better (than chemically grown) aroma, luster, and keeping quality in leaf coriander at village *Ballon-ka-guda* in Udaipur district. In that whole village vermicompost was produced and used for coriander production. Farmers told that they get 1.5-2.0 times higher price in vegetable market and for coriander seed as well.

Therefore, organic farming is the only option for low cost quality production.

There are same other issues which compel to opt organic farming for the betterment of society and farmers as well.

1. *Low and reducing supply of fertilizers*: To some extent nitrogenous fertilizers and most part of other fertilizers are imported from various countries. Not only these imported fertilizers but also supply of nitrogenous fertilizers is decreasing due to changing international scenario. Moreover, most of the fertilizers companies give priority to irrigated areas Punjab, U.P., Haryana, Maharashtra etc. for supply and the major spice cultivation area *i.e.* semi arid and arid regions are remained short supplied as happened this year also. Therefore, to reduce dependency on imported fertilizers and recurring problem of short supply in rainfed areas, opting organic farming is the only solution.
2. *Ethical need-grow medicine not poison*: Most of the spices like cumin, fennel, ajwain, fenugreek etc. are important ingredient of ayurvedic, allopathic, homeopathic preparations. These all medicines are supposed to give to patient and if these ingredients have residues of pesticides it may be poisonous effect instead of curing the patient (Sharma 1998). Therefore, it is our ethical duty to grown spices and medicinal plant, only organically as social obligation.

With the above discussion it is clear that organic farming is becoming necessary not only for market point of view but also for the welfare of farmer and earth society as a whole.

Organic Farming

Growing environmental hazard, non-sustainability of food production and less safer food quality (bigger concern of affluent society) under chemical dependent conventional (intensive) agriculture are some of the concerns that are being projected in favour of promoting organic agriculture. As explained by Lumpkin (1990), organic agriculture is a production system which avoids or largely excludes the use of synthetic compounded fertilizers, pesticides and growth regulators and livestock additives. To meet the nutrient requirement as well as to manage weeds and other pests etc., it relies on crop rotations, crop residues, legumes, animal manures, green manures, off-farm organic waste, biological pest control, mechanical intervention, etc. than the use of synthetic chemicals to which the conventional farming is centric. Thus, the concept of soil as a living body/system is central to organic agriculture. In a healthy soil the biotic and abiotic components covering organic matter including soil life, mineral particles, soil air and water exists in a dynamic equilibrium and regulate the ecosystem related processes in mutual harmony by complementing and supplementing each other. The population of soil fauna and flora multiplies rapidly in a healthy soil, which in turn sustains the biochemical processes of dissolution and synthesis at a high rate. This rate of soil life and the associated organic transformations enhance the regenerative capacity of the soil and make it resilient to absorb the effects of climatic vicissitudes and occasional failures in agronomic management (Palaniappan, 2004).

Objectives of Organic Agriculture

According to the International Federation of Organic Agriculture Movement (IFOAM) the objectives of organic agriculture are as follows:

- ☆ To produce food of high nutritional quality in sufficient quantities
- ☆ To work in harmony with natural systems rather than seeking to dominate them
- ☆ To encourage and enhance the biological cycles within farming system involving microorganisms, soil flora and fauna, plants and animals
- ☆ To maintain and increase the long term fertility of soils
- ☆ To use, as far as possible, renewable resources in locally organized agricultural systems

- ✩ To work, as much as possible, within a closed system with regard to organic matter and nutrient elements
- ✩ To give all livestock, conditions of life that allow them to perform all aspects of their innate behaviour
- ✩ To avoid all forms of pollution that may result from agricultural techniques
- ✩ To maintain the generic diversity of the agricultural system and its surroundings, including the protection of plants and wild life habitat
- ✩ To allow agricultural producers an adequate return and satisfaction from their work including a safe working environment
- ✩ To consider the wider social and ecological impact of the farming system

Essential Characters of Organic Agriculture

The most important essential characters of organic agriculture are as follows:

- ✩ Maximal but sustainable use of local resources
- ✩ Use of purchased inputs, only as complementary to local resources
- ✩ Increasing crop and animal diversity in the form of polycultures, agroforestry, integrated crop/livestock systems, etc. to minimize risk and improve ecological balance and economic stability
- ✩ Ensuring the basic biological functions of soil-water-nutrient-humus continuum
- ✩ Creating an attractive overall landscape which gives satisfaction to the local people

Thus, to sum up mixed farming, crop rotation and organic cycle optimization are the three key principles of organic agriculture (Sharma, 2001).

Advantages of Organic Agriculture

Some important advantages of organic agriculture are as follows:

- ✩ It creates optimal conditions in soil for improved growth and physiological activities of plants and the higher crop yield
- ✩ It improves soil physical, chemical and biological conditions
- ✩ It improves soil carbon content (carbon sequestration)
- ✩ Utilizing organic wastes as an input minimizes pollution
- ✩ It mitigates environmental degradation
- ✩ It is more resistant to diseases, insects and pests
- ✩ Organically grown produce are believed to be nutritionally superior and good for health
- ✩ Higher demand of organically produced food coupled with its premium price may improve economy of the farmer and the region

Action Plan for Promotion of Organic Farming

Considering the export demand and contribution in the economy of this region; it is need of the hour to do integrated efforts for quality organic production of species. These efforts are need to be done at four level *i.e.* policy, research, market and production of spices. Integration of technologies and programmes and coordination among various agencies is the prime requirement. For example development of package of organic production may not be much effective until and unless promotion policies and good market facilities are not available. Integrated action plan at four level is as follows.

Policy

Policy plays major role in promotion of any programme. Policy in terms of supporting rules and regulations, subsidies, facilities, allocation of budget and personnel etc. can alone is sufficient if executed properly. The best example is Cuba(Latin America), where organic farming was made a national policy and now whole of the country is organic. Similarly some of the state like Uttranchal, NEH states etc. has declared organic state and they are taking lead. Although in India, organic movement was started in early 80s but it got momentum only after 2001 when government of India launched National Programme on Organic Farming (NPOF). Later on most of government agencies have started to give priority to organic farming. However simply giving budget, subsidies etc. may not be sufficient to promote organic farming, as least development has been done in arid and semi arid areas even, they are kept at Priority I and II in NPOF. For better development of organic farming in arid and semi arid zone additional measures need to be taken. They are:

1. *Popularization of organic farming without compulsion of certification*: Most of the farmers in India are very poor and unable to afford the cost of certification. Promoting organic farming with the compulsion of certification has made negative impact on adoption. Instead, at initial stage organic farming should be promoted for improving soil fertility, reducing cost of production and other environmental advantages.
2. *Priority in ongoing rural development programmes*: Organic farming need not to be launched as a separate programme. It would be better if organic farming is given priority in all on going rural development programme like watershed, NREGA, ATMA etc.
3. *Dissemination of organic farming in holistic manner*: Most of the agencies promoting organic farming in piecemeal approach *e.g.*.only vermicompost, only IPM, only INM etc. this makes confusion among the farmers. While organic farming is an integrated approach for nutrient recycling, conservation of natural resources, water conservation, crop rotation/diversification etc. So it must be inclusion of all these aspects which can make a sustainable organic farming in real term.
4. *Integrated efforts of supporting agencies*: Individual agency may not work efficiently for promotion of organic farming; For example KVIC have a scheme of margin money to establish vermin compost unit but they are unable to ensure the use of produce of such unit. Similarly agriculture department, ICAR institutes have wealth of information but unable to provide financial support. Thus there is need of integrated programs by all related agencies.
5. *Encouragement of decentralized input supply*: Encouragement may be given to produce all inputs for organic farming in a decentralized manner at local level so that not only local resources can be utilized but also employment at village level can be generated.
6. *Conservation of water*: With over exploitation of ground water since last 3-4 decades for raising crops mainly Cotton, Sugarcane and in some places Rice, the ground water reached to the dark zone in many parts of Maharashtra, AP, Gujarat, Rajasthan, Punjab, etc. After- 2 decades of experience of demonstration/training on dry land farming it has been realized that farmers did not stopped the practice of intensive agriculture till the ground water exhausted. Some farmers were used saline water and made land permanently barren.

Therefore remedy of this malady is only possible through policy reforms. Some of these measures are:

(*a*) Intensive cropping should be allowed in canal irrigated areas only and drylands should be restricted to integrated farming systems.

(*b*) Use of ground water is allowed for life saving irrigation to the monopoly crops of these regions like spices and medicinal plants. Subsidy should be given only if unavoidable.

7. *Adoption of improved methods of composting*: Majority of the farmers in the drylands apply animal (Table 8.1) and crop waste in undecomposed form to the soil, as a result the availability of nutrients to the plants reduces and also invites several pests. It would be better to apply these materials after composting them with any of the suitable methods (Table 8.2). These methods can be popularized and financially supported under the "Clean Village Scheme" of the central governments.

Table 8.1: Nutrient Availability/Year in the Excretion of Animal and Human Being

Source	*Excreta/Unit/Yr (kg)*	*Urine/Unit/Yr (liter)*	*Available N (Kg)*	*Available P (Kg)*	*Available K (Kg)*
Cow	6000	2000	36	8	16
Bullock	8000	2500	44	11	23
Sheep/Goat	500	160	26.5	3.6	6.3
Human being	110	400	14.3	8.9	7.7

Table 8.2: Nutrient Availability in Different Types of Compost

Composting Method	*Nutrient Availability (kg/ton of compost)*		
	Nitrogen	*Phosphorus*	*Potassium*
FYM	7.5	6.0	12.0
NADEP	10.0	7.0	13.0
Phosphocompost	10.4	55.0	15.0
Biogas slurry	17.5	10.0	10.0
Vermicompost	27.0	17.5	17.5

8. *Awareness and capacity building*: Demonstrations, training, conferences, seminars, farmers fair etc. may be organized to make a general consensus about organic farming.

9. *Subsidy on organic inputs:* Provision of subsidy may be made for organic inputs to make organic produce more competitive.

10. *Promotion of high value crops*: The demand of spices and medicinal plant is increasing when grown organically, so it must be promoted organically in the various agro-climatic regions of the drylands. (Table 8.3).

11. *Development of organic clusters of villages*:Available clusters of villages of watershed programs may be converted into organic cluster of villages by providing technical support. This will be cost effective and make easier the certification process of organic produce.

Table 8.3: Potential High-Value Crops for Organic Farming in Different Agro-Climatic Regions

Agro-Climatic Zone	*Rainfall Average (mm)*	*Suitable Species and Medicinal Crop*
Arid	100–300	Cumin, Senna, Gugool, Isabgol, Guar (gum)
Transitional	300–450	Fenugreek, Cumin, Aloe, Senna
Semi arid	450–750	Fennel, Asvagandha, Fenugreek, Coriander, Lucorice, Aloe, Aonla

Research

Research has been done on various components of organic farming by ICAR/SAUs, yet the research is needed to integrate the efforts and assess their effects. Besides, following aspects of research may be taken simultaneously:

1. Survey and scientific validation of traditional organic farming practices.
2. Use of locally available resources for production of manures and bio pesticides.
3. Assessment of economic and ecological returns from organic vis-a-vis intensive agriculture system.
4. Development of organic farming models for each sub agroecological zone.
5. Organic breeding of spices is a proactive step to accelerate the production and export of organic spices. The major selection parameters for developing organic spice varieties are:.
   (*a*) Rapid early growth
   (*b*) Resistance to pests
   (*c*) Responsive to manure application
   (*d*) Good keeping quality and other quality parameters

Market

Market environment plays major role to motivate producer for opting an enterprise and same is true for farmers. After emerging demand of organic produce from western countries to some extent from domestic market production of organic spices mainly tropical spices like black paper, clove, cardamom etc. has increased. However, major part of arid and semi arid zone which was traditionally organic is yet to come in market as certified quality organic to get market premium. Some of possible measures need to be taken for better market environment and for proper remuneration to farmer are:

1. Development of cooperative organic marketing facility.
2. Encouragement to exporters for bearing expenditure of certification.
3. Development of alternative low cost certification system for domestic market *e.g.* participatory guarantee system.
4. Facilitating organic consumer-producers associations.
5. Development of postharvest processing and packaging facilities at Tahseel level.
6. Identification and registration of geographical indicators in spices is a major area in which little work has been done so far. Once a region is identified and registered as GI of a particular crop it will be a patent like status and gets premium price. However, organic or near natural farming is the only way to maintain the quality of GI.

Production

Although production is the first step to get organic produce but it comes at the end in planning as the policy, market and research aspects should be strengthens before starting production. However, not only for market but also for long term sustenance of soil fertility and farmer's livelihood, basic understanding and principals of organic farming should be applied in the production system. In arid and semi arid region main emphasis need to be given on efficient use and recycling of limited natural resources. Therefore for production and marketing point of view an strategy with following component need to be adopted.

1. Production of inputs (compost, vermicompost, biopesticides etc.) as much as possible at local level.
2. Efficient use of inputs (time, method and quantity).
3. Effective integration of perennials, animals and beneficial organisms in farming system.
4. Adoption of system based production rather than crop based.
5. Improvement in traditional (default) organic system.
6. Continuous experimentation at farm level to understand natural production system and interactions.
7. Doing farming and getting certification as a group effort.
8. Giving importance to quality production rather than quantity only.
9. Harvesting at proper time and cleaning and grading at farm level.

Good Initiatives

Some of the good programmes on promotion of organic farming in relation to arid and semi-arid regions have been started. They are:

1. National Project on Organic Farming was launched in 2004 and arid and semi arid region were placed at priority I and II.
2. Under National Horticulture Mission, clusters of 50 ha organic farms for seed species are being developed in several districts of arid and semi arid regions.
3. Recently a rainfed area development authority has been constituted by government of India that will expected to give priority on organic farming of high value crops like spices.
4. Spice board has also launched several subsidy schemes for organic production of seed spices.
5. Several NGO's started organic group farming of seed spices mainly cumin, coriander etc.

These programme may hopefully will give momentum to organic production.

Conclusion

Organic farming is a holistic production system run with the efficient use and recycling of locally available resources. Some monopoly high value seed spices are having great international demand if produce organically. Organic production of these spices not only boost the economy of this farmers but also sustain the productivity of natural resources. The need is to do research on development of easy and economic technologies, development of processing and marketing infrastructure and financial as well as technical support for quality organic production.

References

Lumpkin, N.H., 1990. *Organic Farming*. Farming Press Books, IPSWICH, U.K.

Lampkin, N.H., 1994. Organic farming: Sustainable agriculture in practice. In: *The Economics of Organic Farming: An International Perspective*, (Eds.) N.H. Lampkin and S. Padel. CAB International, U.K., pp 3–9.

Leclare, J., Miller, M.L., Joliet, E. and Roquelin, G., 1991. Vitamins and mineral content of carrot and celerae grown under organic fertilizer. *Bio. Agric. Hortic.*, 7: 339.

Palaniappan, S.P., 2004. Organic farming as a concept. In: *ISSS Bulletin: Organic Farming*, Indian Society of Soil Science, New Delhi, 22: 8–13.

Sharu, S.R. and Meerabai, M., 2001. Effect of integrated nutrient management on yield and quality of chilli (*Capsicum annum* L.). *Vegetable Science*, 28(2): 184–185.

Sharma, Arun K., 1998. Eco-farming of medicinal plants in the arid tropics of India. *Ecology and Farming*, 4: 24–25.

Sharma, A.K., 2001. *A Handbook of Organic Farming*. Agrobios (India), Jodhpur, 626 p.

Recent Advances in Seed Spices (2011) *Pages* **94–96**
Editors: **Y. Ravindrababu, R.K. Jaiman, K.D. Patel, N.R. Patel and S.B.S. Tikka**
Published by: **DAYA PUBLISHING HOUSE, NEW DELHI**

9

Irrigation: A Key Factor for Seed Spice Crop Production

A.U. Amin, P.K. Patel, B.J. Patel, S.G. Patel and S.P. Patel
Centre for Research on Seed Spices, SDAU, Jagudan, Gujarat

The seed spices are annual herbs, whose dried fruits or seeds are mostly used as spices. The seed spices are used for culinary, confectionery, perfumery, cosmetics and pharmaceutical industries. Thus, these are low volume but high value crops. India is the largest producer, consumer and exporter of spices. Export in terms of in volume and value increase year after year. Therefore, India enjoys a pre-eminent position in spices production in the world and called as "Home of spices". Water is the main constituent of protoplasm and it directly participates in many metabolic processes. Thus, irrigation water provides favourable environment for better growth and higher crop production. Irrigation water is scarce and costly input in arid and semi arid regions of country. Even if all possible sources of irrigation water are tapped, this scarcity cannot be solved. Economical and wise use of water for irrigation is a vital problem for arid and semi arid regions where cultivation of seed spices is becoming more popular due to their ever increasing demand, short duration in nature, less requirements of costly and scarce inputs, well suited in saline and more remunerative as compared to other traditional *rabi* crops *e.g.* wheat, gram, mustard, Isabgul etc. Most important seed spices *viz.*, fennel, coriander, cumin, fenugreek, ajwain and dill seed are well distributed over different agro climatic regions in India. But arid and semi-arid regions of Rajasthan and Gujarat has emerged as "Seed Spices Bowl of India" and together contribute more than 80 per cent of the total seed spices production in the country.

Fennel

Fennel (*Foeniculum valgare* Mill.) is a cold weather long duration crop comes up well in fairly mild climate which used as a spices, pot-herb and in medicinal purposes. With comparison to Fenugreek and Coriander, Fennel recorded higher seed yield and lower sodicity susceptibility index, when sodic water use for irrigation (Yadav *et al.*, 1996). Application of irrigation by drip method at 0.8 IW/CPE ratio under normal or paired row planting system not only gave significantly higher seed yield than surface method but also save water and fertilizer appreciably(GAU, 2000 and SDAU, 2008). The IW/CPE ratio of 0.8 was best for irrigating the *rabi* fennel under light texture soil (Patel *et al.*, 2000). According to Amin (2001) irrigating the crop at 0.8 and 1.0 IW/CPE ratio were equally effective and recorded higher seed, stover and oil yields as compared to 0.6 ratio but quality attributes were not influenced by irrigation treatments. According to Ashok Kumar *et al.* (2002) branching, flowerings and seed fillings growth stages were critical to irrigation, among them, branching and seed filling stages were most sensitive to moisture stress.

Cumin

Cumin (*Cuminum cyminum* L.) is a short duration but risky crop which cultivated in areas where atmospheric humidity is low. It is used in perfumery, beverage and as medicine against digestive and intestinal upsets. According to Patel *et al.* (1992) seed and stover yields significantly increased with increase in IW/CPE ratios from 0.3 to 0.6, however, 0.45 IW/CPE ratio gave maximum water use efficiency. Under light soil texture, application of five irrigations gave higher seed yield and water use efficiency; where as maximum consumptive use of water was recorded with six irrigations (Jangir and Singh, 1996). In North Gujarat condition, application of one or two extra irrigations after 30 days of sowing not only increase yield and net return but also improve seed quality (GAU, 1997 and GAU, 1999). Application of four irrigations (sowing time, 8 to 10, 30 and 50 DAS) with soil mulch was equally effective with five irrigations only.

Fenugreek

Fenugreek (*Trigonella foenum graceum* L.) is an important seed spice, medicinal and green leafy vegetable crop and best suited to tracts of moderate to low rainfall. Bhati (1993) concluded that the test weight and straw yield increased significantly with increase in IW/CPE ratios from 0.4 to 1.0 but seed yield increased significantly up to 0.8 IW/CPE ratio. Application of two irrigations at 45 and 80 DAS gave the highest seed yield, where as only one irrigation at 80 DAS found more economical (Virender *et al.*, 2000). Oliver (2003) concluded that seedling, branching, pod formation and pod development stages were critical for irrigating the crop under middle Gujarat condition.

Coriander

Coriander (*Coriandrum sativum* L.) is one of the oldest spice to be used as a fresh leaves or seeds for flavoring food products and in manufacturing curry powder. According to Tomar *et al.* (2000) branching and capsule formation stages were more critical for irrigation or 0.3 IW/CPE ratio was an optimum for irrigating the crop in heavy textured soil of Rajasthan. Application of irrigation at 0.8 IW/CPE ratio under mini-sprinkler method gave significantly higher seed and stalk yields (GAU, 2001). According to Singh *et al.* (2001) application of four irrigations at 30, 60, 90 and 120 days after sowing gave higher number of umbels per plant and seed yield with better monetary return.

Dill Seed

Dill seed (*Anenthum graveolens* L.) is a long duration, drought tolerance and non risky crop of arid and semi arid regions, which cultivated for spices and medicinal purposes. Seed yield and net profit increased with increase in IW/CPE ratios from 0.4 to 0.6, but further increase in IW/CPE ratios from 0.8 to 1.2 decreased yield and net return (Modi, 1990). According to Nemichand (2002) economically potential seed yield of dill seed was obtained by scheduling the irrigation at 0.6 IW/CPE ratio.

Thus, it can be summarized from the above finding that application of irrigation not only increased seed yield and income but also improve the quality of produce. Water requirement of seed spice crops differ according to agro climatic condition, where they grown. Sodic water is not economical and suitable for irrigating the seed spice crops except fennel. Adoptions of micro irrigation system *i.e.* drip and sprinkler method of irrigation increased the yield and save the water. Scheduling of irrigation on the base of critical growth stages or IW/CPE ratio found more economical under limited or adequate availability of water, respectively.

References

Amin, A.U. and Patel, I.S., 2001. *Gujarat Agricultural University Research Journal*, 26(2): 1–4.

Ashok Kumar, Singh, Ranbir and Chhillar, R.K., 2002. *Indian Journal of Agronomy*, 47(2): 289–293.

Bhati, D.S., 1993. *Indian Journal of Agronomy*, 38(3): 449–452.

GAU, 1997. *Annual Report*, Seed Spices Research Station, Jagudan presented in Agresco of Agronomy and Soil Science Sub Committee.

GAU, 1999. *Annual Report*, Seed Spices Research Station, Jagudan presented in Agresco of Agronomy and Soil Science Sub Committee.

GAU, 2000. *Annual Report*, Seed Spices Research Station, Jagudan presented in Agresco of Agronomy and Soil Science Sub Committee.

GAU, 2001. *Annual Report*, Seed Spices Research Station, Jagudan presented in Agresco of Agronomy and Soil Science Sub Committee.

Jangir, R.P. and Singh, Rajender, 1996. *Indian Journal of Agronomy*, 41(1): 140–143.

Modi, 1990. *Unpublished Thesis*, G.A.U., Sardarkrusahinagar.

Nemi Chand, 2000. *Unpublished Thesis*, G.A.U., Sardarkrusahinagar.

Oliver, 2003. *Unpublished Thesis*, G.A.U., Anand.

Patel, B.S., Patel, K.P., Patel, I.D. and Patel, M.I., 2000. *Indian Journal of Agronomy*, 45(2): 429–432.

Patel, K.S., Patel, J.C., Patel, B.S. and Sadaria, S.G., 1992. *Indian Journal of Agronomy*, 37(1): 209–211.

SDAU, 2001. *Annual Report*, Seed Spices Research Station, Jagudan presented in Agresco of Agronomy and Soil Science Sub Committee.

Singh, Dheeraj, Jat, H.S. and Sita Ram Kantwa, 2002. *Farmers' Forum*, p. 7–9.

Singh, Vijay Pal, Yadav, A.C. and Thakral, K.K., 2001. *Haryana Agriculture University Journal of Research*, 31: 105–110.

Tomar, S.S., Dugarwal, H.S., Sinsinwar, P.S. and Sharma, J.C., 2000. *National Symposium on Agronomy: Challenges and Strategies for New Millennium*, November 15–18 GAU, Campus, Junagadh.

Virender Kumar, Yadav, J.S., Singh, Jagdev and Yadav, B.D., 2000. *Indian Journal of Agronomy Science*, 70(8): 515–517.

Yadav, H.D., Singh, Sultan and Virender Kumar, 1996. *Haryana Agriculture University Journal of Research*, 26: 51–55.

Recent Advances in Seed Spices (2011) *Pages* **97–106**
Editors: **Y. Ravindrababu, R.K. Jaiman, K.D. Patel, N.R. Patel and S.B.S. Tikka**
Published by: **DAYA PUBLISHING HOUSE, NEW DELHI**

10

Integrated Nutrient Management in Seed Spices

M.M. Patel, A.U. Amin, L.J. Desai, Manish M. Patel and S.M. Patel
Department of Agronomy, C.P. College of Agriculture, S.D. Agricultural University, Sardarkrushinagar, Gujarat

Seed spices are high value and export oriented commodity crops, which play an important role in agricultural economy of the country. India is the principal source for supply of seed spices in the global market, though there are number of other countries *viz.* Indonesia, Malaysia, Pakistan, Australia, Spain, Egypt, Tanzania etc., producing and exporting to the international market. India is one of the largest producer, consumer and exporter of quality spices and spice products and has rightly earned the name "Land of spices". Gujarat is the leading state for the production of seed spices. The seed spices are aromatic vegetable products primarily used for seasoning or garnishing food and beverages. They are used in preparation of various value added products *viz.* spice, oil, oleoresins and spices powder. Demand of seed spices increased at the rate of 10-12 per cent per year due to change in market trends, consumer's preferences, health awareness and emergence of super-market. Boosting up crop production is an urgent need to fulfill the requirements of mankind from 0.14 ha land available per capita.

The challenge for agriculture over the coming decades will be meet the world's increasing demand for food in a sustainable way. The overall strategy for increasing crop yields and sustaining them at a high level must includes an integrated approach to the management of soil nutrients along with other complementary measures. Use of chemical fertilizer is the quickest and the surest way for increasing crop production. However, ever increasing price and shortage of chemical fertilizers along with deterioration of soil health by unbalanced and injudicious use of chemical fertilizers necessitated the use of available renewable sources of plant nutrients for complimenting and supplementing the demanded fertilizers. Thus, integrated nutrient management is one of the best approach for judicious use of plant nutrients.

Integrated nutrient management takes in to consideration the nutrient cycle involving soils, crops and livestock, nutrient deficiencies, organic recycling, conjunctive use of organic manures and mineral fertilizers and biological nitrogen fixing potential. If current agricultural practices continue on the existing model, organics may disappear from the scene, and our soils in many areas would soon become deserts or unfit for crop improvement and crop production. The sustained use of organics in crops also increases higher uptake of all nutrients when used along with inorganic fertilizers. Application of organic manures would improve the soil physical properties, increase microbial activity, enhance the response per unit of nutrient added, facilitate slow release of nutrients, reduce nutrient losses and provide a long-term build up of soil fertility.

INM's goal is to integrate the use of all natural and artificial sources of plant nutrients, so that crop productivity increases in an efficient and environmentally benign manner without sacrificing soil productivity of future generations. Declining soil fertility and mismanagement of plant nutrients have made this task more difficult. Balanced application of appropriate fertilizer alongwith its source is a major component of INM. Soil fertility management is the basis for sustainability in every agricultural production system. Creating favourable conditions for soil life and plant growth, nutrient application and soil conservation are important aspects of soil fertility management. Good nutrient management is the basis for sustainable food production.

Crop Response to Integrated Nutrient Management

Cumin (*Cuminum cyminum* L.)

Cumin is one of the risky but more remunerative seed spice crop of arid and semiarid regions, which requires less costly inputs, *i.e.* irrigation, fertilizer and labour. For cumin application of 15-20 tonnes FYM is sufficient under rainfed condition whereas additional 30 kg N is recommended under irrigated conditions (Peter *et al.*, 2000). Crop manured with 5 t FYM ha^{-1} with *Azospirillum* (5 kg/ha) reduced 50 per cent nitrogen requirement (Annual Report,GAU, 2003).

Table 10.1: Effect of Integrated Nutrient Management on Yield of Cumin

Treatments	*Yield (kg/ha)*	
T_1	Inorganic N (100 per cent) + *Azospirillum* (5 kg/ha) + 5t FYM/ha	793
T_2	Inorganic N (75 per cent) + *Azospirillum* (5 kg/ha) + 5t FYM/ha	705
T_3	Inorganic N (50 per cent) + *Azospirillum* (5 kg/ha) + 5t FYM/ha	693
T_4	FYM (5t/ha) + *Azospirillum* (5 kg/ha)	565
T_5	FYM (5t/ha) alone	597
T_6	FYM (10t/ha) + *Azospirillum* (5 kg/ha)	615
T_7	FYM (10t/ha) alone	552
T_8	*Azospirillum* (5 kg/ha) alone	144
T_9	Recommended N and P	678
T_{10}	FYM (5t/ha) + remain N and P in form of inorganic	638
	S.Em.±	48
	C.D. at 5 per cent	141
	C.V. per cent	15.20

Source: Annual Report, GAU, 2003.

Crop fertilized with mustard cake in conjunction with chemical fertilizer increased the yield, quality and net realization (Annual Report, GAU, 2003).

Table 10.2: Effect of Integrated Nutrient Management on Yield and Economics of Cumin

Treatments	*Grain Yield (kg/ha)*	*Oil Content (per cent)*	*Net Realization (Rs/ha)*	*Net CBR*
T_1: F.R.D. in form of castor cake	653	3.04	49070	1: 3.02
T_2: F.R.D. in form of mustard cake	774	3.33	60330	1: 3.53
T_3: F.R.D. in form of F.Y.M.	670	3.14	50890	1: 3.16
T_4: 1/2 R.D. in form of castor cake + 1/2 R.D. in inorganic form.	723	3.13	56585	1: 3.60
T_5: 1/2 R.D. in form of mustard cake + 1/2 R.D. in inorganic form.	869	3.30	70765	1: 4.39
T_6: 1/2 R.D. in form of F.Y.M. + 1/2 R.D. in inorganic form.	769	3.07	61245	1: 3.91
T_7: F.R.D. in inorganic form (N and P)	759	3.11	60383	1: 3.89
T_8: F.R.D. N in inorganic form (N)	746	2.98	59372	1: 3.90
S.Em.±	28	0.07	–	–
C.D. at 5 per cent	78	0.19	–	–

Source: Annual Report, GAU, 2003.

Application of RDN through Mustard cake with fertilizer at the rate of 1:1 recorded the maximum growth and yield attributes as well as yield (869 kg ha^{-1}). Moreover, net returns (Rs. 70,765 ha^{-1}) and benefit: cost ratio (4.39) were also recorded higher by the same treatment. However, fertilizing the crop with inorganic alone or, mustard-cake alone or in conjunction with chemical fertilizer recorded higher oil content but were found statistically at par (Patel *et al.*, 2004).

Table 10.3: Effect of Inorganic N, FYM and Biofertilizer on Yield of Cumin

Treatments	*Seed Yield (kg/ha)*
T_1: Inorganic N (100 per cent) + Azospirillum@ 1.5 kg/ha + 5 t FYM/ha	323
T_2: Inorganic N (75 per cent) + Azospirillum @ 1.5 kg/ha + 5 t FYM/ha	292
T_3: Inorganic N (50 per cent) + Azospirillum @ 1.5 kg/ha + 5 t FYM/ha	256
T_4: Azospirillum @ 1.5 kg/ha + 5 t FYM/ha	243
T_5: 5 t FYM/ha alone	219
T_6: Azospirillum@ 1.5 kg/ha + 10 t FYM/ha	235
T_7: 10 t FYM/ha alone	227
T_8: Inorganic N (100 per cent)	274
T_9: Azospirillum @1.5 kg/ha alone	231
T_{10}: Control	179
CD(P=0.05)	32

Source: Anonymous, 2005.

Combined application of 30 kg N/ha + Azospirillum @ 1.5 kg/ha+ 5 t FYM/ha gave significantly higher seed yield over rest of the treatments except 75 per cent RDN + Azospirillum 1.5 kg/ha + 5 t FYM/ha. (Anonymous, 2005)

Chaudhary *et.al.*(2006) reported that yield and yield attributes, net returns (Rs. 6,357/ha), benefit cost ratio(1.39:1), uptake were recorded significantly higher with combined application of 100 per cent Inorganic N + *Azospirillum* @ 1.5 kg/ha + 5 tonnes FYM/ha.

Fennel (*Foeniculum vulgare* Mill)

Fennel is an important spice and cash crop grows well in fairly mild climate. A cold weather favours higher seed production and it thrives well in drained loamy to loamy clay soil rich in plant nutrients. It can be grown in *kharif* as transplanted crop and in *rabi* as drilled crop.

Patel *et al.* (2003) conducted experiment at Jagudan, they reported that application of RDF gave the maximum fennel seed yield (2456 kg/ha), which was statistically at par with treatments received integrated source on 50:50 basis. Different organic manure (C.C., M.C. and FYM) were equally effective and gave significantly lower seed yield than that of inorganic fertilizer. The highest net profit (Rs./ha 70137) and B:C ratio(4.43) were obtained when RDF applied in the form of chemical fertilizer.

Table 10.4: Fennel Seed Yield and Economic as Influenced by Organic and Inorganic Fertilizers

Treatment	*Yield (Kg/ha)*	*Cost of Production*	*Cost Benefit Ratio*
RD of N in form of C.C.	1996	51900	1:2.89
RD of N in form of M.C.	1915	46545	1:2.27
RD of N in form of FYM	2057	54395	1:3.09
RD of N in form of C.C + inorganic fertilizer @ of 1:1	2264	62827	1:3.83
RD of N in form of M.C. + inorganic fertilizer @ 1:1	2133	56980	1:3.22
RD of N in form of FYM + inorganic fertilizer @ 1:1	2236	62027	1:3.82
RD as inorganic ferti.(N and P)	2456	70137	1:4.43
RD as inorganic ferti. (N only)	2191	61730	1:4.13
CD(P=0.005)	.304	–	–

Note: Purchase price (Rs./tones): castor cake: 2200, mustard cake: 3600, FYM: 200, Nitrogen:8.42 Rs./kg, Phosphorus: 19.25 Rs./kg. Selling price: Rs.35/kg

Source: Patel *et al.* (2003).

Chaudhary and Jat (2004) reported that application of 100 per cent inorganic N (90 kg/ha) + *Azospirillum* + 5 t FYM/ha gave maximum seed yield of 1158 kg/ha which remained at par with 75 per cent N through inorganic source + *Azospirillum* + 5 t FYM and 100 per cent inorganic N alone but proved superior to 5 t FYM + *Azospirillum*, 5 and 10 t FYM alone, *Azospirillum* alone and control. The net return (Rs. 20573/ha) and Benefit cost ratio (2.28) were also found higher under the same treatment.

An application of RD of N and P as integration of organic and inorganic sources were equally effective but produced higher seed yield than organic source alone. A field experiment was conducted on sandy loam soil of Jagudan, reported that integration of organic source *i.e.* FYM (5 t/ha) and biofertilizer (5 kg/ha) reduced the chemical fertilizer need of crop upto 50 per cent (Amin *et al.*, 2007).

Table 10.5: Effect of Inorganic N, FYM and Biofertilizer on Yield and Net Teturn on Fennel

Treatments	*Seed Yield (kg/ha)*	*Net Returns (Rs.ha)*	*B:C Ratio*
Inorganic N (100 per cent) + Azospirillum @ 1.5 kg/ha + 5 t FYM/ha	1158	20573	2.28
Inorganic N (75 per cent) + Azospirillum @ 1.5 kg/ha + 5 t FYM/ha	1090	18453	2.17
Inorganic N (50 per cent) + Azospirillum @ 1.5 kg/ha + 5 t FYM/ha	1023	15753	2.06
Azospirillum @ 1.5 kg/ha + 5 t FYM/ha	933	14868	1.93
5 t FYM/ha alone	907	17058	1.88
Azospirillum@ 1.5 kg/ha + 10 t FYM/ha	1006	15298	1.94
10 t FYM/ha alone	955	20083	1.84
Inorganic N (100 per cent)	1046	15193	2.21
Azospirillum @1.5 kg/ha alone	871	11603	1.97
Control	778	–	1.74
CD(P=0.05)	73	–	–

Source: Chaudhary and Jat (2004).

Table 10.6: Effect of Different Treatments of INM on Seed Yield of *rabi* Fennel Yield

Treatments	*Seed Yield (kg/ha)*
T_1: 100 per cent N as inorganic + Azo. (5 kg/ha)+ 5 t FYM/ha	1581
T_2: 75 per cent N as inorganic+ Azo. (5 kg/ha)+ 5 t FYM/ha	1438
T_3: 50 per cent N as inorganic + Azo. (5 kg/ha)+ 5 t FYM/ha	1170
T_4: FYM (5 t/ha) + Azo. (5 kg/ha)	1105
T_5: FYM (5 t/ha) alone	1044
T_6: FYM (10 t/ha) + Azo. (5 kg/ha)	1245
T_7: FYM (10 t/ha) alone	1225
T_8: Azo. (5 kg/ha) alone	908
T_9: Reco. N and P	1516
T_{10}: T_5 and remain N and P in form of inorganic	1223
T_{11}: T_7 and remain N and P in form of inorganic	1338
C.D. at 5 per cent	195
CV per cent	10.7

Source: Amin *et al.* (2007).

Fenugreek (*Trigonella foenum graceum* L.)

Fenugreek is an important spice, medicinal and green leafy vegetable crop of leguminaceae family, grown in semi arid region with less input costs. The use of symbiotic and asymbiotic nitrogen fixer with phosphatic fertilizer increased the crop yield (1068 kg/ha) and net return (Rs. 3594/ha) remarkably but maximum ICBR was obtained when only biofertilizers (1: 55.50) i. e. *Azotobacter* and PSB were used reported by Parekhiya *et al.* (2000)

Hargilas and Keshwa (2002) observed that application of nitrogen as 25 per cent through urea + 75 per cent through FYM significantly increased the number of pods per plant, seeds per pod and seed yield (16.52 q/ha) over all other treatment combinations of urea and FYM, while the protein content in seed was higher under the treatment received integrated source through 50: 50 basis.

Table 10.7: Effect of Integrated Use of Nitrogen on Yield and Quality of Fenugreek

Treatments	*Seed Yield(q/ha)*	*Protein Content in Seed (per cent)*
Nitrogen at 40 kg N/ha		
100 per cent through Urea	12.2	21.3
75 per cent through Urea + 25 per cent through FYM	14.1	21.3
50 per cent through Urea+ 50 per cent through FYM	15.5	27.7
25 per cent through Urea+ 75 per cent through FYM	16.5	21.4
CD(P=0.05)	1.15	N.S.

Source: Hargilas and Keshwa (2002).

Chaudhary and Jat (2004) recorded the highest fenugreek seed yield(12.56 q/ha), net returns (Rs 9761/ha) and B:C ratio(1.76:1) with 100 per cent inorganic N alone closely followed by 100 per cent inorganic N + *Rhizobium* @ 1.5 kg/ha + 5 t FYM/ha.

Table 10.8: Effect of Inorganic N, FYM and Biofertilizer on Yield and Net Returns (Rs/ha) of Fenugreek

Treatments	*Seed Yield (kg/ha)*	*Net Returns (Rs.ha)*	*B:C Ratio*
Inorganic N (100 per cent) + *Rhizobium* @ 1.5 kg/ha + 5 t FYM/ha	12.52	8414	1.60
Inorganic N (75 per cent) + *Rhizobium* @ 1.5 kg/ha + 5 t FYM/ha	11.76	7146	1.51
Inorganic N (50 per cent) + *Rhizobium* @ 1.5 kg/ha + 5 t FYM/ha	11.77	7264	1.52
Rhizobium @ 1.5 kg/ha + 5 t FYM/ha	11.01	6096	1.44
5 t FYM/ha alone	10.59	5365	1.39
Rhizobium @ 1.5 kg/ha + 10 t FYM/ha	11.78	6232	1.29
10 t FYM/ha alone	11.05	4943	1.33
Inorganic N (100 per cent)	12.56	9761	1.76
Rhizobium @1.5 kg/ha alone	10.27	6014	1.48
Control	9.64	4905	1.39
CD(P=0.05)	9.4	–	–

Source: Chaudhary and Jat, 2004.

Coriander (*Coriandrum sativum* L.)

Coriander is one of the major spice and cash crops of semi arid regions of the state and can be cultivated as a *rabi* crop, which is raised for seed and vegetable purposes. Application of recommended dose of fertilizer through chemical fertilizer along with FYM @ 5 tones ha^{-1} recorded significantly higher seed yield (960 kg/ha) however; use of FYM alone or with fertilizer reduced the fertilizer need of crop reported by (Mallanagouda *et al.*, 1995). Kothary *et al.* (1998) reported that application of RD of NPK(60:30:20 kg/ha) recorded maximum coriander seed yield (14.04 q/ha), which was at par with that of 2.5 t vermicompost + half dose of recommended NPK and 5 t FYM in combination with half dose of recommended NPK.

Table 10.9: Effect of Vermicompost and FYM with and without NPK on Yield of Coriander

Treatments	*Yield (q/ha)*			
	1995-96	*1996-97*	*1997-98*	*Pooled*
5 t vermicompost/ha	8.67	9.42	13.75	10.61
10 FYM	7.42	8.86	10.97	9.08
RD of NPK(60:30:20 kg/ha)	12.00	11.78	18.33	14.04
Half dose of recommended NPK	9.17	9.94	15.56	11.56
Half dose of vermin.(2.5 t/ha)+ half RD of NPK	11.67	11.61	18.19	13.82
Half FYM (5 t/ha) + half RD of NPK	10.25	10.28	17.78	12.77
Farmers practices	8.75	10.17	14.81	11.24
CD(P=0.05)	1.89	1.48	1.33	1.49

Source: Kothary *et al.*, 1998.

Table 10.10: Effect of Integrated Nutrient Management on Growth and Yield of Coriander

Treatments	*Seed Yield (kg/ha)*
Unfertilized control	10.00
RDF @ 90:80:50 kg/ha	12.46
25 per cent RDF + FYM	18.22
25 per cent RDF + *Azospirillum* + PSB	10.52
25 per cent RDF + *Azospirillum* + VAM	10.70
25 per cent RDF + *Azotobacter* + PSB	10.59
25 per cent RDF + *Azotobacter* + VAM	11.55
25 per cent RDF + FYM @ 10 t/ha + *Azospirillum* + PSB	28.14
25 per cent RDF + FYM @ 10 t/ha + *Azospirillum* + VAM	18.65
25 per cent RDF + FYM@ 10 t/ha + *Azotobacter* + PSB	19.00
25 per cent RDF + FYM@ 10 t/ha + *Azotobacter* + VAM	19.18
CD(P=0.05)	7.55

Source: Prabu *et al.*, 2000.

Khoja and Gupta (2004) observed that the integrated supply of 30 kg N/ha through urea + 30 kg N/ha through FYM + Azotobacter gave significantly higher seed yield(15.32 q/ha), net return and B:C ratio over all other combination of sources.

Prabu *et al.* (2000) reported that seed yield of Coriander was significantly higher with 25 per cent recommended dose of fertilizer + FYM @ 10 t/ha + *Azospirillum* + PSB over other combinations of nutrient sources and individual source of nutrient at RDF.

Application of recommended dose of fertilizer with organic sources *i.e.* FYM and *Azospirillum* increased the seed yield and reduced the consumption of fertilizer (AICRPS, 2001).

Chaudhary and Jat (2004) reported that the application of 60 kg N/ha + Azospirillum + 5 t FYM/ha recorded significantly higher seed yield (8.89 q/ha) and maximum net return (Rs. 7732/ha) over rest of the treatments, which was statistically at par with the treatment received 75 per cent through inorganic + *Azospirillum* + 5 t FYM/ha (8.23 q/ha)

Table 10.11: Effect of Inorganic N, FYM and Biofertilizer on Yield (kg/ha) and Net Returns (Rs./ha) of Coriander

Treatments	Seed Yield (kg/ha)	Net Returns (Rs.ha)	B:C Ratio
Inorganic N (100 per cent) + Azospirillum + 5 t FYM/ha	8.89	7732	1.53
Inorganic N (75 per cent) + Azospirillum + 5 t FYM/ha	8.23	6232	1.43
Inorganic N (50 per cent) + Azospirillum + 5 t FYM/ha	7.70	5057	1.36
Azospirillum + 5 t FYM/ha	6.64	2707	1.19
5 t FYM/ha alone	6.27	1807	1.13
Azospirillum + 10 t FYM/ha	7.28	3057	1.20
10 t FYM/ha alone	6.77	1807	1.12
Inorganic N (100 per cent)	8.34	7632	1.58
Azospirillum @1.5 kg/ha alone	6.08	2557	1.20
Control	5.40	882	1.07
CD(P=0.05)	0.68	—	—

Source: Chaudhary and Jat, 2004.

Pariari *et al.*(2007) reported that to maximize the seed yield in Gangetic alluvial plains of West Bendal condition, coriander should be grown with organic and inorganic combination and application of vermicompost has been proved beneficial over control. Maximum coriander seed yield 9.44q/ha with application of chemical fertilizer closely followed by 50 per cent nutrients through chemical fertilizer and 50 per cent nutrient through FYM (9.22 q/ha).

Ajawain (*Trachyspermum ammi* Sprague)

Ajawain is long duration seed spice crop of semiarid region, which is used as an aurvedic medicine also. A field experiments conducted at Ajmer to study the effect of sheep manure, vermicompost and biofertilizer on yield of Ajawain revealed that application of 10 t sheep manure ha^{-1} with

Azospirillium seed treatment improved yield and yield attributes significantly as compared to application of fertilizer or organic manure alone. (Mehta *et al.*, 2007)

Table 10.12: Effect of Sheep Manure, Vermicompost and Biofertilizer on Yield of Ajawain

Treatment	*Umbels/ Plant*	*Umbellates/ Umbel*	*Grains/ Umbellate*	*1000-Seed Weight (g)*	*Grain Yield (q/ha)*
Absolute control	165	9.5	11.8	1.60	10.5
R.F.	176	11.3	12.8	1.74	14.2
B.F. only	170	10.5	12.3	1.65	12.3
BF+R.F.	179	12.5	13.1	1.82	14.5
SM-1 @ 5.0 t/ha	173	10.8	12.2	1.70	12.6
SM-2@7.5 t/ha	176	11.0	12.2	1.73	13.4
SM-3 @ 10.0 t/ha	180	12.5	13.4	1.80	15.3
V. C. -1 @ 2.0 t/ha	172	10.6	12.1	1.68	11.9
V. C. -2 @ 3.0 t/ha	175	11.3	12.3	1.71	13.2
V. C. -3 @ 4.0 t/ha	177	12.5	13.3	1.78	15.0
SM-1@5.0 t/ha+ B.F.	174	11.8	12.5	1.80	13.4
SM-2@7.5 t/ha+ B.F.	182	12.3	12.7	1.85	14.3
SM-3@10.0 t/ha+ B.F.	195	13.5	14.2	1.92	16.4
V.C.-1 @ 2.0 t/ha+ B.F.	180	11.7	12.4	1.74	12.8
V.C.-1 @ 3.0 t/ha+ B.F.	183	12.0	12.6	1.83	14.3
V.C.-1 @ 4.0 t/ha+ B.F.	193	13.3	14.0	1.84	15.2
CD at 5 per cent	2	0.4	0.5	0.06	0.5

Source: Mehta *et al.* (2007).

Dill Seed (*Anethum gravelense* Kurz.)

Dill seed is long duration drought tolerant seed spice crop, which is used in preparation of important soup, sauces, salad and in aurvedic medicines. A field trial conducted at Ajmer to study the effect of sheep manure, vermi-compost and biofertilizer on yield of Dill seed by Mehta *et al.*(2007), concluded that crop manured with 10 t sheep manure ha^{-1} with biofertilizer gave significantly higher yield as compared with fertilizer or organic manure alone.

Table 10.13: Effect of Sheep Manure, Vermicompost and Biofertilizer on Yield of Dill Seed

Treatments	*Grain Yield (q/ha)*	*Treatments*	*Grain Yield (q/ha)*
Absolute control	9.4	RF	13.8
BF only	10.5	BF + RF	14.5
SM-1 @ 5.0 t/ha	12.4	SM-2 @ 7.5 t/ha	13.9
SM-3 @ 10.0 t/ha	14.5	VC-1 @ 2.0/ha	12.2
VC-2 @ 3.0/ha	13.3	VC-3 @ 4.0/ha	14.0
SM-1 @ 5.0 t/ha + BF	13.5	SM-2 @ 7.5 t/ha + BF	14.4
CD	0.5		

Source: Mehta *et al.* (2007).

Conclusion

From the scientific information presented here, it is concluded that adoption of integrated nutrient management practices in seed spices cultivation is low cost and an ecofriendly approach, which not only increase the yield with better quality and profitability, but also reduced fertilizer need of crop and sustain the soil productivity. Thus, adoption of INM reduced cost of production and improved seed quality also. Among the different organic sources, biofertilizer is the cheapest and easily available source for adoption of integrated nutrient management practice.

References

AICRPS, Guntur, 2001. *Annual Report*, pp. 86–87.

Amin, A.U., Patel, I.D., Chaudhary, J.H. and Patel, S.I., 2002. *Annual Report*, Main Spices Research Station, GAU., Jagudan, pp. 1–5.

Anonymous, 2005. All India Coordinated Research Project on Spices, *Twenty Ninth Annual Report*, pp. 47–53.

Chaudhary, G.R., Jain, N.K. and Jat, N.L., 2006. *Indian J. of Agron.*, 51(4): 45–46.

Chaudhary, G.R. and Jat, N.L., 2004. *National Seminar on New Perspectives in Commercial Cultivation, Processing and Marketing of Seed Spices and Medicinal Plants*, Jobner, Rajasthan, March, 25–26, pp. 44–45, 56

Gupta, C.R., Sharma, R.B. and Singh, J., 1999. *Abstracts of Poster Sessions in 4th Agricultureal Sciences Congress*, Held at Madikeri, Karnataka, October 5–6, pp. 76–79.

Hargilash and Keshwa, G.I., 2002. Extended summeries. In: 2nd *International Agronomy Congress*, November 26–30, New Delhi, 1: 305–306.

Jat, N.L., Jain, N.K.and Choudhary, G.R., 2006. *Indian J. of Agron.*, 51(4): 221–222.

Khoja, J.R. and Gupta, A.K., 2004. *National Seminar on New Perspectives in Commercial Cultivation, Processing and Marketing of Seed Spices and Medicinal Plants*, Jobner, Rajasthan, March, 25–26, pp. 33.

Kothary, M.L., Singh, Baldev and Chaodhary, R.L., 1998. *National Seminar on Entomology in 21st Century Biodiversity, Sustainability, Environmental Safety and Human Health*, Held at Udaipur from 30 April to 2nd May.

Mallanagouda, B., Sulkheri, G.S., Hulamani, N.C., Murthy, B.G. and Madalgeri, B.B., 1995. *Current Research*, 24(11): 212–213.

Mehta, R.S., Malhotra, S.K., Vashishtha, B.B. and Meena, S.S., 2007. *Proceeding of National Seminar*, Held at NRCSS, Ajmer, pp. 270–277.

Parekhiya, A.M.; Akberi, A.L. and Andharia, J.H.; 2000. *GAU Res. Journal*, 25(2): 34–38.

Pariari, A.; Imam, M.N. and Chakraborty, S.; 2007. *Proceeding of National Seminar*, Held at NRCSS, Ajmer, pp. 267–369.

Patel, B.S., Patel, I.D., Amin and Chaudhary, J.H., 2003. *Indian Journal of Agronomy*, 48(3): 232–234.

Patel, B.S., Amin, A.U. and Patel, K.P., 2004. *Indian J. of Agron.*, 49(3): 205–206.

Peter, K.V., Srinivasan and Hamza, 2000. *Fertilizer News*, 45(7): 13–28.

Prabu, T., Narwadkar, P.R., Sajindranath, A.K. and Rathod, N.G., 2000. *South Indian Horticulture*, 50(4–6): 680–684.

Recent Advances in Seed Spices (2011) *Pages* **107–118**
Editors: **Y. Ravindrababu, R.K. Jaiman, K.D. Patel, N.R. Patel and S.B.S. Tikka**
Published by: **DAYA PUBLISHING HOUSE, NEW DELHI**

11

Natural Resource Management through Seed Spices and Aromatic Plants

O.P. Aishwath and M.M. Anwar
National Research Centre on Seed Spices, Tabiji – 305 206, Ajmer, Rajasthan

ABSTRACT

Natural Resource Management (NRM) combines economics, policy, and science to study, manage and restore natural resources and ecosystems. NRM help balance the needs of people and economy with the ability of ecosystems to support soil, water, forests, wildlife, fish, and recreational resources. The greatest difficulty for achieving desired results are unrecognized human ignorance acted upon, over exploitation with overconfidence and not considering niche diversification. India has vast area under the different problems of soils and about 167 million hectare suffering from various degradations. These problems are wind and water erosion, water stagnation, salinity and sodicity are most prominent. Other than these problems, some of these soils are poor in fertility and scarcities of total as well as good quality of underground irrigation water make them unsuitable for food grain production. Population pressure on land is increasing day by day and per capita land is shrinking down. Therefore, utilization of under utilized natural resource for the production of seed spices and aromatic plant is a wise step for conserving our natural resources and managing crop biodiversity which ultimately adds to our national economy. The information available on this line are compiled which enunciate that the saline/alkaline soils and water resource could be utilized for the production of fennel, dill, celery, palmarosa, lemongrass, vetiver, jamarosa, citronella, ambrette, chamomile, lavender, mint, occimum and eucalyptus. Almost all the seed spices are fairly tolerable to mild soil acidity and alkalinity while coriander, fenugreek and celery could comes up well even up to 5.5 pH level. For utilization of eroded lands, dill, eucalyptus, *Cymbopogons*, ocimum, lavender and roses are promising. Coriander, cumin, fenugreek, Aromatic grasses, peppermint, davana, lavender and geranium are suitable for nutritionally eroded soil and some of them have association with

VAM leads to effective utilization of available nutrients and water by them. Under submergence or water logging condition, sweet flag, cardamom, clove, kewada, mint and vetiver could be taken successfully with due cares. Seed spices are arid and semi-arid crops and come up with limited water conditions, while dill, ammi, Anise, fennel, papaver crops are fairly tolerant to moisture stress among the seed spices. The aromatic plant recommended for arid areas are ferula, artemisia, sage, lemongrass, guggal and Myrrh. The genera *Rosa*, *Lavandula*, *Mentha*, *Salvia*, *Ocimum*, *Foeniculum*, *Coriandrum*, *Anethum*, *Hyssopus* and *Rhus* can be grown successfully in heavy metal contaminated soil without loss of yield and contamination in essential oil.

Introduction

According to FAO estimates, about 50 per cent total geographical area in India is under various degraded hazards and about 2.1 million ha of land is getting degraded and deforested annually (Sehgal *et al.* 1998). Out of that underground water in arid and semi-arid areas is saline or sodic in nature. While seed spices and aromatic plants may be a boon for these areas, which not only withstand under adverse conditions but also could be productive and helps in mitigating these problems. However, other crops are remain less or unproductive in these areas.

The current export (2005-2006) of seed spices viz coriander, cumin, fennel, fenugreek and celery is worth of Rs 6410, 8800, 1960, 2400 and 1265 lakhs, respectively. Aromatic plants have an equally market potential with the world demand of herbal products growing at the rate of 7 per cent per annum (Aishwath and Tarafdar, 2007). Seed spices and aromatic plants are the natural source of flavour, perfumes and fragrance widely exploited by essential oil industries in the world. Owing to considerably diversity of edaphoclimatic conditions ranging from temperate to tropical and arid to sub-humid that prevails in India and an array of aromatic flora occurs in wild state. Many exotic essential oil bearing plants, including mints, citronella, geranium, patchouli, lavender and chamomile have been naturalized in India. As the country has emerged as one of the main producers of seed spices and some of the essential oils, theses plants are significant foreign exchange earners. Some of the aromatic plants especially grasses can sustain both biotic as well as abiotc stresses, even in the areas unsuitable for conventional crops (Aishwath *et al.*, 2008). Most of the seed spices crops are seasonal crops and comes up well with limited moisture and nutrients. Therefore, under utilized natural resources can be utilized profitably by crop diversification.

1. Utilization of Saline Alkaline Soils Resources for Seed Spices and Aromatic Plants

About 7.2 million ha soils are saline and alkaline out of which 3 million ha is alkali in India. Due to poor physical properties, excessive exchangeable sodium and high pH, most of these lands support a very poor vegetation cover. Most of these areas are under arid or semi-arid regions. This waste area is not utilized properly by the traditional farming. There is lot of scope for utilization of our valuable soil resource by seed spices and aromatic plants for the cultivation of these high-value crops with marginal inputs.

(*i*) Utilization of Saline Soil

Dill can be grown up to EC 3.3 dSm^{-1} (Sharma *et al.*, 2001). In another study on dill with salinity shows that it comes up well with EC 4-5 dSm^{-1} (Rao *et al.*, 2001 and Singh *et al.*, 1995). The beneficial effect of salinity on lemongrass (*Cymbopogon citrates*) and Java citronella (*Cymbopogon winterianus*) was observed at EC 5 dSm^{-1} as compared to lower or its higher levels of salinity (Patra and Singh 1995; Pal *et al.*, 1989). Lemon grass is a very hardy crop which grows successfully on cultivable waste land

and the salt affected soil having EC up to 11dSm^{-1}. The essential oil of German chamomile (*Matricaria chamimilla*) popularly known as blue oil can withstand up to salinity 12.0 dSm^{-1} without leaving any adverse effects on yield. Palmarosa comes up well in various textural types of soils. The salinity upto 5.5 dSm^{-1} found beneficial with respect to its productivity. Based on the pot culture study, there was no adverse effect on yield and quality of palmarosa observed upto EC 11.5 and beyond that soil salinity may adversely affect the yield and quality of palmarosa. Figure 11.1 enunciates that vetiver, jamarosa and citronella could withstand up to salinity 12, 12.5 and 5.5 EC dSm^{-1}, respectively (Singh and Anwar, 1985, Patra and Singh, 1995). In aromatic trees, various species of eucalyptus have wide range of salinity tolerance *i.e.* 3.0 to 20.0 EC dSm^{-1} (Figure 11.2) as reported by Gupta and Gupta (1997). Some of the aromatic plants are salt includer type salinity tolerant has potential for hyper accumulation of salt resultant helps in reclaiming the saline soil such as palmarosa is an hyper salt accumulator (Aishwath and Pal, 2004). Palmarosa grown on the soil having EC 4.8 dSm^{-1}, after two years of cropping, the EC of the soil reduced down to 0.64 dSm^{-1} (Prasad and Singh, 1998).

(*ii*) Utilization of Alkaline Soil

Fennel performs well on heavy textured soils and can also be cultivated on salt affected black soil with salinity of 8-10 dS m^{-1} (Singh, 1999). Fennel and coriander grown on partially reclaimed sodic soil having Exchangeable Sodium Per cent (ESP) 20 with various N and P levels yielded well and large accumulation of Na in fennel stover indicating its tolerance to sodicity. Reduction in soil pH and ESP and increase in exchangeable calcium indicates its capability for reclaiming sodic soils (Garg *et al.*, 2000). Dill (*Anethem graveolens*) can tolerate the pH up to 8.6 (Farooqi, and Sreeramu, 2001). It has been reported that aromatic grasses such as, palmarosa (*Cymbopogon martinii* Roxb. Wats.) and lemon

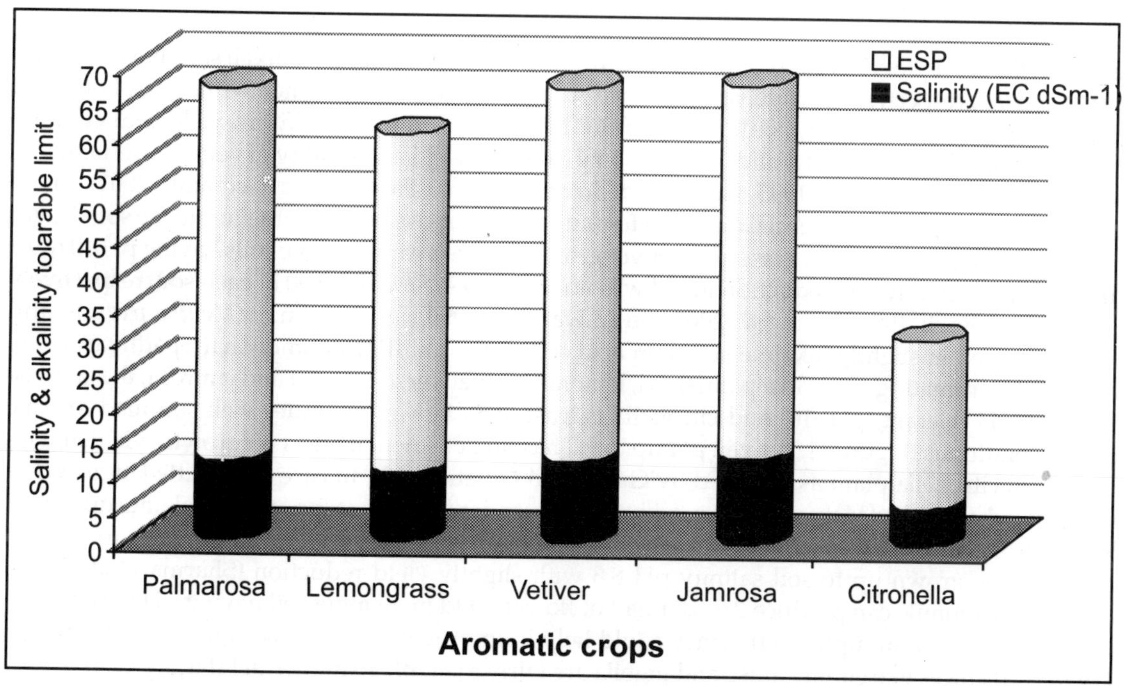

Figure 11.1: Adaptability of Aromatic Crops with Salinity and Alkalinity

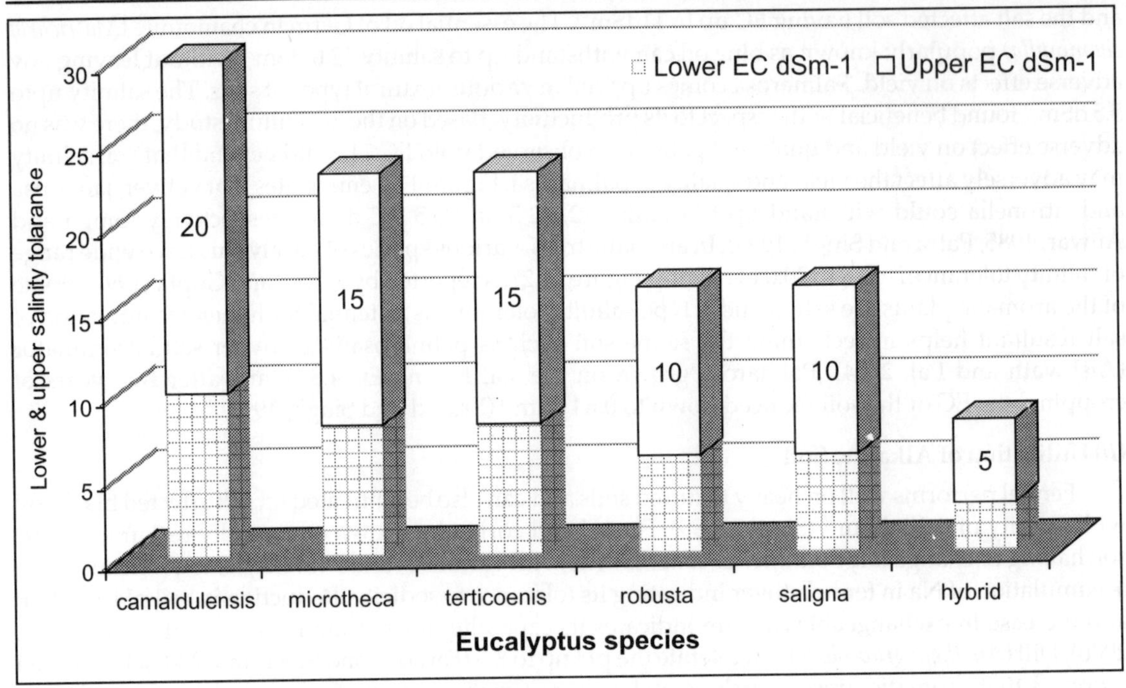

Figure 11.2: Salinity Tolerant Limit of Various Species of *Eucalyptus* sps

grass (*C. flexuosus* Steud. Wats.) could successfully be grown on moderately alkali soils having pH up to 9.0 and 9.5, respectively, while vetiver (*Vetiveria zizanioides* L. Nash) which withstands both high pH and stagnation of water, could successfully be grown without significant yield reduction on highly alkali soils (Figures 11.1 and 11.3). In some cases the yield and quality of vetiver was better on sodic soil (Patra and Singh, 1995, Gupta and Gulati, 1965 and Farooqi and Sreeramu, 2001). These grasses not only produce essential oils used for industrial purpose but also ameliorate the soil (Dagar *et al.* 2004). Palmarosa, lemon grass and vetiver grown for two years on sodic soils having pH 10.6, 9.8 and 10.5, respectively, the reduction in pH was noticed as 9.4, 8.95 and 9.50 in each soil, respectively (Prasad and Singh, 1998; Patra *et al.*, 2002 and Anwar *et al.*, 1996). Japanese mint (*Mentha arvensis*) also comes up well in slightly alkaline soil than acid sedentary soil (Prasad and Chattopadhyay 1999). Among the aromatic grasses like palmarosa, lemongrass, vetiver, citronella and jamrosa, the jamrosa is a high pH tolerant (pH=10) and citronella is least (8.5). However, under well supplied organic matter conditions (Figure 11.3), even pH up to 9.7 has no detrimental effect on growth, yield and oil content in citronella plant as reported by Gupta and Gulati (1965). In case of amrette (*Abelmoschus moschata*), chamomile (*Matricaria chamomilla*), and Lavender (*Lavandula species*) can tolerate upper pH level as 8.6, 9.0and 8.4, respectively (Farooqi, and Sreeramu, 2001). An another study it was found that dill could be grown up to soil salinity pH 8.8 with slightly yield reduction (Sharma *et al.*, 2001). German chamomile can produce 2.5-3.0 t ha^{-1} of flower yield in alkaline soil having pH 9.5 and also have the high sodium uptake efficiency could help in improvement of sodic soil (Patra and Singh, 1995). Seed spices like cumin, anise and nigella are fairly tolerable to higher pH (Figure 11.4).

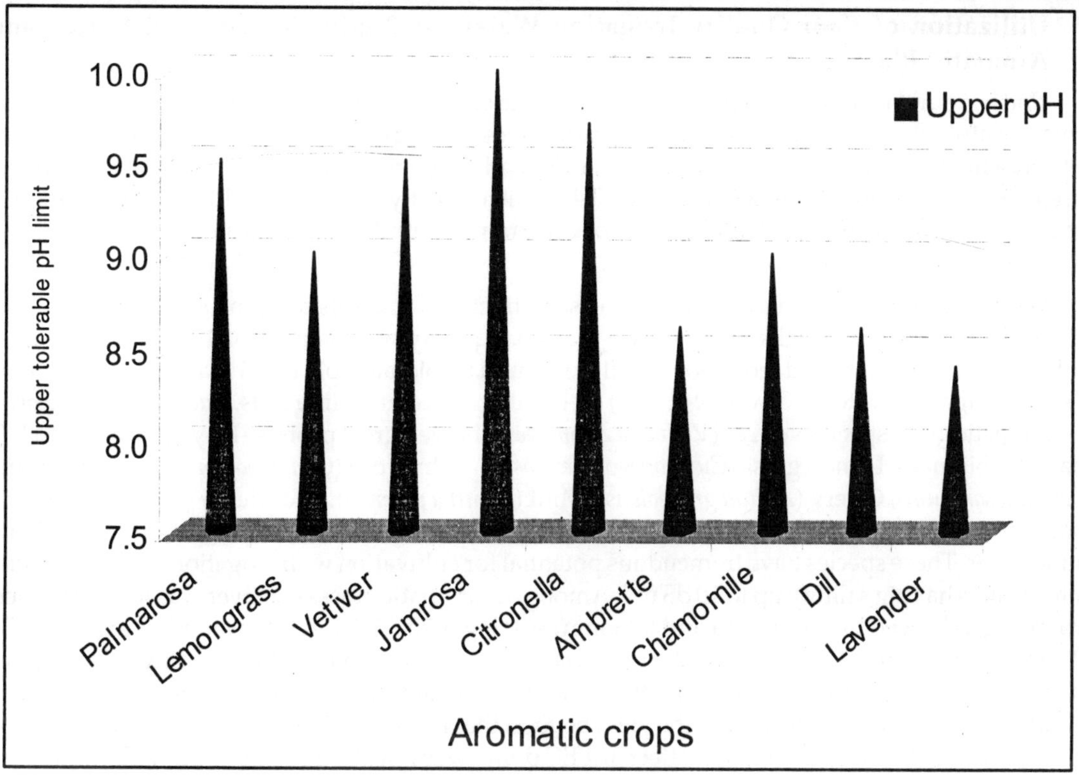

Figure 11.3: Upper Permissible Limit for Adaptability of Aromatic Crops with pH

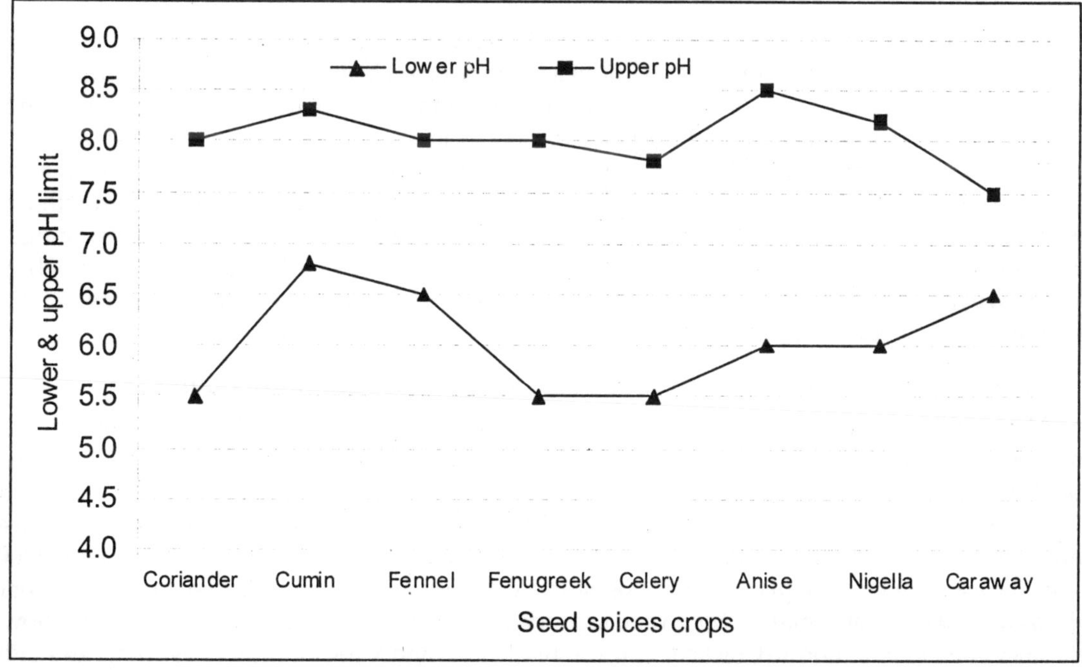

Figure 11.4: Adaptability of Seed Spices Crops with Various pH Ranges

2. Utilization of Poor Quality Irrigation Water for Production of Seed Spices and Aromatic Plants

Underground water always contains some soluble salts. Continuous use of such water may leave some harmful affect the crop growth and soil properties depending upon the quality and quantity of salts present. Accumulation of salt in soil profile depends upon the texture, structure and the depth of the soil (Kelly, 1963 and Aishwath and Pal, 2004), which directly affects the yield and composition of the crops (Aishwath and Pal, 2000). The major constraints of underground water uses for irrigation are (*i*) Salinity and (*ii*) Sodicity.

Fennel can be cultivated profitably in the salt affected black soils by using saline water of 4 dS m^{-1}. It is also revealed that the water of sub-marginal quality with salinity of 4 dS m^{-1} can be used without any significant yield reduction in dill. It is found to tolerate a pH of 8.6 (Rao *et al.*, 2001) Usage of saline irrigation water (ECiw 8-10 dS m^{-1}) on highly calcareous saline soils, *Eucalyptus tereticornis* and aromatic grass species vetiver (*Vetiveria zizanoides*) showed great promise by yielding 1.19 to 1.73 t ha^{-1} root biomass. Lemon grass (*Cymbopogon flexuosus*), Palmarosa (*Cymbopogon martinii*), Ocimum (*Ocimum sanctum*), Celery (*Apium graveolens*), Mint (*Mentha piperita*), and Metha (*Trigonella foenum-graecum*) were irrigated either with saline or canal water alone or with the alteration of canal and saline water. These species have tremendous potential for cultivation with irrigation of underground saline waters having salinity up to 10 dS m^{-1}. Among the aromatic grasses, vetiver was found the most promising, followed by palmarosa and lemongrass (Dagar 2007). Irrigation water having water EC 10.0 and EC 16.0 dSm^{-1} does not have adverse effect on herb, essential oil yield and quality of lemongrass and palmarosa, respectively (Patra *et al.*, 1992). German chamomile is also tolerant to irrigation water salinity up to 8 dSm^{-1} and decreased the yield beyond that level of salinity (Patra and Singh 1995). In *Cymbopogon martinii* the detrimental effect of 4 EC dSm^{-1} was not observed in sandy soil, while same EC level was detrimental for palmarasa on sandy clay loam and sandy loam soils (Aishwath *et al.*, 1990 and Aishwath and Pal 2004).

The irrigation water applied in sandy loam and sandy clay loam soil having SAR 9.1 leaves adverse effect on the yield of palmarosa without any effect on oil content. However, at same level of SAR no adverse effect observed on yield of palmarosa in sandy soil (Aishwath and Pal, 2004). Up to seven numbers of irrigations with sodic (RSC 10.0) water, there was no adverse effect on growth and yield of ambrette, palmarosa and lemongrass. At the initial stage of establishment of kewada has problem with this sodic water, but after 3-4 years of its establishment no adverse effect on growth was observed with sodic water (Aishwath, 2005a). However, some of the aromatic plant like ginger, lemongrass and jasmine are sensitive for sodic water (Aishwath, 2005b). The adverse effect of high SAR (10.0) of irrigation water did not appear on growth and yield palmarosa. However beyond that level the growth, yield and uptake of the N, P, K, Ca and Mg reduced (Pal *et al.* 1993).

3. Management of Acid Soils through Seed Spices and Aromatic Plants

Almost all the seed spices are fairly tolerable to mild soil acidity and alkalinity while coriander, fenugreek and celery could comes up well even up to 5.5 pH level (Figure 11.4). It has also been reported that coriander comes well with the soil pH 6.5-8. Cumin can be grown with pH ranging from 6.8 to 8.3. While, fennel could be grown in heavy clay to sandy loams soils with pH of 6.5-8.0. In case fenugreek, it is cultivated on well drained loams or sandy soils and the acceptable pH rang is 6.0-8.0. The heavy loam or clay with pH 6.0-7.5 is most suitable. Celery thrives well in acid soil but not beyond pH 5.5 and rich in organic matter. Micro and secondary nutrient are essential otherwise plant shows deficiency symptoms as Boron- Cracked stem, Ca–Black heart and Mg-Chlorosis. Anise requires sandy loam to clay loam soil and pH 6.0-8.5 is well suited to this crop. For nigella, well drained sandy to heavy clay soil acid to alkaline pH is suitable for optimum plant growth. Caraway requires neutral to slightly alkaline soil having pH 6.5-7.5 is most suitable. It shows that these crops have wide adaptability with respect to soil reaction could be fitted into any adverse soil conditions (Aishwath *et al.*, 2008).

4. Utilization and Management of Eroded Lands Through Seed Spices and Aromatic Plants

Out of total land area, about 50 per cent is under degraded and cultivated land subjected to serious hazards of water, wind, shifting cultivation, water-logging, salinity and alkalinity and changing water courses. (Narayana and Babu 1983). While seed spices crops and aromatic plants can combat against these problems of erosion on sloppy land. The cultivation of seed spices and aromatic plants like *Anethum sowa* and *Ocimum kilimandscharicum* recommended in degraded habitat orchards as inter-crop (Gupta *et al.*, 1998 and Narayana, 1998). Aromatic grasses on sloppy lands have wide potential for the extraction of oil. In Cymbopogon species, *Cymbopogan martinii* produced oil yield of 68.3 kg/ha/year followed by *C. winterianus*, 41.3kg/ha/year. At many a places, Cymbopogons are utilizing for dual purpose to control the gully erosion and oil extraction in Uttar Pradesh, Madhya Pradesh, Rajasthan and Gujarat (Verma and Chinnamani, 1998). Similarly, Eucalyptus globules and Eucalyptus robusta utilized in humid regions for dual purpose to control the soil erosion and stabilization of gully erosion as well as essential oil extraction (Ramamohan Rao, 1998). Lavender (*Lavandula officinale*) and rose (*Rosa domascena*) could also be grown in sloppy and erodes lands or terraces in hill areas (Shiva *et al.*, 2002).

5. Utilization of Nutritionally Eroded Lands Through Seed Spices and Aromatic Plants

Fast mineralization of soil organic matter and release of nutrients in soil leads to losses of nutrients through water run off in arid and semi-arid areas. It has been estimated that India is loosing about 8.2 million tonnes of nutrients annually (Singh 1998). In these areas major portion of soil is composed of sand (>85 per cent) resulting lower retention power of nutrient, high infiltration rate (9-16 cm hr^{-1}) and poor moisture storage capacity. Lower fertility and moisture stress which offer a limited choice for increasing crop production under varying rainfall situation.

Coriander, cumin, fenugreek, Aromatic grasses, peppermint, davana, lavender and geranium are suitable for nutritionally eroded soil and some of them have association with VAM leads to effective utilization of available nutrients and water by them. It is well known fact that grasses can squeeze the soil by extracting the nutrients with their complex root system. Here, vast potential of aromatic grass could be harvested by utilizing nutritionally eroded soil with limited water (Aishwath *et al.*, 2008). Some of the seed spices and aromatic plants have lower requirement of nutrient can be grown on lower fertile soil (Table 11.1). It has also been reported that VAM simulate plant growth by physiological effects other than by enhancement of nutrient uptake or by reducing the severity of diseases caused by soil borne pathogens (Dehne 1982). The responsiveness of VAM to the host specific seed spices and aromatic plant is given in Table 11.2. Recently, possibilities have been explored for the cultivation of rose scented geranium (*Palargonium sps*) in semi-arid conditions by multiple cuttings/harvesting (Kothari *et al*, 2004).

6. Management of Arid/Dry and Rainfed Areas Through Seed Spices and Aromatic Plants

The arid regions of the world occupy very extensive areas of the continental land masses and constitute nearly one-half of the land surface of the Earth. The major portion of the African continent and a great part of the Asian continent are arid.

Table 11.1: Nutrient Removal by Some Seed Spice and Aromatic Crops

Crop	N Uptake (kg ha^{-1})	P Uptake (kg ha^{-1})	K Uptake (kg ha^{-1})
Menthol mint	109	28	129
Citronella	181	33	255
Lemongrass	160	32	194
Palmarosa	430	62	339
Geranium	180	24	137
Coriander	30	4	36
Davana	179	26	171
Japanese mint	109	28	129
Pappermint	42	–	–

Source: Prakasa rao, 1992 and 1993.

Table 11.2: Per cent Colonization of VAM Fungi in Some Seed Spices and Aromatic Plants

Crop	Per cent Colonization	Crop	Per cent Colonization
Basil	63	*Mentha arvensis*	46
Citronella	75-91	*Mentha piperita*	100
Coriander	32	*Mentha spicata*	100
Cymbopogon citratus	80	*Mentha citrata*	40
Cymbopogon flexuosus	72	Musk	83
Cymbopogon jwarancusa	50	Palmarosa	41-98
Cymbopogon khasianus	76	Patchouli	43
Davana	86	Rose-scented geranium	22-59
Fenugreek	92	Sacred basil	77
Lavender	80	Vanilla	57
Lemongrass	21-75	Vetiver	42

Source: Aishwath and Tarafdar, 2007.

Vegetation in the dry zones is very sparse. The general landscape is desolate and barren. Herbs appear only during a short period of the year when the conditions become favorable. The shrubs, herbs and trees have various means of storing water. These adaptations include increased ability to store water in their succulent stems or leaves; thickening of the leaf cuticle or reduction of leaf surface or the entire absence of leaves to lower the transpiration rate, and the ability to survive as a seed through many years of aridity. Drought – resisting plants may be divided into two groups: Succulents and Ephemerals. Seed spices are arid and semi-arid crops and come up with limited water conditions, while dill, ammi, Anise, fennel, papaver crops are fairly tolerant to moisture stress among the seed spices. The aromatic plant recommended for arid areas are ferula, artemisia, sage, lemongrass, guggal and Myrrh. Recent research on *Ammi visnaga* and *A. majus*, *Pimpinella anisum*, *Foeniculum vulgare*, *Ferula asafetida*, *F. foetida*, Artemisia proved to be adaptive in dry areas. Some artemisias are of medicinal interest and a few are prized for their volatile oils. The chief source of santonin, in the U.S.S.R., is

A. cine Berg. Growing in abundance in Russian Turkestan and Persia. Other species which have been found to contain santonin include *A. mexicana* Willd., A. neo – mexicana Woot, and *A. wrightii. A. grey*, found in America and *A. gallica* Willd. in East Germany, France, England and Scotland, and A. maritime L., a fairly widely distributed species spread from England to as far east as Chinese Mongolia. The content of santonin in the American and English species is too low to warrant commercial extraction. *A. fragrans* Willd. and *A. parviflora* Roxb. occurring in Afghanistan have also been reported to contain santonin. *A. maritime* growing in certain areas of Kashmir (India) and Kurram (Pakistan) has been found to contain substantial amounts of santonin (1–2 per cent), which is commercially extracted for aromatic and medicinal purpose. Lemongrass or Indian Verbena (*Cymbopogon citrates, C. schoenanthus* and *C. nardus*) southern parts of the eastern Egyptian desert and which yields a volatile oil are adaptive to water stress. Myrrh is still used to some extent in pharmacology as an aromatic stimulant; fenugreek (*Trigonella foenumgraecum* L.) contains diosgenin content could serve for the demi-synthesis of cortisone derivatives and sex hormones are come up well under water stress (Chopra *et al.*, 2003).

7. Utilization of Water Logged Areas through Spices and Aromatic Plants

There has been no accurate estimate of the area afflicted in India under water logging. Framji (1974) reported that nearly 3.5 million hectares areas become waterlogged with the rise in water table in the root zone. Recently, Sehgal *et al.* (1998) reported about 20 million hectares area is under flooding causes water logging.

Most of the seed spices are susceptible for the water logging condition. However, water logging or submerged conditions can be utilized for the production of *Acorus calamus*, while, cardamom (*Elettaria cardamomum*) is best suited in swampy areas for its ideal production. Clove can withstand for a shorter period under water logging conditions. In case of kewada (*Pandanus species*), it is highly tolerable to water logging conditions for a longer period. The mint species *Mentha arvensis* is moderately tolerable to water logging for 8-10 days provided low atmospheric temperature. However, vetiver is capable of prolong water stagnation and also yielded with full potential (Shiva *et al.*, 2002).

8. Utilization of Metal Contaminated Soils and Environment Protection through Seed Spices and Aromatic Plants

With our industrial progress, we are also progressing to contaminate our soil and water resource by various metals and heavy metals causes serious problem to the living ecosystem and their environment. Aromatic crops can provide economic return and metal-free final product, the essential oil. The study in Bulgaria using coriander, sage, dill, basil, hyssop, lemon balm, and chamomile grown at various distances from the smelter revealed that aromatic crops may not have significant phytoremediation potential, but growth of these crops in metal contaminated agricultural soils is a feasible alternative. Herbage essential oil yields of basil, chamomile, dill, and sage were reduced when they were grown closer to the contaminants. Metal removal takes place from the site with the harvestable plant parts was as high as 180 g ha^{-1} for Cd, 660 g ha^{-1} for Pb, 180 g ha^{-1} for Cu, 350 g ha^{-1} for Mn, and 205 g ha^{-1} for Zn. It also reported that high concentrations of heavy metals in soil or growth medium did not result in metal transfer into the essential oil. Of the tested metals, only Cu at high concentrations may reduce oil content (Valtcho *et al.*, 2008). No detectable amount of Cd, Cu, or Pb was found in the oils plant species Peppermint, basil, and dill. These can be grown in soils enriched with Cd, Pb, and Cu medium without risk for metal transfer into the oils, and without significant alteration of essential oil composition that may impair marketability (Valtcho *et al.*, 2006). Zheljazkov and Jekov (1995) investigated by Furnace Atomic Absorption (GFAA) and Inductively Plasma (ICP) techniques. The content of Cd, Pb, Cu, Mn, Zn and other trace elements determined in essential oils and plant extracts from the genera *Rosa, Lavandula, Mentha, Salvia, Ocimum, Foeniculum, Coriandrum, Anethum, Hyssopus* and *Rhus*. On the other hand the heavy metal content in oils from different regions were compared to heavy metal content in some oils obtained from species grown on heavily polluted soils and under severe atmospheric pollution. Results, obtained from the two apparatuses were very similar. They found that the concentration of the most hazardous heavy metals in all of the tested oils and plant extracts was very low, near the detection limits of the used apparatuses. It was concluded that most of the essential oil and medicinal crops could be successfully grown on heavy metal polluted soils and under atmospheric pollution as substitutes for some other edible crops.

Reference

Aishwath, O.P., 2005a. Appraisal of underground water in four consecutive years at NRC MAP, Boriavi (Gujarat) and impact of its irrigations on medicinal and aromatic plants. *Ecology, Environ. and Conserv.*, 11: 373–380.

Aishwath, O.P., 2005b. Coefficient of variation and correlation coefficient in underground water quality parameters in and adjoining municipal area of Boriavi, Gujarat. *Pollution Res.*, 24: 9–17.

Aishwath, O.P. and Pal, B., 2000. Performance of palmarosa (*Cymbopogon martinii* var Motia) under various soil texture and salinity levels. *Indian Perfumer*, 44: 285–290.

Aishwath, O.P. and Pal, B., 2004. Mineral composition of palmarosa (*Cymbopogon martinii*), soil EC and pH as influenced by saline water irrigation in texturally different soils. *J. Guj. Soc of Agron. Soil Sci.*, 4: 42–47.

Aishwath, O.P., Pal, B. and Singh, A.K., 1990. Effect of soil texture and water salinity on the growth, dry matter production, nutrients uptake and oil content of palmarosa (*Cymbopogon martinii*). *J. Agri. Sci. Res.*, 32: 45–47

Aishwath, O.P. and Tarafdar, J.C., 2007. Role of organic farming in medicinal and aromatic plants. In: *Organic Farming and Sustainable Agriculture*, (Eds.) Tarafdar *et al.* Scientific Publisher, Jodhpur, pp 135– 185.

Aishwath, O.P., Vashishtha, B.B. and Malhotra, S.K., 2008. Scope of production of aromatic plants in arid and semi-arid regions. In: *Proc. National Workshop on Spices and Aromatic Plants*, Held on 6–7 February at ARS (RAU) Mandor, Jodhpur, Rajasthan, pp. 10.

Anwar, M., Patra, D.D. and Singh, D.V., 1996. Influence of soil sodicity on growth oil yield and nutrient accumulation in vetiver (*Vetiveria zizanioides*). *Ann. Arid Zone*, 35: 49–52.

Chopra, L.C., Abrol, B.K. and Handa, K.L., 2003. *Medicinal Plants of the Arid Zone*. Today and Tomorrow Printers and Publishers, pp. 11–55.

Dagar, J.C., 2007. Alternative land uses in areas with water scarcity. In: *On-farm Land and Water Management: A Compilation of National Training Course*, pp. 154–165.

Dagar, J.C., Tomar, O.S., Kumar, Y. and Yadav, R.K., 2004. Growing three aromatic grasses in different alkali soils in semi-arid regions of northern India. *Land Degradation and Development*, 15 : 143–151.

Dehne, H.W., 1982. Interaction between vesicular arbuscular mycorrhizal fungi and plant pathogens. *Phytopathol.*, 726: 1115–1119.

Farooqi, A.A. and Sreeramu, B.S., 2001. Ambrette, Chamomile, Dill, Lavender, and Palmarosa. In: *Cultivation of Medicinal and Aromatic Plants*, University Press (India) Ltd., Hyderabad, pp. 295–423.

Framji, K.K., 1974. *International Commission on Irrigation and Drainage Bull.*, pp. 36–41.

Garg, V.K., Singh, P.K and Katiyar, R.S., 2000. Spices and aromatic plants. In: *Proc. Centennial Conference on Spices and Aromatic Plants*, 20–23 September, Calicut, Kerla, India, pp. 133–138.

Gupta, R. and Gulati, B.C., 1965. Salt tolerance in *Cymbopogon nardus* (Lin) Rendle (Citronella cylon type) possibilities of commercial cultivation in *ushar* lands. In: *Proc. Sem. Sea Salt and Plants*, held at Bhavnagar, pp. 100–106.

Gupta, R.K., Bhardwaj, S.P. and Mittal, S.P., 1998. Management of degraded lands in the Himalayas. In: *Technologies for Wasteland Development*, (Eds.) I.P. Abrol and V.V. Dhurva Narayana. Information Div., ICAR, KAB, New Delhi, pp. 196–208.

Gupta, S.K. and Gupta, I.C., 1997. Crop tolerance to saline conditions. In: *Management of Saline Soils and Water*. Scientific Publishers, Jodhpur, India, pp. 99–117.

Kelly, W.P., 1963. Use of saline irrigation water. *Soil Science*, 96: 305–391.

Kothari, S.K., Vijay Kumar, Y., Bhattacharya, A.K. and Ramesh, S., 2004. Harvest management in rose-scented geranium (*Pelargonium* sps.) for higher essential oil yield and superior quality under semi-arid tropics. *Archives of Agron. and Soil Sci.*, 50: 593–600.

Narayana, V.V. Dhurva, 1998. Land degradation due to water erosion. In: *Technologies for Wasteland Development*, (Eds) I.P. Abrol and V.V. Dhurva Narayana. Pub. Information Div. ICAR, KAB, New Delhi, pp. 190–213.

Narayana, V.V. Dhurva and Babu, R., 1983. Estimation of soil erosion in India. *J. Irri. Drain. Engg., ASCE*, 109: 419–434

Pal, B., Singh, A.K., Singh, P.K. and Singh, R.P., 1993. Effect of saline water on herb, oil yield and composition of palmarosa (*Cymbopogon martini* Var Motia). *Indian Perfumer*, 37: 267–269.

Pal, B., Singh, K.K., Jadaun, S.P.S. and Parmar, A.S., 1989. Effect of water salinity on herb and oil yield and composition of some *Cymbopogon species*. *Indian Perfumer*, 33: 196–199.

Patra, D.D., Anwar, M. and Singh, D.V., 1992. Herb yield essential, oil yield and nutrient accumulation by palmarosa under saline water irrigation. In: *Abstract of Technical Paper Presented at the Conference on Symposium on Environmental Soil Science*, Canadian Society of Soil Science, August 8–13, Alberta, Canada, pp 36.

Patra, D.D. and Singh D.V., 1995. Utilization of salt affected soil and saline sodic irrigation water for cultivation of medicinal and aromatic palnts. *Current Res. on Med. and Arom. Pl.*, 17: 378–381.

Patra, D.D., Prasad, A., Anwar, M., Singh, D., Chand, S., Ram, B., Katiyar, R.S. and Kumar, S., 2002. Performance of lemongrass cultivars intercropped with chamomile under sodic soil with different levels of gypsum application. *Comm. Soil Sci. Pl. Analy.*, 33: 1707–1721.

Prakasa Rao, E.V.S., 1992. Fertilizer management in medicinal and aromatic crops. In: *Non-traditional Sectors for Fertilizer Use*, (Ed.) H.L.S. Tandan. Fertilizer Development and Consultation Organization, New Delhi, India, pp. 73–93.

Prakasha Rao, E.V.S., 1993. Nutrient management in some important tropical aromatic crops present and future. *Indian Perfumer*, 37: 35–39.

Prasad, A. and Chattopadhyay, A., 1999. Growth and yield of Japanese mint (*Mentha arvensis*) in acid sedentary and slightly alkaline soil. *Indian Perfumer*, 43: 9–14.

Prasad, A. and Singh, D.V., 1998. Effect of organic amendments on yield of palmarosa (*Cymbopogon martinii* Roxb. Wats) var Motia and chemical properties of sodic soil. *Indian Perfumer*, 42: 21–26.

Ramamohan Rao, M.S., 1998. Approaches to management of denudated lands in peninsular, India. In: *Technologies for Wasteland Development*, (Eds.) I.P. Abrol and V.V. Dhurva Narayana. Pub. Information Div. ICAR, KAB, New Delhi, pp. 232–244.

Rao, G.G., Nayak, A.K., Chinchmalatpure, A.R. and Khandelwal, M.K., 2001. Yield of dill (*Anethum graveolens*) on saline black soils of different unirrigated farm sites of Bhal area in Gujarat. *Indian Journal of Agricultural Sciences*, 71: 711–712.

Rao, G.G., Nayak, A.K., Chinchmalatpure, A.R. Singh, R. and Tyagi, N.K., 2001. Resource characterization and management options for salt affected black soils of agro-ecological region of Gujarat state, pp. 48–53.

Sehgal, J.L., Saxena, R.K. and Pofali, R.M., 1998. Degraded soil: Their mapping through soil surveys. In: *Technologies for Wasteland Development*, (Eds.) I.P. Abrol and V.V. Dhurva Narayana. Pub. Information Div. ICAR, KAB, New Delhi, pp. 1–20.

Sharma, P.C., Mishra, B., Singh, R.K. and Singh, Y.P., 2001. Variability in the response of spinach, fenugreek and coriander to alkalinity and salinity stresses. *Indian Journal of Plant Physiology*, 6: 329–333.

Shiva, M.P., Lehri, A. and Shiva, A., 2002. Introduction, Calamus, Cardamom, Clove Kewada Mint and Vetiver. In: *Aromatic and Medicinal Plants*. International Book Distributors, Dehra Dun, India, pp. 67–307.

Singh, D.V. and Anwar, M., 1985. Effect of soil salinity on herb and oil yield and quality of some *Cymbopogons species*. *J. Indian Soc. Soil Sci.*, 33: 362–365.

Singh, R., 1999. Lavaniya jal aur mrda mai sounf ke kheti. *Phal Phool*, October–December, pp. 30–35.

Singh, R., Bhargava, G.P. and Singh, R., 1995. Response of safflower (*Carthamus tinctorius*) and dill (*Anethum graveolens*) to soil salinity. *Indian Journal of Agricultural Sciences*, 65: 442–444.

Singh, R.P., 1998. Land degradation problems and their management in semi-arid tropics. In: *Technologies for Wasteland Development*, (Eds.) I.P. Abrol and V.V. Dhurva Narayana. Pub. Information Div. ICAR, KAB, New Delhi, pp. 125–136.

Valtcho, D.Z., Lyle, E.C. and Baoshan, X., 2006. Effects of Cd, Pb, and Cu on growth and essential oil contents in dill, peppermint, and basil. *Environmental and Experimental Botany*, 58: 9–16.

Valtcho, D.Z., Lyle, E.C., Baoshan, X., Niels, E.N. and Andrew, W., 2008. Aromatic plant production on metal contaminated soils. *Science of the Total Environment*, 395: 51–62.

Verma, V. and Chinnamani, S., 1998. Rehabilitation of revine lands. In: *Technologies for Wasteland Development*, (Eds.) I.P. Abrol and V.V. Dhurva Narayana. Pub. Information Div. ICAR, KAB, New Delhi, pp. 263–272.

Zheljazkov, V. and Jekov, D., 1995. Heavy metal content in some essential oils and plant extracts. In: *Proc. International Symposium on Medicinal and Aromatic Plants*, held on 1 November at Amherst, USA, Published in Acta Horticulturae, 2006, pp. 426.

Recent Advances in Seed Spices (2011) *Pages* **119–122**
Editors: **Y. Ravindrababu, R.K. Jaiman, K.D. Patel, N.R. Patel and S.B.S. Tikka**
Published by: **DAYA PUBLISHING HOUSE, NEW DELHI**

12

Cumin Wilt and its Management

S. Gangopadhyay[1] and R. K. Jaiman[2]
[1]Professor and Head, Department of Plant Pathology, College of Agriculture, Rajasthan Agricultural University, Bikaner
[2]Assistant Research Scientist, Centre for Research on Seed Spices, SDAU, Jagudan, Gujarat

Cumin (*Cuminum cyminum* L.) is one of the major seed spice crop grown in India. Besides India, it is cultivated in Iran, Turkey and in Syria mainly for exports. The crop is also cultivated in Egypt, Afghanistan, Pakistan, North America and Chile. It is widely used as spice and for medicinal purposes all over the world. Rajasthan and Gujarat are the two major cumin growing states of the country. The area under cumin cultivation in India during last couple of years ranged from 3.5 to 4.0 lakh hectares and the production ranged 1.5 to 2.0 lakh tonnes. In Rajasthan, it is mainly grown in the districts of Jalore, Jodhpur, Barmer, Nagour, Pali, Sirohi and Ajmer. In Gujarat, the crop is primarily cultivated in the districts of Ahmedabad, Basankantha, Mehsana, Patan, Jamnagar, Rajkot, Surendranagar and Junagadh. India contributes about 70 per cent of total world production of cumin. In India, the highest ever production was recorded in 2001-02, which was 2.06 lakh tonnes from a total area of 5.26 lakh hectares. More than 80 per cent production of cumin is internally consumed (Anonymous, 2008). The crop is very much weather sensitive and so it gets affected by slightest change in weather.

Fungal diseases like wilt, blight and powdery mildew are serious problem in this crop and cause considerable yield losses in seed yield in most of the cumin growing areas of India. Out of these diseases, wilt caused by *Fusarium oxysporum* (Schlecht) f. sp. *cumini* (Prasad and Patel, 1963) is almost an endemic problem in most of the cumin growing areas in India and causes substantial yield losses. The disease is also prevalent in cumin growing countries of the world.*viz.*, Iran, Turkey, Argentina, Israel, Russia and Turkey. The losses due to *Fusarium* wilt depends on inoculum density of the pathogen in soil and environmental conditions favourable for the disease. Seed yield losses varied to the extent of 65 per cent (Mathur and Mathur, 1964). Some aspects of biology and management of the disease will be discussed in the present paper.

Biology

The disease usually appears in patches and the plants may get infected throughout the growing season. Initially drooping of tips and leaves is observed. Later on, the infested plants show yellowing of leaves followed by wilting and drying of entire plants. Infection at early stages of growth results in complete plant mortality. *Fusarium oxysporum* f. sp. *cumini* is a soil-borne pathogen and it produces micro-conidia, macro-conidia and chlamydospores abundantly. Chlamydospore is resistant to environmental stresses and can survive for long time in absence of host. The seed-borne inocula play an important role in spread of the disease to new areas. It is commonly observed that the severity of the disease progresses fast if the crop is cultivated continuously for more than two to three seasons (Israel and Lodha, 2004). The eradication of this pathogen from infested soil is difficult. The fungus is known to produce toxins like Fusaric acid which is responsible for its pathogenesis in host crops. Limited information is available on biology and epidemiology of *Fusarium* wilt in cumin.

Disease Management

Fusarium wilt is difficult to control because of ecological fitness of the pathogen and susceptibility of cumin crop to biotic and abiotic stresses. In fact, there is a need for efficient disease control strategies that is compatible with the environmentally sustainable crop production system. The integrated wilt management strategies involves selection and application of effective and harmonious range of control strategies. The objective of the programme should be acceptable in economic terms to farmers and at the same time caùse least disturbance to the environment and minimum hazardous effect on non-target microbes, plant and animal health.

Cultural Practices

Cultural practices are considered as first line of defense and are helpful in reducing the inoculum level which slow down or delay the epidemics. Crop rotation effectively controls those pathogens that survive in soil or crop residues. The population of *F. oxysporum* f. sp. *lycopersicae* increases in soil if cumin is grown continuously. *F. oxysporum* although considered as root inhabitant, the chlymodospores produced can survive for long time. Hence, the dynamics of *Fusrium* population, in non -host crops included in rotation has to be studied using suitable selective enumeration technique.

Host Resistance

Resistant cultivar being non-disruptive to environment, is most desirable option for disease management. Unfortunately, resistance is lacking in cumin germplasms. The shortcomings of resistant cultivar is that the selections may lack adequate agronomical traits. The use of resistant cultivar is considered as one of the important component of integrated disease management. Since the genes for vertical resistance is lacking in cumin accessions and the pathogen exhibits high variability and extensive survival capacity, certain other parameters related to nature of host-pathogen interactions be included along with the routine observations of disease severity.

Plant Nutrition and Composts

Plant nutrition is an important consideration in enhancing the crop productivity. A well balance supply of soil nutrients results in healthy and vigorous plants. Potassium generally inhibit disease development and counteract the disadvantages of excess nitrogen. Zinc is known to suppress *Fusarium* wilt disease in cotton. More studies are required on influence of nutrients on pathogenesis process and growth of *Fusarium* in relation to cumin wilt.

Addition of composts into soils is an important practice both under conventional and organic farming. Besides improving soil fertility it helps in suppressing the pathogens by enhancing the population of antagonistic and beneficial microbes in soil. Organic substrates like farm yard manure, mustard cake, neem and groundnut cake reduces *Fusarium* population in soil (Diyora and Khandar, 1995). In fact, soil amendments with organic matter leads to diverse soil environment which is less favourable for the growth of soil-borne pathogens. Studies have revealed that disease control capacity of bioagents like *Trichoderma* spp. *Pseudomonas fluorescens*, against *Fusarium* wilt of cumin can be increased by the addition of farm yard manure to soil (Gangopadhyay, 2006). The interaction of bioagents and organic amendments in soil is an important area which require more attention for effective management of root diseases.

Use of Bioagents

Use of bioagents *viz., Trichoderma* spp, *Pseudomonas fluorescens, Bacillus subtilis* etc. are safe and environment friendly component of sustainable disease management. A large number of bioagents have been screened against *F. oxysporum* f. sp. *cumini* under laboratory and green house conditions. Talc based preparation of *T. harzianum* used as seed treatment at 6 g/kg seed plus soil application at 5 kg per hectare along with farm yard manure 10 tonnes per hectare effectively controlled *Fusarium* wilt of cumin under field conditions. Survival of *P. fluorescens* in soil was higher than *Trichoderma* spp (Gangopadhyay, 2006; Chawla, 2008). Since cumin wilt is primarily soil-borne and also transmitted through seed and may infect cumin at any stage of crop growth, selection of biogents having tolerance to fungicides is required for integrated disease management. This will ensure high level of protection at the seed germination and early growth stages of plants by the fungicide followed by protection of root system by the biological component. Several commercial formulations of the bioagents are available in India and abroad. These formulations are commonly used as seed treatment which results in development of antagonist population in rhizosphere protecting the crop from soil-borne pathogens. The mechanism of antagonism are mycoparasitism, antibiosis, competition, siderophore production and induction of host resistance. There is great scope of using the biocontrol agents successfully in integrated disease management programme against *Fusarium* wilt of cumin. The bioagents must survive in soil for longer period for effective suppression of *F. oxysporum* f. sp. *cumini.*

Use of Botanicals

Efficacy of some commercial formulations of botanical extracts and essential oils *viz.*, clove, neem, pepper, mustard, etc are reported as possible alternatives to soil fumigation for control of *F. oxysporum* f. sp. *chrysanthemi*. Significant reduction of population density of *F. oxysporum* was recorded in treated soil under green house conditions (Bowers and Locke, 2000). Similarly, the antifungal effect of some medicinal plants like *Adhatoda vasica, Azadirachta indica, Ocimum sanctum* etc against *F. oxysporum* f. sp. *melongenae* has been reported (Siva *et al.*, 2008). The application technique, feasibility and commercial viability of potential botanical pesticides need further attention.

Use of Chemicals

Seed dressing fungicides *viz.*, carbendazim, thiophanate-methyl provide protection against *Fusarium* wilt in cumin (Agnihotri and Sharma, 1987; Ram Gopal, 2002). Continuous use of same fungicide may result in development of resistant strain. Hence, new chemicals should be tested periodically. Since fungicidal seed treatment alone can not provide protection for longer period, the combined use of bioagents and fungicide has been suggested for management of soil-borne diseases including *Fusarium* wilt (Kaur and Mukhopadhyay, 1992). Further, the fungicides used should be least harmful to beneficial microflora.

Disease Diagnosis and Prediction

In order to make the disease management more effective and accurate, timely diagnosis of plant diseases is required. This is particularly important when the crop suffers attack of more than one pathogens- not necessarily both be soil-borne. There is a need to standardize quick diagnostic method that can detect the presence of more than one pathogens in the host. Scouting of fields to locate and identify infection foci and hot spots or larger infection patches in a particular area shall help in determining the early actions for preventing further build up and spread of pathogen inocula. Periodical monitoring of pathogen propagules in soil can be done with suitably standardized selective enumeration method.

References

Anonymous, 2008. *Anand Rathi Commodity Research.*

Agnihotri, J.P. and Sharma, A.K., 1987. Efficacy of fungicides in inhibiting the growth of *Fusarium oxysporum*. f. sp. *cumini* and in reducing the wilt of cumin. *Indian Cocoa Arecanut and Spices Journal*, 10: 85–87.

Bowers, J.H. and Locke, J.C., 2000. Effect of botanical extracts on the population density of *Fusarium oxysporum* in soil and control of *Fusarium* wilt in the greenhouse. *Plant Disease*, 84: 300–305.

Chawla, N., 2008. Studies on biological control of Cumin wilt caused by *Fusarium oxysporum* (Schlecht) Snyd. and Hans. f. sp. *cumini* Prasad and Patel. *Ph.D. Thesis*, Rajasthan Agricultural University, Bikaner.

Diyora, P.K. and Khandar, R.R., 1995. Management of wilt of cumin (*Cuminum cyminum*) by organic amendments. *J. Spices and Aromatic Crops*, 4: 80–81.

Gangopadhyay, S., 2006. Studies on biocontrol potentiality of microbial antagonists against important soil-borne diseases and insect pests of Cumin and Chickpea. *Final Report ICAR Adhoc Project.* College of Agriculture, Rajasthan Agricultural University, Bikaner.

Israel, S. and Lodha, S., 2004. Factors influencing population dynamics of *Fusarium oxysporum*. f. sp. *cumini* in the presence and absence of cumin crop in arid soils. *Phytopathol Mediterr.*, 43: 3–13.

Kaur, N.P. and Mukhopadhyay, A.N., 1992. Integrated control of chickpea wilt complex by *Trichoderma* and chemical methods in India. *Tropcal Pest Management*, 38: 372–375.

Mathur, B.L. and Mathur, R.L., 1964. Studies on wilt disease of cumin caused by *Fusarium oxysporum*. f. sp. *cumini*. *Indian J. Agric. Sci.*, 34: 131–137.

Ram Gopal, 2002. Studies on Management of Cumin Wilt (*Fusarium oxysporum* (Schlecht) Snyd. and Hans. f. sp. *cumini*. *M.Sc. Thesis*, Rajasthan Agricultural University, Bikaner.

Siva, N., Ganesan, S., Banumathy, N. and Muthuchelian, 2008. Antifungal effect of leaf extract of some medicinal plants against *Fusarium oxysporum* causing wilt disease of *Solanum melogena* L. *Ethnobotanical Leaflets*, 12: 156–163.

Recent Advances in Seed Spices (2011) *Pages* **123–153**
Editors: **Y. Ravindrababu, R.K. Jaiman, K.D. Patel, N.R. Patel and S.B.S. Tikka**
Published by: **DAYA PUBLISHING HOUSE, NEW DELHI**

13

Management of Diseases in Seed Spices

R.L. Patel[1], R.K. Jaiman[2], N.R. Patel[2], K.D. Patel[2], D.K. Joshi[2] and Y. Ravindrababu[2]
[1]Department of Plant Pathology, C.P. College of Agriculture, SDAU, Sardarkrushinagar, Gujarat
[2]Centre for Research on Seed Spices, SDAU, Jagudan – 382 710, Dist. Mehsana, Gujarat

ABSTRACT

India is a largest consumer of the spices as well as a predominant country in spices production in the world. Therefore, it is known as "The land of spices". The seed spices *viz.*, cumin, fennel, coriander, fenugreek, ajwain dill seed are important firstly because of their demand for domestic consumption and secondly for export earning. These are grown generally as rabi crops in cool and dry climate.

The seed spices are aromatic vegetable products mostly used in whole or ground from for imparting flavour, aroma and pungency to food. The seed spices possess industrial importance and are used in cosmetic, perfumery and pharmaceutical preparations. There are about 20 seed spices grown in India, the most important among them are cumin, coriander, fennel and fenugreek, which are cultivated conspicuously in larger area. These four crops are categorized as major group of seed spices and remaining crops sowa, ajwain, dill seed and few others are minor group of seed spices, as their cultivation is restricted to only few pockets.

Wilt, root rot, blight and powdery mildew diseases cause severe crop losses in seed spices and have been identified as major production constraints. Wilt of cumin, coriander and root rot of fenugreek are important soil born diseases. Powdery mildew, blight of cumin, fennel, coriander and fenugreek are important foliar diseases that affect these crop. Apart from those fungal diseases, losses caused by plant parasitic nematodes, occurrence and incidence of new disease problem of minor importance *viz.*, sugary and phyllody in fennel. Conductive climate conditions, lack of resistant varieties and non adoption of recommended package of practices by farming communities are the limiting factors for effective disease management in seed spices. In this chapter attempts has been made to review the present status of the knowledge on the occurrence of diseases, their pre-disposition factors and available management strategies are discussed modules to minimize economic losses. Developing different modules of disease management integrating various practices *viz.*, crop rotation, summer ploughing, summer fallowing, soil solarization, organic amendments, time of sowing, seed treatments etc.

Introduction

The seed spices *viz.*, cumin, fennel, coriander, fenugreek, ajwain and dill seed are important for both internal consumption and export. These are generally grown as rabi crops in cool and dry climate. Cumin, fennel, fenugreek, dill seed and ajwain are grown extensively in Gujarat and Rajasthan. Coriander is grown mostly in the southern states *viz.*, Andhra Pradesh, Karnataka and Tamil Nadu apart from Rajasthan and Gujarat. Lack of high yielding strains, improper nutrient management and occurrence of many diseases at various stages of crop growth are the principle constraints to achieve high productivity of seed spices. Number of diseases have been reported and investigated in different research organizations located in the major seed spices producing areas. Efforts have been made in the present chapter to compile available information on the various aspects of important diseases of these valuable foreign exchange earning crops.

Cumin

Occurrence of diseases *i.e.* wilt, powdery mildew and blight along with improper nutrient management are the two major reasons for its low productivity (Dange *et al.*, 1992).

Wilt (*Fusarium oxysporum* f.sp. *cumini* Prasad and Patel)

It is the most important and destructive disease of cumin crop Gaur (1949) first reported the existence of a wilt disease caused by spices of Fusarium from Kekri in Ajmer district of Rajasthan state. Later, the correct identify of the fungus as *F. oxysporum* f.sp. *cumini* was established (Patel *et al.*, 1957). It has also been reported from many other countries (Gaetan and Madia, 1993; Omer *et al.*, 1997). The disease preliminary surveys revealed losses of about 5 to 25 per cent in north Gujarat and 5 to 60 per cent in Rajasthan (Patel *et al.*, 1957; Patel and Prasad, 1963). However, infection may vary from 25 to 40 per cent and may be as high as 70 per cent in some cases (Gaur, 1949). Subsequently, an exhaustive survey carried out in Rajasthan in 1953-54 revealed that wilt incidence ranged from 7.13 to 25.7 per cent and was as high as 60 in some cases. The extent of loss in terms of revenue was nearly four and a half million rupees for the state at that price level (Mathur and Mathur, 1966). In arid districts of Rajasthan, where cumin is extensively cultivated as a cash crop, approximately 40 per cent yield losses were recorded (Lodha *et al.*, 1986). Dange *et al.* (1992) reported 7 to 30.6 per cent losses in Gujarat.

Symptoms

The plants are attacked at all stages of crop growth, but the disease usually appears in patches in the field when the crop is about one month old. The affected plants turn yellow, wither and show characteristic wilt symptoms and dry up.In initial stage, top of the branches droop down and gradually infected plant dry-off. The roots of diseased plant show dark brown margins. In the affected root, vascular bundle appears brown indicating the presence of fungal mycelium.

Symptoms of Wilt (*Fusarium oxysporum* f.sp. *cumini*)

Cumin Wilt

Biology of the Pathogen

The fungus produces abundant aerial mycelium which consists of macro and micro conidia. Micro conidia are single septate, hyline thin walled, elliptical to spherical, single or two celled. The macro conidia are 1-5 septate, hyline thin walled, linear to falcate the ends tapering. Chlamydospores are spherical, smooth and terminal or intercalary. The fungus grows at wide range of temperature, 24°C being optimum and 35°C is the maximum. The pH requirement for growth vary from 4.8 – 8.4 and 6.4 is the optimum (Anon, 1955). Production of pectin methyl esterase (PME), pectin trans-eliminase (PME), polygalacturonase (PG). Polygalacturanase trans-eliminase (PGTE) and cellulose both in vitro and in vivo was reported (Champawat and Pathak, 1990).

Epidemiology

The pathogen is soil as well as seed borne in nature. The fungus is soil borne and root infection is the primary cause resulting in different degrees of root rot leading to varying degrees of foliar yellowing. The disease is a typical vascular wilt and the presence of mycelium and clamydospores in xylem elements was recorded (Anon, 1955; Patel *et al.*, 1957). The pathogen is highly host specific but seedling infection in coriander and carrot was noticed (Mathur and Prasad, 1964). Disease incidence more in light and poor soils and the soil temperature of 54.6-57.0°f in the morning and 61.0-66.7°f in the evening was found conducive for disease development (Mathur and Mathur, 1966). Seed borne nature of the disease was also established (Patel, 1968; Champawat and Pathak, 1990b). The fact that disease was noticed in the plots where previously the crop was not grown (Mathur and Prasad, 1964) supported further the possibilities of disease transmission through seeds. Presence of fungus between seed coat and endosperm was established and the seed infection might occur in the field or during storage (Singh *et al.*, 1972). Presence of sporodochia of the fungus on infected roots was reported (Gaur, 1949: Mathur and Mathur, 1970). The left over infected plant debries in the field and also contaminated or infected seed would be serve as the primary source of inoculum.

Disease Management

The prédominant soil borne nature and longér period of survival of resting structures of Fusarium in the soil have lessened the efforts to manage this disease for profitable cultivation. Many falk songs in Rajasthan and Gujarat depict the plight of resource deficient farmers.

Resistant Cultivars

Patel *et al.* (2005) found varieties Gujarat Cumin-3(GC-3) and Gujarat Cumin-4 (GC-4) have been found resistant against wilt disease. Arora *et al.* (2004) found that none of the variety was found totally resistant to wilt and varieties UC 220 and UC 63 were highly resistant to wilt disease.

Cultural Methods

Rotation with non-host crops has been found to be the most effective way of managing this disease in the same field (Mathur *et al.*, 1967). Rotating cumin fields with mustard and pearl millet in winter and rainy season respectively, has been found to keep incidence of wilt under check. Summer ploughing twice and thrice resultede in 55.5 and 69.1 per cent disease control and consequent increase in yield by 38 and 85.8 per cent respectively (Champawat and Pathak, 1990). Summer fallow of cultivated land also found effective in managing cumin wilt (Mathur *et al.*, 1967).

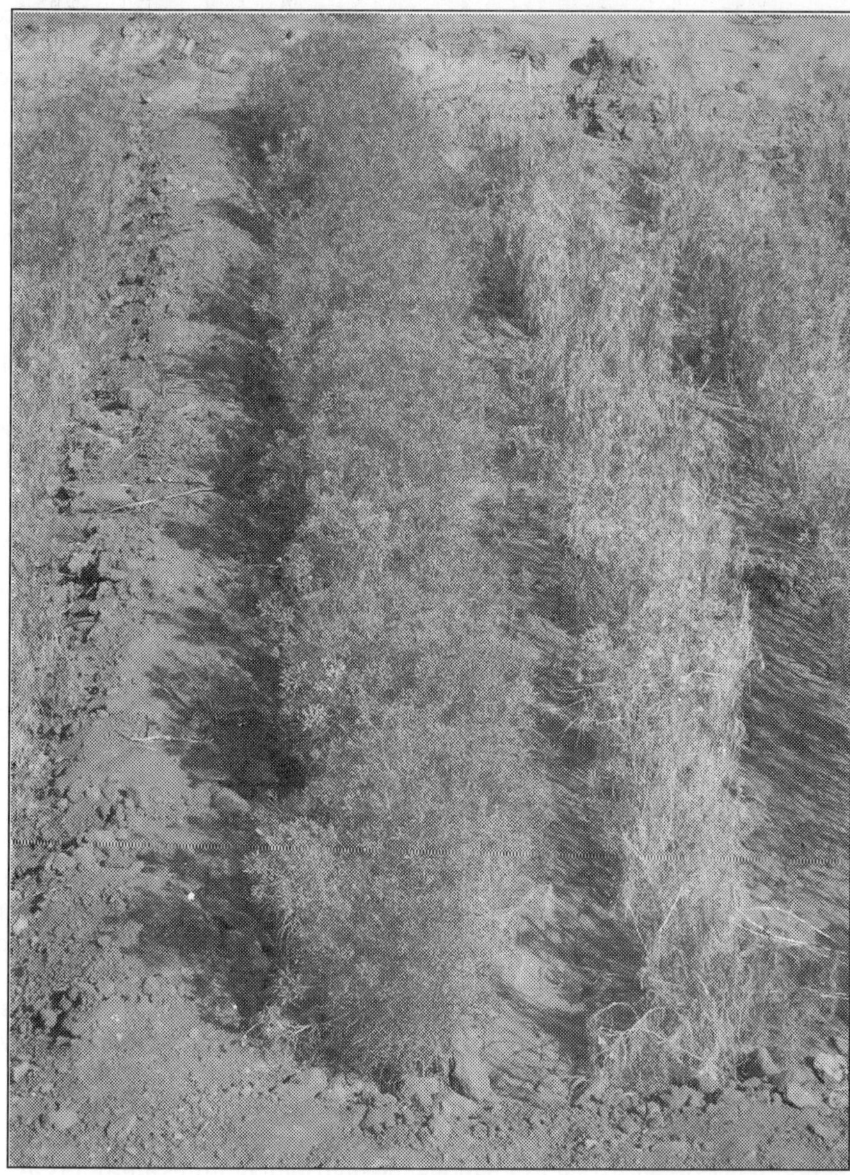

GC-4 (Blight)

Chemical Methods

Incidence of wilt was lowest (24.5 per cent) with 30 kg N/ha followed by 26.8 per cent with 30:10:10 NPK kg/ha (Champawat and Pathak, 1988a). Increased application of potassium decresed wilt incidence. Application of carbofuran 3G@ 66 kg/ha before sowing was most effective in reducing wilt incidence followed by phorate 10G @ 20 kg/ha and one spray of carbendazim (0.1 per cent) or benomyl (0.05 per cent) was effective in reducing wilt incidence in field (Champawat and Pathak,

1991a). A combination of seed treatment with carbendazim @ 3g/kg seed and one spray of carbendazim (0.1 per cent) or benomyl (0.05 per cent) was effective in reducing wilt incidence in field (Champawat and Pathak, 1991a). Soil application of SMDC (Vapam) was effective in checking wilt and increasing plant vigour (Mathur and Mathur, 1967). Seed dressing and soil drenching with methyl thiophanate and carbendazim decreased wilt incidence and increased yield. It was found that herbicide inhibited the mycelium growth of Fusarium (Patel and Patel, 1993). Seed treatment and foliar spray with bavistin or benelate was found effective in reducing the wilt incidence (Agnihotri and Shgarma, 1987). Seed dressing and soil drenching with Emisan-6 (0.02 per cent), Topsin (0.2 per cent) and bavistin (0.2 per cent) decreased disease incidence and increased yield (Anon., 1990). Decreased disease incidence (29.5 per cent) with bavistin seed treatment and increased yield of 203 kg/ha was observed at Jobner compared to 47.9 per cent disease incidence and a yield of 103 kg/ha in untreated control (Anon., 1993). Carbendazim treated seed gave the higher germination and showed lowest wilt incidence (Raheja, 2006). Seed dressing and soil drenching with methyl thiophenate and carbendazim decreased wilt incidence and increased yield (Patel and Patel, 1993). Ghasolia and Jain (2004) showed bavistin reduced pre and post emergence seedling mortality of wilt pathogen and give increased seed germination and vigour index.

Biological Control

Seed dressing with *Trichoderma harzinianum* T2 isolate lowered disease incidence by 65.4 per cent and was found to be more effective than seed treatment with fungicide (Aghnoom *et al.*, 1999). Inoculation of cumin plants with *Glomus fasciculatum*, *G. mosseae* and *Acaulospora laevis* reduced the wilt incidence (Champawat and Pathak, 1991b). Bio control agents like *T. harzianum*, *Aspergillus flavus*, and *A. niger* inhibited the growth of Fusarium in laboratory studies (Patel and Patel, 1998a). Bacterial antagonists, Lactobacilli and *Bifidobacterium adolesentis* inhibited growth of Fusarium in dual culture (Patel and Patel, 1998b). *Aspergillus versicolor* isolated from heated (naturally or solarized) cruciferous residue amended soil was found to parasitize Fusarium in laboratory test and also reduces its population densities in soil (Israel, 2002). Amending soil with *A. versicolor* alone or in combination with *T. harzianum* or Verbisina residues (an obnoxious weed) were significantly better in reducing Fusarium population compared to non-amended control. Seed plus soil treatment with alone resident isolate of *T. harzianum* or combined application as *T. harzianum* + *T. viride* + *P. fluorescence* at both seed and soil level was found to be most promising in suppressing cumin wilt (Raheja, 2006). Baradia and Rai (2006) observed that the treatment of cumin seed with *T. harzianum* @ 8 g/kg seed showed significantly lowest disease incidence offering disease control followed by *T. harzianum* @ 6 g/kg seed and *T. viride* @ 8 g/kg.

Soil Solarization

In arid region, where intense solar irradiations and high temperatures are amply available during summer months, harnessing solar heat has been found effective in minimizing wilt incidence in the field (Lodha and Kumar, 1998). In such regions, two or three summer ploughings decreased the wilt incidence (Champawat and Pathak, 1990b). Solarization of soil by covering it with transparent polyethylene sheet for 15 days greatly reduced the population of Fusarium at 0-30 cm depth (Lodha, 1995). As a results of significant reduction in the incidence of wilt and weed populations and improvements in seed yield were recorded in solarized plots.Combining mustard residues with polyethylene mulching has been found to improve the reduction in populations of Fusarium in hot arid climate (Israel, 2002).

Organic Amendments

Several organic amendments were also found effective in reducing population of Fusarium and wilt incidence. Population of Fusarium was completely eradicated in the mustard cake amended (1 per cent) soil within a period of 30 days after addition (Sharma *et al.*, 1995). Incorporation of this cruciferous amendment also enhanced the population of antagonistic actinomycetes in the soil. Combining summer irrigation with amendment of cruciferous residues during May or June was found highly effective in reducing Fusarium population in the field (Mawar and Lodha, 2002). Amendment of soil with a combination of mustard residues and mustard oil cakes (2.5+0.5 t/ha) was effective in control Fusarium population in soil and incidence of wilt on cumin (Israel *et al.*, 2002). Combining a small dose (0.5 t/ha) of onion, verbisina or mustard oil cake with mustard residues further improved the reduction in Fusarium propagules (Lodha and Isreael, 2005). Champawat and Pathak (1988) observed that lowest wilt incidence and highest yield of cumin was recorded in plots treated with poultry manure followed by mustard cake.Raheja (2006) found mustard cake to be most promising in reducing the wilt incidence followed by groundnut cake, neem cake and caster cake. mustard cakes application at 2t/ha gave lowest wilt incidence (12.3 per cent) and a yield of 180 kg/ha compared to 41.0 per cent wilt incidence and a yield of 100 kg/ha in untreated control (Jain *et al.*, 1990).

Phytoextract

Seed treatment with *Allivum. sativum* (cloves) extract @ 10 ml/kg seed gave higher germination and lowest wilt incidence followed by *Calymma alliaceum* and *A. cepa*.

Blight (*Alternaria burnsii*)

The disease was first reported from Bobbay (Uppal *et al.*, 1937) provience to be caused by *Alternaria sp*. but latter the correct identify the fungus as *A. burnsii* was established (Uppal *et al.*, 1938). It has been observed to be wide spread in all the cumin growing belt of Rajasthan (Mathur, 1949; Joshi, 1955). In Bombay, it appeared usually by the middle of January (Uppal *et al.*, 1938) and in Rajasthan during Febrary and March and becomes serious under humid and cloudy weather. It has also been reported from Pakistan (Shakir *et al.*, 1959). The yield losses may be up to 70 per cent (Patel, 1968).

Symptoms

Generally the blight disease appears at the flowering, however, early infection is not uncommon. The disease is characterized by the appearance of small, isolated, ash colored necrotic areas on the aerial parts of the plants, especially on the tips of the young leaves. These necrotic areas gradually enlarge and coalesce with each other and turn purple, eventually brown and finally black. The succulent leaves and blossoms are more affected. Under wet conditions, the infection spreads to the stem and the whole plant dries up and become black in colour as it is burnt.In severe cases of infection, there will be no seed production. If seeds are produced, they are found shriveled, dark coloured, light in weight and usually non-viable. The quality of the seed is very much reduced.

Pathogen

It has been observed that *Alternaria burnsii* can over summer in exposed plant residues in the field and is capable of withstanding long periods of dry conditions in the laboratory. However, it is not certain that infected plant residues are the primary source of infection. Seeds may play an important role in perennation of the pathogen and as a source of primary infection in new area (Uppal *et al.*, 1938;

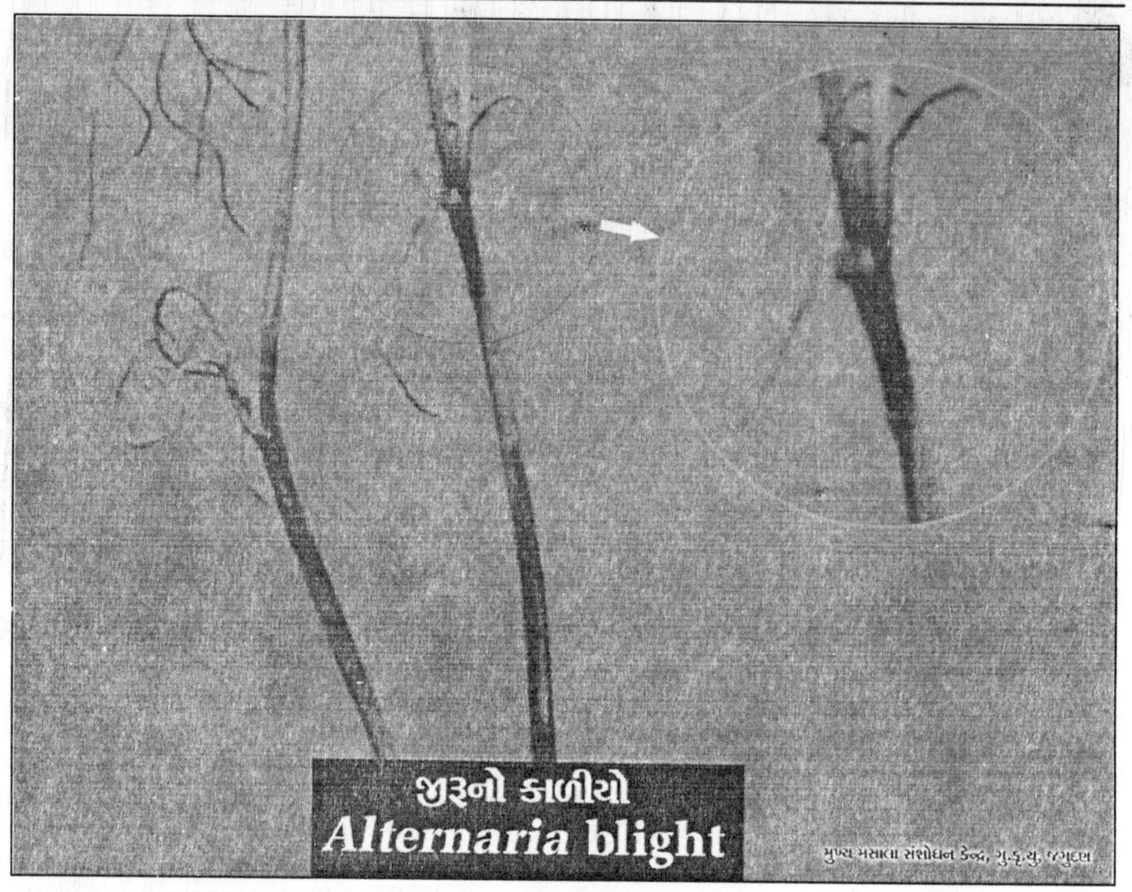

Symptoms of Blight (*Alternaria burnsii*)

Patel and Desai, 1971). Disease development is governed to atmospheric humidity levels and the disease become serious if humid condition continue at the time of flowering (Uppal *et al.*, 1938). The plant become susceptible only after flowering stage. Rain, cloudy humid weather and temperature range 23-28°C are highly favourable for disease development (Gemawat and Prasad, 1972). The reports of the effect of early planting in relation to disease development are contradictory might be related to difference in prevailing weather condition of the particular season and year. While severity was less with early planting (Gemawat and Prasad, 1972), it was reported severe in early planting (Patel and Pallai, 1983). Extended sowing from October to December ensured availability of continuous susceptible host in the areas for longer periods (November- April). This resulted in disease spread to the adjoining areas (Gemawat and Prasad, 1972). The pathogen survives through seed and plant debris and those serve as primary source of inoculum (Patel and Desai, 1971; Gemawat and Prasad, 1972). Secondary inoculum is airborne and spread through wind.

Management

Resistant Cultivars

Available germ- plasm have been screened but none was found resistant.Evaluation of callus of 12 genotypes in vitro revealed resistance in GC-1 only (Shukla *et al.*, 1997). In compression to other varieties, GC-4 showed less disease. Arora *et al.* (2004) found that none of variety was found totally resistant to blight and variety U C 310 was highly resistant to blight.

Chemical Method

Seed borne inoculum was completely eliminated by seed treatment with Agalol, Aureofungin, ceresin wet, Agrosan GN, captan and difoltan. Among these fungicides, cumin, captan, mancozeb, vitavax, plantvax and blitox were found effective (Solanki *et al.*, 1973).Four spraying of carbendazim protected against the infection of *A. burnsii* and gave the highest (0.2 per cent) effective control of cumin blight in Gujarat state. (Mehta and Solanki 1990). Mancozeb 0.2 per cent was effective against cumin blight followed by propiconazole @ 0.025 per cent (Anonymous, 1993).Maximum grain yield and net return was registered in four spray of mancozeb @ 0.25 per cent at 30, 50, 65, and 80 DAS followed by five sprays at 30, 40, 50, 60 and 70 DAS.(Dange *et al.*, 2002). Four sprays of mancozeb @ 0.25 per cent

Chemical Control of Blight by Mancozeb

at 35, 45, 55 and 65 DAS found effective for management of cumin blight (Patel *et al.*, 2007).Minimum blight intensity with highest yield was recorded with three sprays of Difenoconazole were done at an interval of 15 days followed by mancozeb (Bishnoi and Rajpurohit, 2007).Bhatnagar *et al.* (2004) observed the three spray of copper oxychloride (Blitox-50) at the age of 70 days at 7 day interval were found significantly supperior in reducing the CODEX of blight in cumin.Sesame cake was found effective to reduce blight disease up to 35.9 per cent as compared to 59.3 per cent disease intensity recorded in control (Deepak and Choudhary, 2008).

Cultural Method

Late sown crop had comparatively less blight intensity but also yielded less as compared to early sown crop (Baswana *et al.*, 1991). Higher dose of NPK @ 30:20:30 kg/ha resulted in more yield but also the more blight intensity.

Biological Control

Patel (1993) reported the *Pseudomonas* sp. found effective against *Alternaria burnsii* under *in vitro* conditions.

Powdery Mildew (*Erysiphe polygoni* DC)

Powdery mildew is known to occur in Bombay province as early as 1921 (Burns, 1921) and is now reported to occur in all the tracts where cumin is grown extensively. On an average crop loss of about 50 per cent has been reported if the disease occur during flowering stage but losses would be minimum at 10-15 per cent if the disease occurs during late stage of seed formation (Chattopadhyay, 1967). However, appearance of disease under epiphytotic conditions often results in total failure of the crop (Uppal and Desai, 1932). Yield losses of about 5-15 per cent have been reported from Rajasthan due to powdery mildew (Gaur, 1949; Mathur, 1949). The disease has also been reported from Bulgaria also (Dodoft and Naceff, 1936).

Symptoms

The infection take place at any stage of plant growth, but more severe at flowering stage. Small, white or grayish spots appear on the leaves. They gradually enlarge, coalesce and cover the entire surface with whitish mycelial growth and conidia of the fungus. The infection rapidly spreads to stem, flowers and fruits. Under warm and low moist condition, the spread of disease is very rapid and entire plant becomes affected and dry up. In the case of severe attack, the leaves/plants shows ashy white appearance. Seed setting is also adversely affected. The seeds from infected plant are small and light in weight. Late sown crop is severely affected.

Biology of the Pathogen

Being obligate parasite, the mycelium is superficial with haustorial development between the epidermal cells. The fungus is ectoparasitic, spreading on surface of host and sending haustoria into epidermal cells to draw out nutrients. The conidia are ovate to rectangular, hyaline, thin walled, and produced in chains on conidiophore. The perithecia occur either in cluster or scattered on host tissues.

Epidemiology

The disease is noticed during December-January and is favoured by moist warm condition. Temperature appears to be the determining factor in rapid disease development. Disease appears in severe form under at temperature range of 26.5- 35 °C. Occasional showers with warm condition

Cumin Powdery Mildew

during December would favour the disease. Pathogen is obligate in nature. It survives on collateral host from which it provides primary source of infection and secondary infection occurs through air borne conidia. Temperature 20-22 ºC, relative humidity >60 per cent and cloudy weather favors the disease development.

Disease Management

Cultural Methods

Since the disease is known to penetrate through infected seed, hot water and fungicidal seed treatment practiced for wilt might help in ensuring healthy seed. Maintenance of special disease free seed plots was also suggested (Chattopadhyay, 1967). Late sowing should be avoided.

Chemical Control

The disease till now can be managed in the field only by timely spray of fungicides. Uppal and Desai (1932) observed that the disease could be effectively controlled by one application of dusting sulphur @ 28 kg/ha at the time of flowering in January. A subsequent dusting was not usually necessary if the first one was through and timely. However, a second application @ 13.5 kg/ha at the time of seed formation was recommended if weather conditions were favourable for the spread of the disease. Powdery mildew can also be effectively controlled by spraying dinocap (0.1 per cent) (Mathur *et al.*, 1971). Carbendazim (0.1 per cent) or tridemorph (0.05 per cent) have also been found effective. Three sprays of tridemorph as most effective means to manage to powdery mildew (Gohil *et al.*, 1985;1988; Mehta and Solanki, 1990; Mehta *et al.*, 1990).

Coriander

Like cumin, coriander (*Cariander sativum* L.) also suffers from three major diseases viz, wilt, powdery mildew and stem gall causing considerable crop losses.

Wilt (*Fusarium oxysporum* f.sp. *corianderii* Nikem and Joshi):

The disease was reported during 1936 (Anon, 1953). Wilt of coriander caused by *Fusarium oxysporum* f.sp. *corianderii* was first reported by Joshi and Nikam (1953). It is known from Gwalior and Guna districts of Madhya Pradesh and Kota area of Rajasthan. The disease occurs severely in coriander growing tracts of Madhya Pradesh and Rajasthan every year and may cause as high as 60 per cent yield losses (Dange and Pandey, 1994). About 25 per cent yield loss has been reported from Madhya Pradesh. In certain pockets of Madhya Pradesh like Guna district, the average losses were about 40-50 per cent. In Rajasthan, losses as high as 60 per cent have been reported (Chattopadhyay, 1967).

Symptoms

Plants are prone to infection at all stages of crop growth. Root infection leads to droping of the terminal portion of the plants followed by withering and drying up of leaves are the major symptoms. The infected plants ultimately die. Partial infection of the plants results in growth retardation. The leaves of such plants exhibit pinkish yellow to yellow colour. Partial infection leads to sterility in some cases and, if seed set is noticed, such seed retained light and immature. When infection is noticed in entire field, wilting will be sudden. Early infection leads to total failure of the crop (Chattopadhyay, 1967). Pre-emergence death and damping off followed by wilt is also noticed (Mali, 1968).

Biology of the Pathogen

In culture, the mycelium appeas whitish and the substrate shows pinkish colour on oat meal agar and old lilac colour on steamed rice. Micro conidia are scattered and are hyline,ovoid, ellipsoid or reniform. Macro conidia appear in sporodochia. The fungus grows at a temperature range of 12-35 °C and best growth was observed at 19.5 °C.

Epidemiology

Temperature range 24-27 °C was found optimum for disease development and even though wilting was seen even at 20-29 °C. Soil moisture content of 50-60 per cent was found favourable for the disease incidence (Chattopadhyay, 1967) and wilt increased with increase in soil moisture. Wilt incidence was high at a temperature range of 28-33 °C and was low at 20-24 °C. Mortality of the plants was high at soil pH range of 5.8-6.9. Symptoms development was gradual in alkaline soils (Srivastava, 1972). The pathogen remained viable in the seed for five months and could survive for long in soil. Soil borne inoculum appears to be the primary source of infection (Srivastava, 1967).

Disease Management

Application of organic amendments and fungicidal treatments were the adopted methods of disease management. Use of disease free seed, crop rotation and summer ploughing has been suggested as disease management practices (Bhati *et al.*, 1985).

Organic Amendments

Organic soil amendments reduced the disease incidence. Disease incidence was less at higher concentration of amendments (Srivastava and Sinha, 1971). Considerable reduction of disease under field conditions by amendments of soil with oil cake and adjusting the soil pH at 8.2 has been reported by Srivastava (1972). Application of amendments with low N and high P and K (40:80:80) gave the lowest wilt incidence of 8.3 per cent compared to high N and low P and K (80:40:40) with wilt incidence of 59.7 per cent (Srivastava, 1967, 1972).

Cultural Control

The pathogen is specific to coriander, therefore crop rotation may result in reduced wilt incidence. It has been found that sowing in the first and second weeks of November reduces the wilt incidence.

Biocontrol

Varying degree of antagonistic activity of four strains of Streptomyces on Fusarium isolates of coriander was noticed (Mathur, 1963). However, their in vivo activities are not known.

Chemical Control

Seed treatment with Agrosan GN was found effective in wilt control (Nikam *et.al.*, 1959). Seed dressing and spraying twice with bavistin or benelate at the age of one and two months reduced wilt incidence upto 60 per cent (Jain and Agarwal, 1978).

Host Resistance

UD-20 has been reported resistant to wilt and varieties Sindhu, CS-287, Co-3, Swathj, Sadhana, Rcr-41, RD44, Gujarat Coriander-1 and 2 were reported tolerant (Edison *et al.*, 1990). Gwalior No. 5365 has been reported resistant (Singh Banafar and Singh, 1985). Variety M 55365 was found tolerant to five isolates of Fusarium (Mathur, 1963).

Stem Gall (*Protomyces macrosporus* Unger)

This is a soil borne dsease. It was first reported in India by Butler (1918) but was not considered serious. However, it was later considered as important (Mundkar, 1949). It is noticed in all tracts where coriander is grown. The pathogen has been observed on many other plants of umbelliferae in India and Europe.In Gwalior (M.P.), an approximate loss of 23 per cent was reported by Gupta (1951). Similarly in U.P. a yield loss of 9.6-33.8 per cent was reported by Singh *et al.* (1984).

Loss in Quality

The hypertrophied leaves and fruits loose their flavour. The fat loss in fruit was about 15 per cent (Gupta, 1956a). The diseased fruits loose their flavour and also their fat, protein, carbohydrates and aminoacids (Gupta, 1964; Prasad, 1983). In diseased leaves and fruits ash, magnesium, sodium and silica contents increased, while calcium, manganese, phosphorus, potassium and iron decreased (Gupta, 1975). Increased concentration of reducing sugar and aminoacids and decreased content of non-reducing and total sugars in the diseased tissues was reported. Increased levels of invertase, protease acid, phosphatase, peroxidase and polyphenols have been reported (Goel *et al.*, 1983).

Symptoms

The disease manifests itself in the form of small tumour-like hypertrophied swelling on all the aerial parts of the affected plants like, stem, petiole, flower stalks and veins of the leaves. The swelling are usually elongated and varying according to the severity of disease. The galls are soft and fleshy when young later becoming hard and woody, as they grow old. The infected plants are stunted. The hypertrophied seeds fail to germinate and also unfit for the use. The hyperplasy of the outer cortical cells restricted to those in the neighborhood of intercellular hyphae results in the formation of tumours. Often, the thin-walled parenchymatous cortical cells may replace the collenchyma in the hypertrophied tissue. The infection becomes systemic in the host stem before or during the pre-flowering period, incluing organogenic changes in flowers and fruits due to hypertrophy and hyperplasy. Galls had high concentration of amino acids and reducing sugars and low concentration of non-reducing total sugars.

Biology of the Pathogen

The organism could be cultured on potato dextrose agar (Mukhopadhyaya and Pavgi, 1964). The mycelium is intercellular and branched, producing thick walled, globose to ellipsoid chamydospores with three-layred membrane, a thick brownish exospore and thin meso and endospores. Chlamydospores germ in water forming sporangia and spores. The released spores from the sporangium are connected in pairs and later they fuse (Gauman and Dodge, 1928). High acid phosphatase activity was noticed in chlamydospores (Meheswari and Chaturvedi, 1983). Considerable variability of the pathogen was noticed (Gupta and Sinha, 1963; Narula and Joshi, 1963). Chlamydospores showed 10 per cent germination after 6 years and lost the viability after 7 years of storage (Mukhopadhyaya and Pavgi, 1971).

Epidemiology

Inoculum potential, microclimate and crop growth phase influenced the susceptibility of the plants. The disease was severe in heavy and irrigated soils. Low temperature, abundant dew with higher day temperature has been reported to predispose the plant to infection. The disease was severe in low lying heavy and ill drained soils (Mukhopadhyaya and Pavgi, 1971). Infection was minimum (20.8 per cent) at a soil pH of 4.6 and increased with rise in pH upto 7.4 with maximum infction (91.6 per cent). Fertilizer application of N resulted in disease reduction and consequent nutritional status influenced the disease incidence. Increased levels of manganese increased disease. At lower dose of P disease reduction was noticed but increased with higher dose of P (Gupta, 1976).

Disease Management

Available information on the effect of fertilizer application, regulation of soil pH and drainage need to be adopted to reduce the disease incidence.

Cultural Control

Butler (1918) reported that disease incidence can be reduced by removal and destruction of gall bearing parts of the plant. It has been observed that early and late sown crop contracted less incidence of stem gall than normal sown crop (Lakra, 2000).

Resistant Cultivars

Naqvi (1985) found cultivars, *viz.*, UD-03, UD-41, UD-46 and UD-48 resistant to stem gall. Tripathi *et al.* (1998) reported cultivars G-C-88-8, G-5365-91, Pant-1 and UD-20, moderately resistant against stem gall. The variety NP-95 was least affected (Das, 1971). Shujalpur local, NP-92, NP-95, Delhi local, Synthetic North, Kota local and P-107 were tolerant (Gupta and Sinha, 1973). Rajendra, Swathi, Rcr-41 and UD-20 were found resistant (Edison *et al.*, 1991).

Chemical Control

Seed treatment with thiram (0.25-0.4 per cent), tetrachloronitro amisole (0.5 per cent), captafol (0.2 per cent) were found effective in checking chlamydospore germination and in improving seedling vigour (Lakra, 2000). Spray of streptomycin and carboxin 30 days after sowing were also effective in the control of disease. Penicillin and streptomycin spray (400ppm) at three weeks interval upto the age of 13 weeks reduced disease intensity by 85.4 and 87.9 per cent respectively (Gupta and Sinha, 1963). Seed and soil treatment with thiram at 250g/100kg seed gave better protection (Nene *et al.*, 1966). Foliar spray with thiram (0.2 per cent) captafol (0.2 per cent) or carboxin (0.1 per cent) at fortnightly intervals gave better protection (Bhardwaj and Shresta, 1985).

Powdery Mildew (*Erysiphe polygoni* DC)

Erysiphe polygoni is the most common pathogen causing powdery mildew in India. However the occurrence of *E. chicaracearum* and *E. umbelliferaum* has also been reported (Hirata, 1966). *E. umbelliferaum* has also been reported from Afghanistan, Burma, Iran and Russia. Khan and Kamal (1962) reported *Leveillula taurica* from Pakistan. *L. taurica* from *Capsicum sp.* has been found to infect coriander and other hosts in India (Pawar *et al.*, 1985). The disease occurs in varying intensities in almost all coriander growing areas of the country. The disease affects the quality of leaves, seeds and causes 15-20 per cent yield losses (Srivastava *et al.*, 1971).

Symptoms

The disease manifests first as small, white circular patches on young parts of stem and leaves. With time these patches increase in size, often coalesce to cover large areas of leaf surface. The infected leaves get distorted and reduced in size. Early infection does not permit seed formation. However, if infection occurs at a late stage, seed formation may take place but such seeds will be small and shriveled, thus affecting the yield and quality (Kalra *et al.*, 1995).

Epidemiology

Infection is noticed during February-March. Dry season and high temperature favoured disease development (Butler, 1918). However, importance of humid condition for disease development in Rajasthan has been reported (Mathur, 1949).

Disease Management

Similar to powdery mildew of cumin, disease management so far adopted is mostly through agrochemicals.

Cultural Control

Early sown crop during October showed less disease incidence compared to late sown crop (Keshwal *et al.*, 1979). Early planting of coriander has been found effective in managing powdery mildew (Kalra *et al.*, 2000)

Chemical Control

Sulphur dusting once when the crop in about two months old and again after fruit setting was found effective (Nikam *et al.*, 1959). Two rounds of spray with the fungicides *viz.*, cosan, sultaf, karathane, elosal, thiovit and morocide were effective in checking the disease (Srivastava, 1971). Three round of spray with sultaf (0.25 per cent) was effective in checking the disease (Keshwal *et al.*, 1979). Sulphur fungicides *viz.*, wettable sulphur (0.6 per cent) and karathane gave maximum control of the

Coriander Powdery Mildew

disease and increased yield (Raju *et al.*, 1982). Sulphur dusting thrice at 25 kg/ha followed by three sprays of Dinocap (0.05 per cent). The disease intensity was 21.1 and 37.4 per cent and yield was 1700 and 1600 kg/ha respectively, an increase of 57 and 56 per cent respectively (Parkhia *et al.*, 1990). Karathane (0.1 per cent) has been found to be effective by Vinayagamoorthy *et al.* (1985). Karathane (0.1 per cent) or bavistin (0.1 per cent) or wettable sulphur (0.25 per cent) first at the time of initiation of the disease followed by second spray ten days later control the disease (Khader and Vedamuthu, 1989). Seed treatment with thiram, thiophnate-methyl or captafol were found effective in reducing powdery mildew severity and in increasing the seed yield (Ali *et al.*, 1998). Kalra *et al.* (2000) tested different spray schedules of dinocap and demonstrated that disease control varied with cultivars differing in maturity periods. Weekly application of fungicides provided greater disease control and substantial increase in yield compared with bi-weekly applications in late maturing varieties.

Host Resistance

Varieties Gujarat Coriander 1 and 2, Rcr-41, Sadhana, Co3, Sindhu and UD-20 were found tolerant while Swati, an early maturing variety, escaped disease (Edison *et al.*, 1991). Cultivars CS-362, CH-15 and CS-52 had the lowest disease incidence (19-24 per cent) under Central Indian conditions (Keshwal and Khatri, 1998).

Blight or Anthracnose (*Colletotrichum gloeosporiodes*)

The blight disease caused by *Colletotrichum capsici* was first reported from Kota district of Rajasthan (Sehgal *et al.*, 1965) subsequently this blight caused by *C. gloesporiodes* (*Glomerella cingulata*) was reported from Anantpur of Andhra Pradesh (Masses and Rao, 1969). The fungus is pathogenic to other hosts too. It produces thermostable non specific toxin which is toxic to many other hosts.

The disease affects mainly the inflorescence of plants. Flower initiation and early seed setting stages are susceptible stage of the plant (Bhatta and Hiremath, 1990). It infects young pedicels at juncture of flowers due to which flower become pale yellow and drop. In severe disease, flowers generally do not open and get killed. In advanced disease stage whole of the inflorescence and leaves get affected, inhibiting seed formation. In severe infection of inflorescence, the shriveled, light weight seeds are formed.

The disease can be managed by spraying bayleton (triadimefon) @ 0.05 per cent, foltaf and cumin L @ 0.3 per cent at flowering stage of crop (Bhatta *et al.*, 1988; Naik *et al.*, 1986).

Besides this, *Ramularia corianderi* has been reported to cause severe blight especially in wet weather condition. Seed treatment with thiram followed by spraying zineb resulted in effective disease management (Savenko and Pinkovskii, 1981).

Stem Rot (*Sclerotinia sclerotium, Rhizoctonia solani* and *Macrophomina phaseolina*)

The disease has been found to be most severe in rainy seasons mainly affecting the seedlings leading to pre-emergence mortality. Post emergence infection has been observed to occur on lower parts of hypocotyls. A brown erotic ring appears in this region that ultimately leads to collapse of the seedling.

Pathogen

A very mild form of stem rot caused by *Sclerotinia sclerotiorum* (Lib.) de Bary has been reported by Mehta *et al.*, 1946. Another stem rot caused by *Rhizoctonia solani* Kuhn was first reported by Mall (1968). A similar disease of coriander caused by *Macrophomina phaseolina* also been reported by Mahor *et al.* (1982).

Bacterial Diseases

A number of bacterial diseases caused by three genera *viz.*, *Erwinia*, *Pseudomonas* and *Xanthomonas* have been found to occur in coriander growing region of India, Argentina, Brazil and Germany.

Erwinia and *Xanthomonas*

Chakravarti and Rangrajan (1966) reported bacterial soft rot to leaves caused by *E. aroiatae*. A bacterial sudden yellow death and soft rot was reported from Brazil (Romeiro *et al.*, 1994). Biochemical and other studies suggested that bacterium was most closely related to *Eriwinia carotovora* sub sp. *betavasculorum*.

Chemical Control

Seed treatment with 0.4 per cent formalin, 0.1 per cent panogen, 0.1 per cent mercury sublimate and dry hot air at 53 0C for 30 minutes gave the best control of the disease.

Pseudomonas

Bacterial blight caused by *P. syringae* pv *coriandricala* is a major disease of coriander (Refshauge and Nayudu, 2001).

Symptoms

The disease is characterized by dark brown discoloration of blossoms and umbels, water soaked and brown spots on leaves and stems, seed decay and wilting. The blackening of leaf veins and surrounding tissues is the most typical symptoms of the host for multiplication and transport leading to infection of many tissues and potentially host death. The pathogen is host specific.

Virus Diseases

The different virus diseases in the crops have been reported in different countries. 'Western celery mosaic virus' from Britain (Walkey *et al.*, 1970), Feathery red vein virus transmitted by aphid, hyadophis from USA (Misari and Sylvaster, 1983) and coriander chlorotic lesion virus from Hissar, India (Poonam Dhawan and Rishi, 1992) have been reported in low to moderate form on coriander. A viral disease caused by carrot red leaf virus has been reported from southern India (Prakasam *et al.*, 1988). The virus was not transmitted by sap inoculation or by seeds from affected plants.

Phyllody

Phyllody of coriander in the range of 0.5 to 7 per cent has been reported from different crop growing areas of Tamil Nadu (Prakasam *et al.*, 1988).

Coriander Phyllody

Fenugreek

Powdery mildew, downy mildew, root rot and leaf spot are some of the known diseases of fenugreek (*Trigonella foenum-graecum* L.).

Powdery Mildew (*Erysiphe polygoni* DC and *Leveillula taurica* (Lev.) AM)

Erysiphi polygoni has been reported to infect fenugreek from France, germany, India, Iran, Italy, Yogosalavia, Sudan, America and Lithuania, while *Leveillula taurica* from India, Israel, Sudan, Pakistan and *Oidium* sp. from Egypt, Portugal and Israel (Hirata, 1966). Occurrence of *E. polygoni* has also been reported from Canada (Zimmer, 1984). Gupta *et al.* (1982) found *E. trifolie* infecting fenugreek at Delhi. It is a serious disease in all the crop areas and appears during February-March at seed formation stage. The disease appears on leaves and stem and seriously affects the quality of leaft vegetable and reduced yield and seed quality. Saxena and Tripathi (1985) studying the epidemiology of the powdery mildew caused by *L. taurica* observed that the disease is more severe in mid December and is favoured by 60-70 per cent R.H. at 15-25 °C temperature.

Symptoms

Powdery mildew is characterized by the appearance of white floury patches on both the surface of leaves and other aerial parts of plants. Symptoms of *L. taurica* developed as mycelial patches on the surface of leaves. *L. taurica* survive from one season to another through cleistothecia present on the debris of infected crop (Chattopadhyay and Maiti, 1990). Cleistothecia of *L. taurica* appears on leaf surface at a later stage as small black specks immersed in the mycelial tissue. In cleistothecium acsi are formed which contain ascospores.

Management Strategies

Resistant Cultivars

Of the 44 lines screened against powdery GC-39 and UM-32 found free from fungal infection while GC-7, GC-20 and UM-34 were categorized as resistant (Prakash and Saharan, 1999). Collection No. 21-76-326 has been reported to be most resistant (Saxena *et al.*, 1984). Variety Prabha has been reported to be resistant (Sharma and Bhati, 1985). Varieties Rajendra Kanti, RMT-1 and Lam Selection-1 have been reported to be moderately resistant/tolerant (Edison *et al.*, 1991).

Cultural Control

In early sowing (Oct. 20-30), the disease appeared in the last week of January but further spread was checked due to low temperature and relative humidity (Sharma, 1999). However, in late sown condition (Nov., 20-30), the crop remained in flowering or pod formation stage thus environmental conditions were congenial for disease spread, maximum seed yield was also recorded from Oct. 30 sown crop.

Chemical Control

Two application of cecosal, cosan, thiovit, sultaf, karathane-WD or Morestan controlled the disease and increased yield by 220-326 kg/ha (Maish *et al.*, 1970). High volume spray with Gandhok gol (0.5 per cent) or cosan (0.2 per cent) at 20 days intervals efficiently controlled the disease (Mathur *et al.*, 1972). Dusting of 300 mesh sulphur @25 kg/ha done trice at 15 days interval was found to be effective in controlling the disease (Edison, 1989). Dange *et al.* (2003) observed that significantly least powdery mildew (22.4 per cent) was recorded in sulphur dust @25kg/ha followed by 0.4 per cent wettable

Powdery Mildew of Fenugreek

sulphur (25.6 per cent), 0.05 per cent hexaconozole (26.3), 0.02 per cent tridimefon (28.4 per cent) and 0.005 per cent penconzole (29.8 per cent). Higher pooled yield was found in sulphur dust treatments 852.5 kg/ha as compared to 495.2 kg/ha in control. Singh (2003) reported that spray of bayleton (0.05 per cent and 0.025 per cent) were most effective for the management of the disease followed by karathane (0.1 per cent) which was at per with sulphur dust. Effective control was obtained with triazole fungicides compared to conventional fungicides (Dhruj *et al.*, 2000).

Downy Mildew (*Perenospora trigonellae* Gaum)

The disease was first reported from Maharashtra (Uppal *et al.*, 1935). Infected leaves are early recognized by the presence of yellow patches on the upper surface. Presence of cottony grayish-violet mycelium is seen on the lower surface. The conidiophores emerge through the stomata. The disease evidently penetrates through ospores. It can be controlled by spraying of 0.2 per cent difolatan. Accessions like UM38, UM69, UM79 etc. have shown tolerance to the mildew and gave higher yields in Tamil Nadu (Edison, 1989).

Leaf Spot (*Cercospora traversiana* Sacc)

Occurrence of *Cercospora sp.* from Israel (Ming and Solel, 1959) of *C. traversiana* from North America (Leppik, 1960), Hungary (Varos and Nagy, 1972) and from Canada (Zimmer, 1984) has been reported. Zimmer (1984) observed that *C. traversiana* causes severe defoliation and affects stem and pods and possesses potential to be serious constraint to the production of fenugreek. Varos and Nagy (!972) observed the disease causing severe losses in Hungary. In India, the occurrence of *C. traversiana* has been reported from Punjab (Chahal *et al.*, 1975), Maharashtra (Sarode and Kadam, 1979), Uttar Pradesh (Singh and Pavgi, 1979), Delhi (Gupta *et al.*, 1982) and Madhya Pradesh (Khan *et al.*, 1984).

Symptoms

Symptoms of the disease are spots observed on the leaves, stems, young pods and seedlings, damaging the plants before maturity.

Chemical Control

The disease could effectively be controlled, seed yield increased and soil borne inoculum reduced by hexaferb, dithane-M-45 (Mancozeb) and aureofungin (Singh and Pavgi, 1979). Seed treatment with systemic fungicides can reduce seed borne infection and disease incidence. Among spray fungicides, benlate has been found most promising (Vorus and Nagy, 1972).

Collar Rot/Root Rot (*Rhizoctonia solani* Kuhn.)

The fungus is soil borne and was first reported from Karnataka (Hiremath *et al.*, 1978). This is also known from Rajasthan (Shuka, 1978), Gujarat (Gupta, 1980) and Tamil Nadu (Vanayagamurthy *et al.*, 1985).

Symptoms

It causes lesions mostly at the collar regions. Water soaked lesions appear that ultimately cause damage by rotting of the epidermal or vertical tissues at the collar regions causing damping off. As the disease advances, the stem becomes constricted at the base and the plants collapse. The sclerotia of the fungus can survive in the soil for several months, depending upon the temperature and moisture. Initial infection starts from the pre-infested soil. The disease is destructive in high humid and warm environmental conditions,

Disease Management

Carbendazim (0.1 per cent) both as seed dressing and soil drenching twice, effectively controlled the disease and increased yield (Prasad and Hiremath, 1985; Prakasam *et al.*, 1990a; Rajan *et al.*, 1990a). Captan is effective both as a soil mix and as soil drench while brassicol (PCNB) is superior as seed dressing (Hiremath *et al.*, 1978). Soil amendment with neem cake at 1t/ha and farm yard manure was also found effective in reducing the root rot and increasing the yield (*Rajan et al.*, 1990a). Soil application of neem cake at 1500kg/ha alone or in combination with Trichoderma reduced root rot and increased yield (Anon, 1993). Out of the 10 lines tested in IET trial the Acc-464, 113 and 231 showed lowest disease incidence (<5 per cent). In a CYT trial involving 14 lines HM-57 showed disease incidence of 2.1 per cent in kharif and 1.8 per.cent in rabi during 1992 (Annn., 1993).

Fennel

The crop suffers mainly with blight, root rot and powdery mildew.

Powdery Mildew (*Leveillula taurica* (Lev.) Am. Var. *languinosa* Salm)

It is caused by *Leveillula taurica* (Lev.) Am. Var. languinosa Salm, which appears in February and March in Rajasthan occasionally causing severe losses. Dusting of sulphur is recommended for its control (Chattopadhyay and Maiti, 1990). Variety IS-171 was least affected (Gidanavaer and Krishna Murthy, 1979).

Leaf Spot or Blight (*Cercospora foeniculi* Magn.)

The blight caused by *Cercospora foeniculi* Magn. has been reported first time from Pusa, Kashmir and other parts of India and Pakistan (Sydow and McRae, 1928; Prasad *et al.*, 1961; Kamal and Khan, 1962). This disease caused widespread damage in Rajasthan resulting in substantial yield reduction due to destruction of the foliage (Chattopadhyay, 1967).The fungus causes brown to black lesions on leaves, stems, peduncles and seeds. Severely infected plants give a blighted appearance. It is favoured by high temperature and humidity. If the plants are affected in the early stage, seed formation does not take place otherwise seeds are shriveled and blackened (Prasad *et al.*, 1961). Infected plant debris may constitute an important source of infection for the development of the disease. Severe epiphytotics were reported during 1955-56 in Ajmer due to prevailing warm temperature and low rainfall (Joshi, 1958).

Disease Management

Bordeaux mixture spray at monthly intervals from January checked the disease (Joshi, 1958). Spraying with suspension of colloidal sulphur (0.25 per cent) in water during February-March was recommended for the control of the disease (Chattopadhyay, 1967).

Ramularia Blight (*Ramularia foeniculi* Sib)

The disease is severe in Rajasthan and Gujarat and appears during the month of January. The temperature between 20 to 24 °C and RH >75 per cent favours disease severity.

The initial symptoms of the disease are minute angular, brown, necrotic areas on all the aerial parts of plant. Such spots enlarge and coalesce to form large necrotic areas. These necrotized spots get covered with grayish-white fungus growth. In advanced disease stage such symptoms appear on stem, peduncle umbels and leaves leading to drying of leaves. Plant infected at early flowering stage produces only a few shriveled, blackened seed (Prasad *et al.*, 1961). Such seeds are having less market value due to poor germination.

Three sprays of benomyl @ 0.02 per cent, tridemorph 0.1 per cent and dithane M 45 0.2 per cent or carbendazim 0.2 per cent starting from flowering stage at an interval of 20 days have been suggested for management of disease (Chaudhary and Patel, 1987; Lakra, 1993).

Alternaria Blight (*Alternaria alternata*)

The disease caused by *Alternaria alternate* has been reported to cause heavy losses (Deshpande and Sehgal, 1964). The disease was reported for first time from Kota (Rajasthan) and subsequently from other fennel growing areas (Chaudhary and Patel, 1987).Infection is confined mostly to the inflorescence. The buds become pale yellow and lose turgidity. Infected buds gradually become light brown and dry up. They are shed prematurely. The whole inflorescence and pedicel are affected within two weeks. Seeds may also be attacked and killed in severe cases (Chattopadhyay and Maiti, 1990).This disease appears at flowering particularly in persistent cloudy weather conditions. Cool

Ramularia Blight of Fennel

weather and unusual rainfall favours the disease epidemics (Cox, 1969). The disease can be managed effectively by spraying mancozeb (dithane M-45) at 20 days intervals after disease appearance.

Drooping Off/Stalk Rot/Root Rot (*Sclerotinia sclerotiorum* (Lib.)

The disease caused by *Sclerotinia sclerotiorum* was first reported from Tonk, Rajasthan, in 1965-66 (Sehgal and Agarwal, 1971). The fungus also cause stalk rot (Gaetan *et al.*, 1997). The pathogen also causes rotting near soil surface (Sahgal and Agrawal, 1971).

Phyllody

A Phyllody disease was noticed to occur in this crop during December-2005 at research farm of Main Spices Research Station, SDAU, Jagudan. During last two years, the disease severity was observed

Phyllody of Fennel

Phyllody of Fennel

in increasing trend. Incidence of the disease was recorded about 1 to 10 per cent at this station and also in Sabarkantha and Mehsana districts farmer's field. For confirmation of this disease, a sample of diseased plant with healthy plant was send at Indian Institute of Spices Research, Calicut and they confirmed by PCR analysis that this disease is Phyllody and caused by phytoplasma. Vector of this disease is leaf hopper. This disease started as purple/copper colour of leaves instead of green colour. Fully or partially plants converted in vegetative growth and no flower and seed setting was found. Same symptoms were also observed in other seed spices like cumin, coriander, dill and ajwain. During survey of Sabarkantha and Mehsana districts, this disease was observed high incidence where sesame (Til) crop grown around fennel field and less incidence was observed where pulses crop are grown in surrounding areas of fennel. For confirmation of this results a trial was conducted at Jagudan and same trend was observed.

References

Anonymous, 1953. *ICAR Scheme on Studies in the Wilt Disease of Coriander in Madhya Bharat*. Indian Council of Agricultural Research, New Delhi.

Anonymous, 1955. *Annual Report of the Scheme for Wilt Disease of Jira in Rajasthan*. Indian Council of Agricultural Research, New Delhi.

Anonymous, 1993. All India Coordinated Research Project on Spices, *Annual Report for 1992–93*. National Research Centre for Spices, Calicut..

Agnihotri, J.P. and Sharma, A.K., 1987. *Indian Cocoa Arccanut and Spices Journal*, 10: 65–67.

Aghnoom, R., Falahati-Rastegar, M. and Jafarpour, B., 1999. Comparison of chemical and biological control of cumin wilt (*Fusarium oxysporum* f. sp. *cumini*) in laboratory and greenhouse conditions. *Iranian Journal of Agricultural Sciences*, 30: 619–630.

Ali, S.A., Pathak, R.K. and Saraf, R.K., 1998. Fungicidal seed treatment in corinader. *Advances in Plant Science*, 11: 303–304.

Baswana, K.S., Jalali, I. and Thakral, K.K., 1991. Effect of sowing dates and fungicides on seed quality, yield and disease incidence of cumin. *Indian Cocoa, Arecanut Spices J.*, 14: 155–157

Bharadwaj, L.N. and Shrestha, S.M., 1985. *Agriculture, Ecosystems and Environment*, 13: 319–323.

Bhatnagar, K., Sharma, B. and Kant, U., 2004. *J. Mycol. Pl. Pathol.*, 34: 359–360.

Butler, E.J., 1918. *Fungi and Diseases in Plants*. Thacker Spink and Co., Calcutta, pp. 359–361.

Chakravati, B.P. and Rangarajan, M., 1966. Occurrence of *Erwinia aroideae* on two hosts in India. *Plant Disease Reporter*, 50: 701–702.

Champawat, R.S. and Pathak, V.N., 1988a. Role of nitroghen, phosphorus and potassium fertilizers and organic amendments in cumin (*Cuminum cyminum*) wilt incited by *Fusarium oxysporum* f. sp. *cumini*. *Indian Journal of Agricultural Sciences*, 58: 728–730.

Champawat, R.S. and Pathak, V.N., 1988b. Soil application of different insecticides and nematicides for the control of wilt (*Fusarium oxysporum* f. sp. *cumini*) of cumin. *Indian Journal of Plant Protection*, 16: 195–196.

Champawat, R.S. and Pathak, V.N., 1990. Management of cumin wilt by summer ploughing. *Indian Cocoa Arecanut and Spices Journal*, 13: 23–26.

Champawat, R.S. and Pathak, V.N., 1990b. Studies on seed microflora of cumin. In: *Proc. National Seminar on Seed Spices*, (Eds.) C.K. Goerge, S. Edison, E. Velappan and R. Naidu. Spices Board, Ministry of Commerce, Government of India, Cochin, pp. 592–599.

Champawat, R.S. and Pathak, V.N., 1991a. Effect of fungicidal seed treatment on wilt disease of cumin. *Journal of Turkish Phytopathology*, 20: 23–26.

Champawat, R.S. and Pathak, V.N., 1991b. Interaction between vesicular arbuscular mycorrhizal fungi and *Fusarium oxysporum*. *Proc. Indian Nat. Sci. Acad.*, 57: 59–62.

Chattopadhyaya, S.B. and Maiti, S., 1990. *Diseases of Betelvine and Spices.* ICAR, New Delhi, pp. 160.

Chattopadhyay, S.B., 1967. *Diseases of Plant Yielding Drugs, Dyes and Spices*. ICAR, New Delhi, 100 pp.

Chahal, S.S., Rawal, G.S. and Rawal, H.S., 1975. *Proc. Nat. Acad. Sci., India*, 45B: 276–280.

Dange, S.R.S., Pandy, R.N. and Savalia, R.L., 1992. Diseases of cumin and their management: A review. *Agricultural Reviews*, 13: 219–224.

Dange, S.R.S., Patel, R.L., Patel, S.I. and Patel, D.S., 2003. *J. Mycol. Pl. Pathol.*, 33: 502.

Das, A.M., 1971. Studies on stem gall disease of coriander incited by *Protomyces macrosporus*. *Plant Breed. Abstr.*, 943: 857.

Deshpande, A.L. and Sehgal, S.P., 1964. Blight of *Foeniculum vulgare* Mill caused by *Alternaria tenuis*. *Science and Culture*, 31: 203–204.

Dhruj, I.U., Akbari, L.F., Khandari, R.R. and Jadeja, K.B., 2000. Field evaluation of fungicides against powdery mildew of fenugreek. *J. Mycol. Pl. Pathol.*, 30: 98–99.

Edison, S., 1989. *Proc. First National Seminar on Seed Spices*, Jaipur, 1: 195–196.

Edison, S., Johny, A.K., Nirmal Babu, K. and Ramadasan, A., 1991. *Spices Varieties: A Compendium of Morphological and Agronomic Characters of Improved Varieties of Spices in India*. National Research Centre for Spices, Calicut, Kerala, India.

Gaur, M.M., 1949. Plant protection work in Ajmer–Merwara in 1948 Plant Diseases. *Plant Protection Bulletin*, Govt. India, 1: 20–21.

Gaumann, E.A. and Dodge, C.W., 1928. *Comparative Morphology of Fungi*. McGraw Hill Book Co. Inc., New York.

Gupta, J.H., 1980. *J. Mycol. Pl. Pathol.*, 10: 103.

Gupta, R., Mohendra, P. and Mukerji, P.G., 1982. *Nova Hedwigia*, 36: 675–680.

Gupta, J.S., 1956a. Chemistry of diseased fruits of *Coriandrum sativum* L. affected by *Protomyces macrosporus* Ung. *Indian Phytopath.*, 9: 217.

Gupta, J.S., 1964. Influence of stem gall disease on fruit quality in coriander. *Agra Univ. J. Res.*, 13: 55–58.

Gupta, J.S. and Sinha, S., 1963. Some therapeutic control trials on the stem gall disease of coriander. *Indian Phytopath.*, 16: 106.

Gupta, J.S. and Sinha, S., 1973.Varietial field trials in the control of stem gall disease of coriander. *Indian Phytopath.*, 26: 337–340.

Gupta, R.N., 1976. Impact of nitrogen, phosphorus and manganese on the stem gall disease of coriander. *Acta Botanica Indica*, 4: 30–35.

Gupta, R.N., 1973. Longevity of chlamydospores of coriander gall fungus. *Indian Phytopath.*, 26: 581–582.

Gemawat, P.D. and Prasad, N., 1972. Epidemiological studies on Alternaria blight of *Cuminum cyminum*. *Indian J. Mycol. Pl.Pathol.*, 2: 65–67.

Gidnavar, V.S. and Krishnamurthy, K., 1979. IS–711 variety of fennel has a possibility as a pure crop. *Curr. Sci.*, 8: 111.

Goel, A.K., Kumar S. and Tayal, M.S., 1983. Biochemical alteration induced in *Coriandrum sativum* L. by *Protomyces macrosporus* Ung. *Indian Bot. Rept.*, 2(1): 62–64.

Gohil, W.P., Patel, A.T. and Jain, S.M., 1985. Field evaluation of fungicides for control of powdery mildew of cumin. *Indian Phytopath.*, 38: 783–784.

Hirata, 1966. *Host Range and Geographical Distribution of the Powdery Mildew*. Faculty of Agricultural Niigata Univ., Niigata, pp. 472.

Hiremath, P.C., Ponnappa, K.M., Janardhan, A. and Sundaresh, H.N., 1978. *Pesticides*, 12: 30–31.

Israel, S., 2002. Effect of amendments on wilt of cumin caused by *Fusarium oxysporum* f.sp. *cumini* Prasad and Patelin Aridisols. *Ph.D. Thesis*, J.N.V. University, Jodhpur.

Jain, K.L. and Agrawal, J.M., 1978. Control of Fusarium wilt of coriander by systemic fungicides. *Indian J. Mycol. Pl. Pathol.*, 8: 101–102.

Joshi, L.K. and Nikam, B.G., 1953. *Studies in the Wilt Disease of Coriander in the Madhya Bharat*. Govt. Central Press, Gwalior, pp. 20.

Joshi, N.C., 1958. *Agri. Univ. J. Res. (Sci.)*, 7: 87–91.

Kalra, A., Parameshwaran, T.N., Ravindra, N.S. and Dimri, B.P., 1995. Effect of powdery mildew (*Erysiphe polygoni*) on yields and yield components of early and late maturing coriander (*Coriandrum sativum* L.). *Journal of Agricultural Science*, 125: 395–398.

Kalra, A., Parameshwaran, T.N., Ravindra, N.S., Rao, M.G., Sushil-Kumar and Kumar, S., 2000. Effect of planting date and dinocap application on the control of powdery mildew and yields of seed and seed oil in coriander. *Journal of Agriculture Science*, 135: 193–197.

Kamal, M. and Khan, S.A., 1962. Identity of *Cercospora foeiculi P. magnus* causing infection of *Foeniculum vulgare* in Indo-Pakistan sub-continent. *Biologia*, 8: 61–62.

Keshwal, R.L. and Khatri, R.K., 1998. Reaction of some high yielding varieties of coriander to powdery mildew. *J. Mycol. Pl. Pathol.*, 28: 58–59.

Khader, M.D. Abdul and Vedamuthu, P.G.B., 1989.

Khan, S.A. and Kamal, M., 1962. *Pakistan J. Sci. Res.*, 5: 41.

Khan, A.J., Sattar, A. and Thussain, A., 1984. *Indian J. Pl. Path.*, 2: 201.

Keshwal, R.L., Choudhary, P.C. and Singh, K., 1979. *Pesticides*, 13: 25.

Lakra, B.S., 2000. Management of stem gall of coriander (*Coriandrum sativum*) incited by *Protomyces macroporus*. *Indian Journal of Agricultural Sciences*, 70: 338–340.

Lodha, S., 1995. Soil solarization, summer irrigation and amendments for the control of *Fusarium oxysporum* f. sp. *cumini* and *Macrophomina phaseolina* in arid soils. *Crop Protection*, 14: 215.

Lodha, S. and Kumar, A., 1998. Harnessing solar heat to manage cumin wilt. *Indian Farming*, 47(12): 23–24.

Leppik, E.E., 1960. *Pl. Dis. Reptr.*, 44: 440.

Mahor, R., Naqvi, S.M.H. and Gupta, 1982. A new fungal disease of coriander in India. *Acta Botanica India*, 10: 323.

Maheshwari, D.K. and Chaturvedi, S.N., 1983. Histochemical localization of acid phosphatase in two fungus galls. *Indian Phytopath.*, 36(1): 167–170.

Mall, O.P., 1968. Studies on wilt disease of coriander with special reference to variability of the pathogen. *Indian Phytopathology.*, 21: 379–388.

Mathur, B.L. and Mathur, R.L., 1966. Fusarium wilt of cumin: Influence of certain factors on disease incidence. *Proceedings of National Academy Sciences*, 36: 33–38.

Mathur, B.L. and Prasad, N., 1964. Studies on wilt disease of cumin caused by *Fusarium oxysporum* f. sp. *cumini*. *Indian Journal of Agricultural Sciences*, 34: 131–137.

Mathur, B.L., Sankhla, H.C. and Mathur, R.L., 1967. Influence of cultural practices on cumin wilt incidence. *Indian Phytopathology*, 20: 32–35.

Mathur, R.L., 1949. Plant protection notes and news. *Plant Protection Bulletin*, Govt. India, Jaipur, 1: 30.

Mathur, R.L., Masih, B. and Sankhla, H.C., 1971. Evaluation of fungicides against powdery mildew disease of cumin (*Cuminum cyminum* L.) caused by *Erysiphe polygoni*. *Indian Phytopathology*, 24: 796–798.

Mathur, B.L. and Mathur, R.L., 1967. *Pl. Dis. Reptr.*, 51: 630–632.

Mathur, B.L. and Mathur, R.L., 1970. *Proc. Nat. Acad. Sci., India*, 38: 49–52.

Mathur, R.L., 1963. Studies on the Fusarium wilt of coriander in Rajasthan. *Ph.D. Thesis*, Univ. of Rajasthan, Jaipur.

Mathur, R.L., Mathur, B.N. and Sharma, B.S., 1972. *Pl. Dis. Reporter*, 43: 1051.

Masih, B., Sankhla, H.C. and Mathur, R.L., 1970. *Rajasthan J. Agric.* 1: 18–20

Mawar, R. and Lodha, S., 2002. Integration of brassica amendments and summer irrigation for the control of *Macrophomina phaseolina* and *Fusarium* in hot arid region. *Phytopathologia Mediterranea*, 41: 45–54.

Mehta, P.R., Singh, B. and Bose, S.K., 1946. Some new host of *Sclerotinia sclerotiorum* (Lib.) de bary. *Current Science*, 15: 171–172.

Mehta, K.G. and Solanki, V.A., 1990. A status report on plant pathological work carried out Main Spices Research Station, Jagudan, Dist. Mehsana, Gujarat.

Mehta, K.G., Patel, D.B. and Solanki, V.A., 1990. Effect of fungicidal sprays on powdery mildew disease of cumin. *Indian Cocoa, Arecanut and Spices Journal*, 13: 144–145.

Moses, G.J. and Govinda Rao, P., 1969. Coriander anthracnose caused by *Glomerella cingulata*, a new record from India. *Indian Phytopath.*, 22: 145.

Mukhopadhyay, A.N. and Pavgi, M.S., 1971. *Annual Phytopath. Soc., Japan*, 37: 215–219.

Mundkar, B.B., 1949. *Fungi and Plant Diseases*. McMillan and Co. Ltd., London.

Naqvi, S.A.M.H., 1985. Varietal screening of coriander against stem gall in relation to disease intensity and crop loss. *J. Mycol. Pl. Pathol.*, 12: 270–276.

Narula, P.N. and Joshi, M.G., 1963. Differential response of coriander strains to stem gall. *Sci. Cult.*, 29: 206.

Nene, Y.L., 1966. *Mycopath. Mycol Appl.*, 29: 142–144

Omer, E.A., Nofal, M.A., Lashin, S.M. and Haggaq, W.M.E., 1997. Effect of some growth regulators on the growth parameters and oil content of cumin with disease incidence under two types of soil. *Egyptian Journal of Mycology and Plant Pathology*, 12: 270–276.

Parakhia, A.M., Naliyadhar, M.V. and Vaishnav, M.V., 1990. Chemical control of powdery mildew of coriander in Gujarat. In: *Proc. First National Seminar on Seed Spices*, (Eds.) C.K. George, S. Edison, E. Elappan and R. Naidu. Spices Board, Ministry of Commerce, Govt. of India., Cochin, pp. 569–579.

Patel, R., 1968. Alternaria blight of cumin (*Cuminum cyminum* L.) and its control. *M.Sc. Thesis*, S.P. University, Anand.

Patel, P.N., Prasad, N., Mathur, R.L. and Mathur, B.L., 1957. Fusarium wilt of cumin. *Current Science*, 26: 181–182.

Patel, S.M. and Patel, B.K., 1993. Evaluation of herbicidal concentrations against *Fusarium oxysporum* f. sp. *cumini* causing cumin wilt. In: *Proc. of Indian Society of Weed Science International Symposium* Hissar, November 18–20, 2: 131–132.

Patel, S.M. and Patel, B.K., 1998a. Inhibitory effect of different fungal bioagents on *Fusarium oxysporum* f. sp. *cumini* causing cumin wilt. *Annals of Plant Protection Sciences*, 6: 25–27.

Prakasam, V., Vedamuthu, P.G.B., Khader, M.A. and Jeyarajan, R., 1988. A preliminiary report on the occurrence of phyllody in coriander. *Vegetable Science*, 15: 198–199.

Prakash, S. and Saharan, G.S., 1999. Sources of resistance to downy and powdery mildew of fenugreek. *J. Mycol. Pl. Pathol.*, 36: 95–99.

Prasad, N., Mathur, R.L. and Agnihotri, J.P., 1961. Blight of fennel in Rajasthan, I: Morphology and Identity of the pathogen. *Current Science*, 30: 65–66.

Panwas, R.N., Utikar, P.G. and Shinde, P.A., 1985. *J. Maharashtra Agri. Univ.*, 10: 218.

Refshauge, S.J. and Nayudu, M., 2001. Invasion by *Pseudomonas syringae* pv. *coriandricola* is responsible for bacterial blight of coriander. *Plant Science*, 161: 621–625.

Romeiro, R.S., Simoes, A.R., Muchovej, J.J., Oloveria, J.R. and Souza, R.M., 1994. Bacterial sudden yellow death and soft rot of *Coriandrum sativum*. *Plant Pathology*, 43: 944–946.

Sarode, M.S. and Kadau, V.C., 1979. *Maharashtra Vidnyan Mandir Patrika*, 14: 10–12.

Sehgal, S.P. and Agarwal, J.M., 1971. Drooping of fennel (*Foeniculum vulgare*) due to *Sclerotinia sclerotiorum*. *Indian Phytopathology*, 24: 608–609.

Sharma, S.K., Agarwal, R.K. and Lodha, S., 1995. Population changes of *Macrophomina phaseolina* and *Fusarium oxysporum* f.sp. *cumini* in oil cake and crop residue amended sandy soils. *Applied Soil Ecology*, 2: 281–284.

Solanki, J.S., Singh, R.R. and Delela, G.G., 1973. Field evaluation of fungicides in controlling Alternaria blight of *Cuminum cyminum*. *J. Mycol. Pl. Pathol.*, 3: 191–197.

Srivastava, U.S., 1972. Effect of interaction of factors on wilt of coriander caused by *Fusarium oxysporum* Schlecht. Ex. Fr. *F. coriandrrii*. Kulkarni, Nikem and Joshi. *Indian Journal of Agricultural Sciences*, 42: 618–621.

Srivastava, U.S., Rai, R.A. and Agrawal, J.M., 1971. Powdery mildew of coriander and its control. *Indian Phytopathology*, 24: 437–440.

Srivastava, U.S., 1967. Effect of soil conditions in relation to wilt of coriander (*Coriandrum sativum* L.) *Ph.D. Thesis*, Univ. of Rajasthan, Jaipur

Saxena, P., Ahmad, S.T. and Chauhan, S., 1984. Field reaction of methi germplasm against powdery mildew. *Indian J. Mycol. Pl. Pathol.*, 14: 85–86..

Sydow, H. and Mc Rae, W., 1928. Hyphomycetes Indiae Orientalis. *American Cryptogamie. Emt.*, 1: 166.

Tripathi, A.K., Bartaria, A.M., Pandya, R.K., Shasi-Chauhan and Chauhan, S., 1998. Evaluation of coriander cultivars against stem gall disease. *Flora and Fauna– Jhansi*, 4: 98.

Uppal, B.N., Patel, M.K. and Kamat, M.N., 1935. The fungi of Bombay. *Bull. Dep. Land Reclamation and Agriculture Bombay*, 176(1934): 2.

Uppal, B.N. and Desai, M.K., 1932. Cumin powdery mildew in Bombay. *Bull. Bombay Dept. Agriculture*, 169 p.

Uppal, B.N., Patel, M.K. and Kamat, M.N., 1938. Alternaria blight of cumin. *Indian Journal of Agricultural Sciences*, 8: 9–62.

Voros, J. and Nagy, F., 1972. *Cercospora traversiana* Sacc., a new destructive pathogen of fenugreek in Hungary. *Acta Phytopathologica Academie Scientia Hungarica*, 7: 71–76.

Vinayagamoorthy, A., Mohideen, M.K., Jeyarajan, R. and Prakasham, V., 1985. *South Indian Hort.*, 133: 347–348.

Recent Advances in Seed Spices (2011) *Pages* **154–162**
Editors: **Y. Ravindrababu, R.K. Jaiman, K.D. Patel, N.R. Patel and S.B.S. Tikka**
Published by: **DAYA PUBLISHING HOUSE, NEW DELHI**

14

Management of Insect Pests in Seed Spices

G.M. Patel, B.R. Patel and J.K. Patel
Department of Entomology, C.P. College of Agriculture, SDAU, Sardarkrushinagar, Gujarat

Introduction

There are about twenty seed spices grown in India. Most of the important seed spices them are belong to family Apiaceae (coriander, cumin, fennel etc.). About 0.9 million ha area is under seed spices in the country with annual production of about 5-6 lakh tones (Anonymous, 2003). India is the world's largest producer, consumer and exporter of seed spices. The states, Rajasthan and Gujarat has emerged as Seed Spices Bowl and contribute more than 80 per cent of the total seed spices produced in the country. Cumin, fennel, coriander, fenugreek, ajwain and dill seeds are major seed spices crops of the Gujarat state. The area under cumin crop was 2,32,000 ha. with a production of 1,44,900 mt and fennel crop was grown in 27,300 ha with a production of 43,500 mt in the state during the year 2007–08 (www.gswan.gov.in). These crops are highly sensitive to climatic conditions. Dry and cool weather is favourable for the crops, while high relative humidity and cloudy weather provide congenial conditions for the multiplication of pests.

Heavy attack of insect pests in these crops result in reduction in grain yield and affect the quality of seed spices. The insect pests may cause 5 to 67 per cent reduction in the yield of various spices crops. Important pests of these crops are categorized in to three groups *viz.*, sucking pests, defoliators and stored grain pests.

Sucking Pests

Aphids, thrips, leaf hoppers, whiteflies and mites are reported to infest seed spices crops.

Aphids

Various species of aphids infest seed spices crops. The coriander aphid (*Hydaphis coriandri* (Das) and cotton aphid (*Aphis gossypii* Glower) are major pests of fennel, coriander, cumin and dill seeds. The population of aphids attains peak level during the month of January in Gujarat (AGRESCO, 1996). *Aphis craccivora* infests fenugreek. Apart from *H. coriandri, Myzus persicae* also infests cumin. The coriander aphids *Hyalopterus pruni* Geoffroy, *Brevicoryne coriandri* and *Acrophosiphan pisum* are reported on fenugreek and other seed spices. The aphids, *H. coriandri* remained active during February to April with peak activity in the first week of March in fennel (Patel, 2008).

Thrips

Thrips flavus was found to infests fennel and dill seeds. *T. tabaci* and *Sirtothrips dorsalis* are pests of coriander, cumin, and Psyllium crops. Both nymphs and adults scrap the epidermis of leaves and tender parts and suck oozing cell sap. The infested leaves turn brown. Excessive feeding may lead to curling of leaves and stunted growth of the plants. Long dry spell favours fast multiplication of this pests.

Two species of mite *viz., Petrobia latus* and *Tetranychus telarius* are found on coriander, cumin and fennel crops. White flies (*Bemesia tabaci*) and leaf hopper also damage these crops in early stage under favourable environmental conditions. Stink bugs (*Agnoscelis nubile F.* and *Anezara viridule*) may attack cumin and coriander crops.

Management of Sucking Pests

- ☆ Install yellow sticky traps for monitoring incidence of aphids/white flies @ 10 traps/ha.
- ☆ Late sowing should be avoided. Coriander sown during last week of October to first week of November may record minimum damage caused by aphids.
- ☆ Clip the heavily infested plant parts which help in reducing aphid infestation in the early stage.
- ☆ Keep the crop free from weeds either by chemical or manual weeding.
- ☆ Apply organic manure and avoid excess use of nitrogenous fertilizers.
- ☆ Apply organic amendments *viz.*, castor or neem cake @ 2 t/ha.
- ☆ For the management of aphids and leaf hoppers in fenugreek, seed treatment with thiamethoxam 70 WS @ 2.8 g/kg seed (CBR 1:2.12) was effective in coriander, seed treatment with thiamethoxam 70 ws @ 4.2 g/kg seed (CBR 1:4.94) was found effective against aphids (AGRESCO, 2007).
- ☆ The insecticidal seed treatments of thiamethoxam 70 WS@ 4.2g/kg seed (ICBR 1:9.22) and imidacloprid 70WS @ 10g/kg seed (ICBR 1:2.58) is recommended for the control of thrips in cumin crop (AGRESCO, 2007).
- ☆ The insecticidal seed treatments of thiamethoxam 70 WS@ 4.2g/kg seed (ICBR 1:5.29) and imidachloprid 70WS @ 10g/kg seed (ICBR 1:4.99) proved most effective in controlling the aphid in dill seed crops (AGRESCO, 2007).
- ☆ Avoid irrigation at the time of active seed filling stage to minimize incidence of aphids and other sucking pests.
- ☆ Conserve the predators like Coccinellids, Chrysopids and syrphid fly etc. The Coccinellid predators (*Coccinella semptpunctata* and *Menochilus sexmaculatus*) were reported in fennel ecosystem and played vital role in checking the population of aphids (Patel, 2008).

Table 14.1: Effect of Different Treatments on Aphid Infestation in Cumin with Economics

Sl.No.	Treatments	Infestation (Per cent) Pre-Treatment	Reduction in Aphid Infestation (Per cent) at DAS		Cumin Yield (Kg/ha)	Increase in Yield over Control (Kg/ha)	Cost of Treatment (Rs/ha)	Realization Over Control (Rs/ha)	Net Gain (Rs/ha)
			7	10					
1.	Carbosulfan 25 EC @ 1250 ml/ha	61.71 (51.77)*	24.11 (29.41)	19.33 (29.08)	372.5	210.0	465.0	9450.0	8985.0
2.	Carbosulfan 25 EC @ 2500 ml/ha	64.00 (53.13)	53.69 (47.12)	45.52 (42.43)	389.0	226.5	847.5	10192.5	9345.0
3.	Carbosulfan 25 EC @ 625ml/ha	61.65 (51.74)	65.29 (53.90)	57.45 (49.28)	474.0	311.5	1612.5	14017.5	12405.0
4.	Thiamethoxam 25WG @ 10 gm/ha	64.37 (53.35)	26.77 (30.94)	15.27 (23.00)	328.0	165.5	682.5	7447.5	6764.5
5.	Imidacloprid 17.8 SL @ 100 ml/ha	62.63 (52.32)	21.97 (27.97)	15.04 (22.82)	369.0	206.5	402.5	9292.5	8890.0
6.	Control	63.78 (53.00)	–	–	162.5	–	–	–	-
	C.D. at 0.05	NS	2.04	2.16	70.36				

SDAU, S.K. Nagar. *Source*: Patel *et al.*, 2005.

DAS: Days After Spraying.

Figures in parentheses are retransformed values.

Table 14.2: Efficacy of Methyl-O-Demeton Against Cumin Aphid Sprayed through CDA and Knapsack Sprayer

Treatments	*Per cent Umbels Infested by Aphids*		*Yield (Kg/ha)*
	3 DAS	*7 DAS*	
S_1: CDA sprayer	36.97	40.15	511.88
S_2: Knapsack sprayer	77.80	79.42	282.09
S.Em.±	1.88	1.92	17.95
C.D. at 5 per cent	5.52	5.65	5279
ST_1: Sticker	54.09	55.47	453.44
ST_2: No sticker	60.67	64.11	340.53
S.Em.±	1.88	1.92	17.95
C.D. at 5 per cent	5.52	5.65	52.79
I_1: Insecticde	47.21	48.95	416.72
I_2: No insecticide	67.55	70.63	377.25
S.Em.±	1.88	1.92	17.94
C.D. at 5 per cent	5.52	5.65	NS
C.V. per cent	13.09	12.84	18.08

SDAU, S.K. Nagar. *Source*: Patel *et al.*, 2005.

DAS: Days After Spraying.

- Augmentative release of *Coccinella septumpunctata* @ 5000 beetles/ha against the pest, may help to suppress aphid population.
- Methyl-O-demeton applied through Controlled Droplet Applicator (CDA) sprayer was found more effective than the application with manually operated knapsack sprayer in cumin crop and it was observed that addition of sticker (Sandowit) in insecticidal spray fluid increased the effectiveness of methyl-O-demeton (Patel *et al.*, 2005).
- In the experiment conducted at Sardarkrushinagar, dimethoate 0.03 per cent, imidachloprid 0.06 per cent and carbosulfan 0.05 per cent proved effective against *H. coriandri* on fennel (Patel, 2008).
- The results of field experiment conducted at Sardarkrushinagar during 2001-02 and 2002-03 indicated that application of carbosulfan 25 EC @ 2500 ml/ha found effective in reducing aphid infestation in cumin crop however, its lower dose *i.e.* 1250 ml/ha was also effective and economical. The residues of insecticide were below detectable level (*i.e.* < 0.1 ppm) in the grain sample collected one month after spraying (Patel *et al.*, 2005).
- Spraying of monocrotophos 0.04 per cent (ICBR 1:18.46) or endosulphan 0.07 per cent (ICBR, 1:12.56) or quinalphos 0.05 per cent (ICBR, 1:12.16) twice at 15 days interval is recommended for the control of *A. gossypii* in cumin (AGRESCO, 2003).

Table 14.3: Effect of Different Insecticidal Seed Treatments on Leaf Hopper Population of Fenugreek

Sl.No.	Treatment	Jassids/3 Leaves				Yield kg/ha	CBR
		8 WAG	9 WAG	10 WAG	11 WAG		
1.	Imidacloprid 70WS @ 5g/kg seed	1.00 (0.49)	1.43 (1.56)	1.65 (2.22)	1.69 (2.35)	1610	1:0.33
2.	Imidacloprid 70WS @ 7.5g/kg seed	0.93 (0.36)	1.31 (1.22)	1.42 (1.52)	1.50 (1.76)	1687	1:0.51
3.	Imidacloprid 70WS @ 10g/kg seed	0.85 (0.23)	1.08 (0.66)	1.06 (0.62)	1.15 (0.83)	1798	1:0.70
4.	Thiamethoxam 70 WS @ 2.8g kg seed	0.81 (0.15)	1.14 (0.79)	1.22 (0.99)	1.33 (1.26)	1748	1:2.12
5.	Thiamethoxam70WS @ 4.2g kg seed	0.81 (0.16)	0.96 (0.43)	0.87 (0.26)	1.09 (0.69)	1827	1:1.98
6.	Dimethoate 30 EC @ 0. 03 per cent on need base	1.08 (0.66)	1.78 (2.68)	2.28 (4.69)	2.51 (5.78)	1570	1:0.42
7.	Control (water spray)	1.15 (0.82)	1.76 (2.59)	2.28 (4.72)	2.45 (5.52)	1570	–
8.	Control (Untreated)	1.12 (0.76)	1.82 (2.82)	2.28 (4.69)	2.53 (5.89)	1552	–
	S.Em.±for Treat.	0.04	0.05	0.05	0.05	45.73	
	C.D. at 5 per cent for Treat.	0.11	0.14	0.14	0.16	132.5	
	S.Em.±for year	0.02	0.02	0.02	0.03	22.87	
	C.D. at 5 per cent for year	0.05	0.07	0.07	0.08	66.23	
	S.Em.±for YXT	0.05	0.07	0.07	0.08	64.67	
	C.D. at 5 per cent for YXT	NS	NS	NS	NS	NS	
	C.V. per cent	9.47	8.60	7.46	7.45	6.71	

JAU, Junagadh. *Source*: AGRESCO, 2007.

$\sqrt{x} + 0.5$ transformations used.

Data in parentheses are retransformed values.

WAG: Week after germination.

Defoliators and Leaf Miner

Helicoverpa sp. attacks cumin, coriander, fenugreek and fennel crops. Cutworm (*Agrotis ipsilon*) and beet army worm (*Spodoptera exigua*) were recorded on dill seed crop. Cabbage looper (*Trichoplusiani*) and leaf miner (*Liriomyza sativa* and *L. trifolii*) are also pests of seed spices. These pests sporadic in nature and found only in suitable environmental conditions. Heavy attack of *Helicoverpa* at flowering and seed setting stage of cumin and coriander crops caused severe damage.

Management

- ☆ Deep summer ploughing will check the pest by exposing their resting stage to sun and insectivorous birds.
- ☆ Collect the larvae of cut worm and leaf defoliators and kill them.

Table 14.4: Effect of Different Insecticidal Seed Treatments on Thrips Population of Cumin

Sl.No.	Treatment	Average Thrips/5 Plants/Week				Yield kg/ha	CBR
		3 WAG	6 WAG	9 WAG	Mean Pooled		
1.	Imidacloprid 70WS @ 5g/kg seed	2.76 (7.61)	5.14 (26.42)	7.14 (51.03)	5.15 (26.52)	337	1:0.78
2.	Imidacloprid 70WS @ 7.5g/kg seed	2.50 (6.26)	4.58 (20.94)	6.95 (48.32)	4.80 (23.06)	385	1:1.49
3.	Imidacloprid 70WS @ 10g/kg seed	1.55 (2.41)	3.78 (14.29)	7.03 (49.42)	4.18 (17.46)	482	1:2.58
4.	Thiamethoxam 70 WS @ 2.8g/kg seed	2.38 (5.66)	3.97 (15.74)	7.00 (48.97)	4.54 (20.60)	446	1:7.72
5.	Thiamethoxam70WS @ 4.2g/kg seed	1.66 (2.76)	3.58 (12.85)	6.87 (47.22)	4.05 (16.41)	553	1:9.22
6.	Dimethoate 30EC 0.03 per cent on need base	3.25 (10.56)	5.28 (27.89)	5.64 (31.83)	5.18 (26.83)	332	1:1.73
7.	Control (water spray)	3.13 (9.81)	5.18 (26.83)	7.08 (50.06)	5.35 (28.62)	308	–
8.	Control (Untreated)	3.27 (10.67)	5.35 (28.61)	7.20 (51.78)	5.46 (29.83)	311	–
	S.Em.±for Treat	0.12	0.47	0.53	0.24	30	
	C.D. at 5 per cent for Treat	0.35	1.57	NS	0.69	88	
	S.Em.±for year	0.06	0.24	0.27	0.12	15	
	C.D. at 5 per cent for year	0.18	0.79	0.89	0.34	44	
	S.Em.±for YXT	0.17	0.18	0.23	0.34	43	
	C.D. at 5 per cent for YXT	NS	0.51	0.66	NS	NS	
	C.V. per cent	11.59	6.58	5.76	12.19	19.00	

JAU, Junagadh. *Source*: AGRESCO, 2007.

$\sqrt{x}$ transformations used.

Data in parentheses are retransformed values.

- ☆ Destroy the egg mass and gregarious young larvae of *Spodoptera*.
- ☆ Conserve parasites and predators and mass release of *Trichogramma spp* against lepidopterous pests.
- ☆ Spray HaNPV @ 300 LE/ha during evening hours when young *Helicoverpa* larvae are noticed.
- ☆ Install bird perchers @ 20-25/ha to promote predation of moths and larvae by birds.
- ☆ Install 8 pheromone traps each of *Helicoverpa* and *Spodoptera spp*/ha.
- ☆ If needed, apply endosulphan 0.07 per cent or DDVP 0.025 per cent along with stickers.

Storage Pests

Fennel and coriander seeds are infested by seed midge (*Systole albipennis* Walker) in the field at maturity. The female midge lay eggs in the embryo of grain and developing larvae feeds remaining inside the grain. Larvae pupate inside the grain and adults emerge from the stored grains by making hole. Oil content and germination power in damaged seed spices are reduced. The infestation of this pest is reported up to 40 per cent in fennel, 30 per cent in coriander, 27 per cent in cumin and 10 per cent in ajawain. In fennel, the damage of *S. albepennis* Walker was found lowest in January (5.4 per cent) and highest in March (20.7 per cent). Cigarate beetle (*Lasioderma sericorne* Fabricious) and Drug store beetle (*Stegobium sp.*) are also reported to damage stored spices.

Table 14.5: Effect of Different Insecticidal Seed Treatments on Aphid Population of Dill Seed

Sl.No.	Treatment	Average Thrips/5 Plants/Week				Yield kg/ha	CBR
		7 WAG	10 WAG	12 WAG	Mean Pooled		
1.	Imidacloprid 70WS @ 5g/kg seed	0.83	1.87	1.60	1.52	1032	1:1.16
2.	Imidacloprid 70WS @7.5g/kg seed	0.67	1.53	1.37	1.27	1120	1:1.38
3.	Imidacloprid 70WS @ 10g/kg seed	0.27	0.67	0.70	0.59	1246	1:1.69
4.	Thiamethoxam 70 WS @ 2.8g/kg seed	0.50	1.13	1.13	0.98	1174	1:4.99
5.	Thiamethoxam70WS @ 4.2g/kg seed	0.17	0.37	0.47	0.44	1323	1:5.29
8.	6. Dimethoate 30 EC 0.03 per cent on need base	1.00	1.07	0.70	0.82	1065	1:1.23
7.	Control (water spray)	1.00	2.43	2.00	1.87	926	1:0.06
8.	Control (Untreated)	1.00	2.53	2.10	2.06	921	-
	S.Em.±for Treat	0.09	0.23	0.25	0..11	43	-
	C.D. at 5 per cent for Treat	0.27	0.77	0.84	0.36	123	-
	S.Em.±for year	0.05	0.12	0.13	0.02	21	-
	C.D. at 5 per cent for year	0.14	0.39	NS	0.056	62	-
	S.Em.±for YXT	0.13	0.17	0.17	0.055	60	-
	C.D. at 5 per cent for YXT	NS	0.48	0.50	0.16	NS	-
	C.V. per cent	19.37	14.69	17.02	13.84	9.00	-

JAU, Junagadh. *Source*: AGRESCO, 2007.

WAG: Week After Germination.

Management

Control measures of insect pests of stored grains is divided in to two groups *viz.*,

1. Prevention of infestation
2. Control of established infestation

Prevention of Infestation

- ☆ The grains should be dried thoroughly in sunlight, to bring down moisture content below 9-10 per cent.
- ☆ Damaged grains should be sieved before storage to remove broken grains to avoid infestation of secondary pests in the store.

Table 14.6: Efficacy of Different Treatments Against *H. coriandri* in Fennel Crop During Rabi 2008

Treatments	*A.I./Plant at Different Intervals of Spray*				*Yield (Kg/ha)*
	3 DAS		*7 DAS*		
	Before Spray	*Mean*	*Before Spray*	*Mean*	
Dimethoate @ 0.03 per cent	1.61 (2.10)	1.04 ·(0.61)	1.54 (1.90)	1.39 (1.46)	765.38
Imidacloprid @ 0.06 per cent	1.65 (2.23)	1.03 (0.58)	1.60 (2.06)	1.43 (1.58)	724.31
Endosulfan @ 0.07 per cent	1.53 (1.86)	1.16 (0.88)	1.49 (1.73)	1.54 (1.90)	577.50
Acetamiprid @ 0.0065 per cent	1.63 (2.16)	1.11 (0.78)	1.51 (1.80)	1.53 (1.85)	500.50
Carbosulfan @ 0.05 per cent	1.49 (1.73)	1.07 (0.68)	1.50 (1.76)	1.47 (1.67)	632.94
Verticillium lecani @ 2 g/l	1.48 (1.70)	1.27 (1.13)	1.47 (1.66)	1.59 (2.06)	468.16
Tobacco decoction @ 1.00 per cent	1.53(1.86)	1.22 (1.02)	1.50 (1.76)	1.59 (2.04)	392.70
NSKE @ 5.00 per cent	1.69 (2.36)	1.25 (1.1)	1.61 (2.10)	1.58 (2.01)	403.48
Neem oil @ 0.50 per cent	161 (2.10)	1.25 (1.09)	1.56 (1.96)	1.61 (2.10)	341.36
Untreated control	1.57 (1.96)	1.34 (1.32)	1.53 (1.86)	1.67 (2.32)	224.84
SEm (±)	0.045	0.015	0.049	0.016	0.249
CD (5 per cent)	NS	0.044	NS	0.046	0.741
CV (per cent)	4.96	2.92	5.62	2.32	13.22

SDAU, S.K. Nagar. *Source*: Patel *et al*., 2008.

Outside the parentheses are $\sqrt{x + 0.5}$ transformed values while values in parentheses are retransformed value.

DAS: Days After Sowing.

- ☆ Clean up the godown/warehouses inside and outside before storage of new produce.
- ☆ All the cracks, crevices and holes present in the floor walls and ceiling of the godown should be filled up with cement and white washed.
- ☆ The ceiling and walls of godown may be sprayed with malathion 0.5 per cent.
- ☆ Old bags should be impregnated with deltamethrin 0.025 per cent or malathion 0.25 per cent before they are filled up again.
- ☆ Old bags should be fumigated with ED/Ct mixture at the rate of 1 lit./100 bags.
- ☆ Do not store the different commodities in a store.
- ☆ Remove old stock of grains from the godown/store before storage of new stock.
- ☆ Use of methoprene @ 5 to 10 ppm against cigarette beetle infestation in storage was found effective.
- ☆ Methyl-O-demeton 0.025, monocrotophos 0.04 and thiometon 0.03 per cent are the effective insecticides against the midge infesting fennel. These insecticides should be applied at the flowering stage to minimize the midge damage during seed setting and its carry over in storage.

Control of Established Infestation

When stored seed spices are infested by pests, it can easily be controlled by following methods.

- ☆ Infested grains should be spread in a thin layer on cemented floor during hot period of day to kill the pests.
- ☆ Fumigation of infested grains with ED/CT mixture @ 0.5 kg/metric tonnes.
- ☆ Aluminium phosphide @ 3 tablets/tonne for an exposure period of 3 days was found to be highly effective against all the storage pests. Aluminium phosphide should be used under strict supervision of experts of pest control operators.

References

AGRESCO, 1996. *Proceeding of 31st PPSC of Agricultural Research Sub Committee Meeting of GAU* held at Navasari, 8–9 April, pp. 11.

AGRESCO, 2003. *Proceeding of 38th PPSC of Agricultural Research Sub Committee* held at AAU, Anand, April 12–13, pp. 7.

AGRESCO, 2007. *Proceeding of 3rd combined AGRESCO Meeting of PPSC of SAUs* held at AAU, Anand, April 12–13, pp. 7.

Patel, B.R., Patel, G.M., Shukla, N.P. and Patel, S.R., 2005. Comparative efficacy of different doses of carbosulfan (Marshal 25 EC) in comparison to some newer molecules against cumin aphid. In: *Abstracts of National Symposium on Stress Management* held at SDAU, Sardarkrushinagar during April 11–13, pp. 94.

Patel, B.R., Patel, G.M. and Patel, I.S., 2005. Impact of sprayer and sticker on efficacy of methyl-O-demeton against cumin aphids. In: *Abstracts of National Conference on Applied Entomology,* held at MPUAT, Udaipur during September 26–28, p. 8–9.

Patel, S.A., 2008. Biology and ecosafe management of fennel aphid, *Hydaphis coriandri* (Das) in fennel crop. *M.Sc.(Agri.) Thesis,* S.D.A.U., Sardarkrushinagar.

Recent Advances in Seed Spices (2011) *Pages* **163–165**
Editors: **Y. Ravindrababu, R.K. Jaiman, K.D. Patel, N.R. Patel and S.B.S. Tikka**
Published by: **DAYA PUBLISHING HOUSE, NEW DELHI**

15

Nematodes Diseases of Seed Spices and their Management

Ashok D. Patel*
Professor of Nematology, Department of Nematology, B.A. College of Agriculture, Anand Agricultural University, Anand – 388 110, Gujarat

Spices have a major role in influencing the course of history and civilization of all over the world. They even today play an important role in the national economy. India is considered as the "Home of Spices" from ancient times. It produces a large array of spices with an annual output of 2.5 million tones. India enjoys a pre-eminent position in spices production in the world. It is also the largest consumer of spices with the domestic needs requiring more than 90 per cent of spices production. Although nearly 60, out of the recorded 107, spices are cultivated in Gujarat, Rajasthan, TN and Karnataka. Only 16 *viz.* black pepper, cardamom, ginger, turmeric, large cardamom, clove, chilli, garlic, saffron, celery, cumin, coriander, fennel, fenugreek, ajwain and suwa are economically important. Among these, cumin, coriander, fennel, fenugreek, ajwain and suwa are classified as seed grain spices. They are mainly cultivated in Gujarat, Rajasthan with a sizeable acreage in states of MP, UP, AP, Haryana, Punjab and Bihar. India produces (Vashishtha and Malhotra, 2004-05) seed spices of 564.72 thousand tones from 850.28 thousand hectors.

Phytonematodes are tiny, round-bodied, unsegmented, bilaterally symmetrical, pseudocoelomatic, triploblastic, colour less, vermiform invertebrate bisexual animal and generally not visible with naked eyes. They range from 0.25 mm to 19 mm in length and from 0.01–0.1 mm in body width. About 3,000 nematodes (10 per cent of 30,000 so far identified) feed on crop plants. Apart from several abiotic

* E-mail: patelad2003@yahoo.co.in.

stresses, spice crops suffer from several diseases and insect pests including nematodes resulting in enormous loss both qualitatively and quantitatively. Among these, damage caused by plant parasitic nematodes is a serious in spice production (Koshy and Geetha, 1992). Sasser and Freckman (1987) estimated 12.3 per cent yield loss in agricultural crops incurring monitory losses of about US $ 77 billion annually. Jain *et al.* (2007) worked out monitory loss to the tune of 21,068.73 million rupees in 24 important crops under All India Coordinated Research Project on Plant Parasitic Nematodes with Integrated Approach, for their control, ICAR, New Delhi. Patel *et al.* (1995) reported 28.0 per cent loss in fennel grain by root-knot nematode, *Meloidogyne javanica* pt. 2. Their economic importance is now well established in many spice crops. However, nematological investigations are concentrated mainly on major spices and to a lesser extent on seed spices.

Root-knot nematodes (*Meloidogyne* spp.) considered as one of the most important global pests attacking almost all the spice crops. Of more than 98 species of *Meloidogyne* so far reported in the world. *M. incognita, M. javanica, M. hapla* and *M. arenaria* are key nematode species widely distributed throughout the country. Beside root-knot nematodes, reniform nematode, *Rotylenchulus reniformis,* stunt nematodes, *Tylenchorhynchus vulgaris* and lesion nematodes, *Pratylenchus* spp. are also reported to attack spices. The typical symptoms of nematode injury can be seen on above ground and below ground plant parts. Foliar symptoms are stunting, yellowing or bronzing of the foliage, unthriftiness, typical nutrient deficiency, premature wilting and slow recovery on improved soil moisture condition. Below ground symptoms include profuse root galling, stunting, root decay and darkened roots.

There are well established management strategist including prevention, physical, cultural, biological, chemical, host resistance and legal procedures.

Soil application of neem cake @ 1000 kg/ha + Phorate @ 1 kg/ha proved effective to reduce root-knot disease and there by increase 65.7 per cent fennel (Patel *et al.*, 2005).Jain and Paruthi (2008) reported fennel varieties SF1, GF2, HF33, HF39 and PF35 tobe poor host for *Meloidogyne incognita.* Similarly Patel *et al.*(1990) also reported fennel varieties FS 256, Guj.-1, ICA 2, UF1, 31,35,90 and 112, Fenugreeck varieties EC 4314, 16392, 26177, IC 689, 1978, 2010,3693,5686 and Cumin MC Sel. 9 to be resistant to the root-knot nematodes. Cumin varieties, GC 1, 2, 3, Coriander, GCO 2, Fennel, GF 1 and GF 2, Ajwain GA 1 and Dilseed GD 1 were recorded highly resistant to reniform nematode by Ghelani and his coworkers in 2003.

References

Ghelani, Y.H., Patel, A.D. and Patel, D.J., 2003. Host status of released varieties of spices against reniform nematode, *Rotylenchulus reniformis.* In: *Proceedings: New Perspectives in Spices, Medicinal and Aromatic Plants*, pp. 375–376.

Jain, R.K., Mathur, K.N. and Singh, R.V., 2007. Estimation of losses due to plant parasitic nematodes on different crops in India. *Indian J. Nematol.*, 37: 219–220.

Jain, R.K. and Paruthi, I.J., 2008. Host status of different seed spices against root-knot nematode (*Meloidogyne incognita*). *Indian J. Nematol.*, 38: 121–122.

Koshy, P.K. and Geetha, S.M., 1992. Nematode pests of spices and condiments. In: *Nematode Pests of Crops*, (Eds.) D.S. Bhatti and R.K. Walia. CBS Publishers and Distributor, Delhi, pp. 228–238.

Patel, D.J., Patel, H.R. and Patel, C.C., 1990. Susceptibility of spices to root-knot nematodes. In: *The Proceedings of First National Seminar on Seed Spices*, pp. 256–265.

Patel, D.J., Patel, S.K. and Patel, H.V., 1995. Yield loss estimation in fennel (*Foeniculum vulgare* Mill.) due to root-knot nematodes. *Journal of Spices and Aromatic Crops*, 4: 84–85.

Patel, S.K., Patel, H.V., Patel, A.D. and Patel, D.J., 2005. Integrated management of root-knot nematode, *Meloidogyne javanica* (Treub) Chitwood in fennel (*Foeniculum vulgare* Mill). *Journal of Spices and Aromatic Crops*, 14: 152–154.

Sasser, J.N. and Freckman, D.W., 1987. A word perspective of Nematology: The role of the society. In: *Vistas on Nematology*, (Eds.) J.A. Veech and D.W. Dickson. Pub. By Society of Nematologist, U.S.A., p. 7–14.

Recent Advances in Seed Spices (2011) *Pages* **166–168**
Editors: **Y. Ravindrababu, R.K. Jaiman, K.D. Patel, N.R. Patel and S.B.S. Tikka**
Published by: **DAYA PUBLISHING HOUSE, NEW DELHI**

16

Status of Pesticide Residues in Seed Spices in India

P.G. Shah*, Kalpana Diwan and M.F. Raj
AINP on Pesticide Residue, ICAR Unit-9, Anand – 388 110, Gujarat

Plant protection chemicals, commonly known as pesticides are inevitable in the present day agriculture. Judicious use of these chemicals has not only protected the crops from various pests and diseases, but has saved millions of lives by controlling vector-borne diseases such as typhus and malaria. For the country like India, where the food production is trailing population growth, there is an urgent need to safeguard the crops and animals against pests. Thus, the use of agrochemicals is justified at field as well as postharvest stages. Unfortunately, pesticides are one of the largest groups of poisons deliberately introduced in the ecosystem. As these chemicals are designed to kill or control the pests, their side effects to the flora and fauna could seriously hamper the ecological balance.

The first generation of pest control agents, consisting largely of botanicals and inorganic substances containing copper, lead, sulfur, or arsenic, received little regulatory interest, and thus there was little monitoring of the products applied or of the residues remaining in/on them. Visual monitoring could be done for some materials, such as Paris Green (blue–green due to copper content), or lead arsenate, which often left a white residue on apples and other produce because of the high application rates. Simple gravimetric, titrimetric, or colorimetric methods could be used to quantify residues of many agents, including copper-, arsenic, sulfur and lead-containing products or their derivatives. Thoroughly washing treated fruit and fresh vegetables probably removed most residues since the materials in use were largely nonsystematic and water soluble or water wettable, lessening

* Corresponding Author: E-mail: paresh243@yahoo.co.in.

their hazard to the consumer. This situation changed significantly with the introduction of second-generation pest control agents, largely synthetic organics such as DDT, 2,4-D, and ethyl parathion, from the 1940s on. These chemicals had a number of qualities which invited heightened consumer concern, regulatory attention, and monitoring activity: *e.g.* many chemicals were acutely toxic and/or chronically toxic to humans, domestic animals, and wildlife. Their ability to cause tumors, at least in laboratory animals, was of particular concern.

Regulation was developed in the 1950s and 1960s to include legal limits (tolerances) for residues on foods and in feeds and, with time, in water and air. Enforcing these regulations required analytical methods of ever-increasing sophistication and sensitivity as public concern grew with each new residue-related crisis or toxicology finding. Early efforts at regulating residues confounded the situation. Legislators introduced a 'zero tolerance' concept for pesticides that produced cancer in experimental animals, for agricultural crops or food animal products such as milk and butter.

India is the largest producer, consumer and exporter of spices in the world. Among various spices grown in India, coriander, cumin, fennel, fenugreek, ajawain, dill, celery, anise, caraway and nigella are the important seed spices. These seed spices are gaining importance world over because of their medicinal and culinary usage. The countries like USA, Japan, Canada, Australia and Europe are the important markets for Indian seed spices. However, the post-WTO period has made it mandatory to maintain the minimum standards of their intrinsic and extrinsic quality for international markets. Of late, some consignments of seed spices meant for export from India suffered rejection in the European market due to the presence of pesticide residues.

In 1962, the publication of 'Silent Spring' created world wide awareness about the contamination of food with pesticide residues and the outcry which resulted in increased funds for research in the field of pesticide residue analysis. In fact, most of the advanced countries initiated regular monitoring programme in early eighties. Such studies helped in banning notorious pesticides *e.g.* DDT, BHC, aldrin, etc. India too initiated similar programme under All India Co-ordinated Research Project on Pesticide Residues through ICAR in 1983-84 with 17 centres in different parts of the country. The project has undergone sea-change within 20 years and at present known as All India Network Project On Pesticide Residues with effective 12 centres. Recently, Ministry of Agriculture, Government of India has launched a National Level Monitoring Programme with the help of AINP on Pesticide Residue laboratories and the network has about 30 laboratories all over the country.

As on today, though there is no regular monitoring programme for seed spices for the analysis of pesticide residues, limited work carried out by some laboratories in India have been reviewed hereunder.

During the period 1995 to 1999, market survey of various seed spices for pesticide residues was carried out by AINP on Pesticide Residues, Anand centre. The results revealed low levels of organochlorine group of pesticides in cumin, fennel, dill, ajowan and coriander. This could be due to the fact that though BHC and DDT were either categorized as restricted use or banned, the chemicals were still available illegally at that time. However, such work carried out after the year 2000, did not reveal presence of organochlorine group of pesticides.

Analysis of 18 samples of fennel collected from market places of different cities/towns of Gujarat during 2001-02 revealed presence of only organophosphate group of residues. Among the organophosphates, parathion methyl, chlorpyriphos, quinalphos, dimethoate, monocrotophos, malathion and profenophos were detected in 12, 5, 4, 3, 3, 2 and 1 sample, respectively. Based on the MRLs adopted by Germany 6, 3, 1 and 1 sample exceeded the limits for parathion-methyl, monocrotophos, chlorpyriphos and quinalphos, respectively. In all, out of 18 samples analyzed 15 were found contaminated, wherein 11 exceeded the limits.

In a similar such study carried out for 10 cumin samples, the results revealed contamination of organophosphates and dithiocarbamate in all the samples. From the organophosphate group, phorate, dimethoate, parathion-methyl, malathion, chlorpyriphos, quinalphos, prophenophos and triazophos were detected in 3, 3, 10, 4, 8, 4, 3 and 1 sample, respectively. Based on the MRLs fixed by a major seed spice importing country like Germany, parathion methyl, malathion, chlorpyriphos and quinalphos exceeded their MRLs in 1, 3, 2 and 1 sample, respectively. Though, dithiocarbamate residue (as CS_2) was detected in all the samples, none exceeded its MRL of 3.0 ppm.

Mancozeb, a non-systemic fungicide is widely used for the control of cumin blight. Most of the cumin growers in Gujarat go for prophylactic spray of mancozeb when weather condition is conducive for blight infestation. As a result, sometimes, farmers have to go for frequent foliar sprays till harvest time. The practice might result in high levels of mancozeb residues. The work carried out at Anand Agricultural University, Anand for market as well as export purpose samples revealed contamination of dithiocarbamate residue. To avoid such situation, farmers are advised to maintain safe waiting period between last application and harvest. In one such experiment, conducted at Main Spices Research Station, Jagudan Gujarat, mancozeb was applied as foliar spray @ 1.125 kg a.i. ha^{-1}(recommended) and 2.250 kg a.i. ha^{-1} (double the recommended). The application was carried out twice at 15-day interval and the interval between last spray and harvest was found to be 34 days. The result revealed that the 2.008 and 6.500 mg CS_2 g^{-1} of cumin were detected at recommended and double the recommended doses, respectively. Though, at recommended dose, the residue was well below its MRL, higher dose exceeded the limit.

Conclusion

Considering the strong competition in the world spice market coupled with quality demand in the domestic market, there is an urgent need to address the problems of pesticide residues in seed spices. With banning of chemical like HCH and DDT in agriculture, the problems of organochlorine residues have been minimized. Nonetheless, limited survey carried out in Gujarat revealed contamination of organophosphate group of pesticides. Also the problem of dithiocarbamate residue in cumin is a matter of concern. More such studies are required to carry out farm-gate as well as market level monitoring of seed spices.

Recent Advances in Seed Spices (2011) *Pages* **169–185**
Editors: **Y. Ravindrababu, R.K. Jaiman, K.D. Patel, N.R. Patel and S.B.S. Tikka**
Published by: **DAYA PUBLISHING HOUSE, NEW DELHI**

17

Farm Equipment and Technologies for Mechanization in Major Seed Spices: Prospects and Opportunities

M.M. Pandey[1], Tarun Kapur[2], B.K. Garg[2], and S.K. Rautaray[3]
[1]Director, [2]Principal Scientist, [3]Project Coordinator (UAE)
Central Institute of Agricultural Engineering, Bhopal, M.P.

Introduction

Farm mechanization inputs has been making significant contribution in increasing agricultural production and productivity through timeliness in operations, efficient application of inputs, conserving soil and water resources; and reducing losses, pollution and drudgery of the task involved apart from increase in cropping intensity. Today there is an increasing awareness and concern for appropriate, affordable and energy efficient equipment and technology for cost effective production and processing of crops focused at increased yield, reduced cost of cultivation, prevention of losses and value addition through location specific management practices. National and globas findings have shown that the engineering inputs has increased the crop productivity by 10-15 per cent, besides the productivity of other factors of production such as land, water and labour, reduced the cost of production by 20-25 per cent and also reduced the field, transport, processing and storage losses to the order of 10 per cent.

Several studies on mechanization coupled with the use of improved crop inputs have shown direct increase in yield due to timely completion of operations in most favorable periods by use of appropriate equipment for improved seedbed preparation, proper placement of seed and fertilizer, efficient application of irrigation water, timely adoption of plant protection measurers and mechanized harvesting and threshing of crops. The farmers have realized these advantages and mechanization is in increasing trend in many parts of the country. However the type of equipment suitable for cultivation of seed spices and the rate of introduction of new equipment is to be considered with multidisciplinary approach specific to the site needs.

Agriculture is season specific and time bound. It is the livelihood of 65 per cent of India's population. 60 per cent of our cultivable land being rainfed is critical as delay in field operations cause loss of yield. India's mechanization is based on power mix approach with present power availability of 1.3 kW/ha. Use of mechanical power sources (3 million tractor) and the existing draught animal power (Population = 63 million) play complementary role to each other following the approach of selective mechanization. As the improved tools, implements and machinery (From hand hoe to the sophisticated power machinery) compared to the indigenous ones results in increased production and at the same time reduce drudgery, different size and shapes of these have been developed for efficient and expeditious operations under varying conditions. The purpose is to increase the output per human, animal and tractor and thereby ensuring timeliness and minimizing the cost of operations, minimize the energy requirement per unit of crop production, minimize losses, make agricultural production less laborious and reduce drudgery, reduce pollution, meet peak labour requirements such as at seeding, weeding and harvesting, permit an additional crop to increase cropping intensity, bring additional area into production, maximize the efficiency of inputs thereby increasing crop yields, raise farm income and contribute to overall economic development.

Production of Major Seed Spices

The major seed spices (Cumin, fennel, coriander and fenugreek) in India are generally grown in semi arid and arid regions of Gujarat, Rajasthan, Uttar Pradesh and Uttarakhand. The production concentration of these seed spice crops are almost in Gujarat and Rajasthan together contributing more than 70 per cent of the total seed spices produced in the country (Production = 5.70 lakh ton and area = 8.65 lakh ha: 2006-07). Considering the country's domestic requirements and export potential of seed spices (Annual export = 70125 t earning foreign exchange worth Rs 3620/- million against the global demand 1.5 lakh tonne it calls for production and productivity increases with quality upgradation to make those competitive in the global market. The area, production and productivity of selected major seed spices in India and in Gujarat and Rajasthan in given is Table 17.1.

Table 17.1: Area, Production and Productivity of Selected Major Seed Spices in India and in Gujarat and Rajasthan (2005-06) Area: 000 ha, Production: 000 t, Productivity: kg ha $^{-1}$)

Crop	India			Gujarat			Rajasthan		
	Area	Prodn.	Yield	Area	Prodn.	Yield	Area	Prodn.	Yield
Cumin	349.02	158.35	454	213.90	106.10	496	135.11	52.24	387
Fennel	43.67	61.65	1418	38.70	57.40	1483	3.90	3.66	939
Coriander	340.35	223.40	656	-	-	-	136.70	142.40	1042
Fennugreek	32.66	35.71	1093	4.85	5.93	1198	27.29	29.17	1281

Source: http://dacnet.nic.in.

The soil and climatic conditions are generally favorable for seed spice farming in the identified states/areas of the country. Moreover seed spices being high value crops should ensure return to the farmers if the present level of low productivity is overcomed coupled with cost-effective cultivation practices. The seed spices farmers, besides improving the productivity feel under pressure to reduce the cost of cultivation for which mechanization would be the key to achieve the cost-effectiveness.

Concerns Leading to Interest in Seed Spices Production Machinery

The low productivity of seed spices may be attributed to the fact that the farmers normally cultivate these crops on small/marginal lands using locally available seed varieties and following their traditional package of practices without effective weeds and pest control measures. Insufficient protective irrigation, inadequate uses of fertilizer, untimely and delayed cultivation operations are primarily the factors of low productivity. Scarcity of labourers during peak season, high wages, low work rate of labourers with indigenous tools and equipment often delay the farm operations causing loss of yield and increasing the cost of cultivation. The result is low economic return to the farmers. However, opportunity exists to increase the productivity and profitability of seed spices production by mechanization of cultivation operations. Therefore mechanization technology coupled with improved management practices including irrigation may be introduced selectively fitting to the site needs to enhance the capacity of farmers to secure higher returns for their labour and investments besides improving working condition of the farm workers in the field. More area can also be put under these crops by adopting suitable mechanization package.

Mechanization Strategy to Improve Productivity of Major Seed Spices

Several tools and equipment of different size and shape for different crops have been developed and available for land shaping, manure application, seed bed preparation, seeding, weeding, fertilizer application, spraying, irrigation, harvesting and threshing operations. What is needed is to adapt these equipment for seed spices cultivation by suitable modifications as per the requirements placed by the specific crops to make the seed spice farming more energy efficient and cost-effective. Selective mechanization may also be followed for the production operations which are highly labour demanding and weather dependent and where traditional practices have limitations to achieve timeliness of operations. With the availability of desired level of agro inputs: improved seeds, fertilizer and plant protection chemicals the mechanization of cultivation operations would lead to improved productivity and profitability.

The study deals with the identification of existing equipment and technology for different unit operations and adopting those with modifications where ever required to mechanize the cultivation operations for production of major seed spices.

Progress of Mechanization

Agriculture has long tradition of farming using animal energy. The wedge plough so commonly used may not do ideally any of the functions it performs but it is very versatile equipment capable of doing primary and secondary tillage, sowing, and interculture. Traditional local craftsmen have skills to fabricate and maintained the equipment in a self sustaining manner. As a result inspite of rapid growth in mechanization, these traditional equipment evolved in ancient times continue to be in use. With population explosion the traditional agriculture has not been able to suffice the food needs of the people and feed to the livestock. Till modern agriculture practices including farm mechanization were put to practice, agricultural productivity was very low. Education, research and extension has brought transformation and several equipment and technologies have been developed.

Farm Planning and Land Shaping

For modern spice seeds farming proper farm planning is the basic requirement. Land shaping soil/water testing, rain water harvesting and recycling, drainage and recharging of open wells/tube-wells, adoption of micro irrigation systems (sprinkler/drip) have shown that average productivity of field crops was increased by 80 per cent than the traditional farms (yield of corps, t/ha; rice = 2.2, soybean = 0.8, wheat = 1.8, chick pea = 1.0). The cropping intensity was doubled (200 per cent) with increased annual productivity leading to 250 per cent increased annual income compared to the traditional practice of rain dependent single crop per year (Annual net income, Rs/ha = 6000/-).

Land Shaping

Laser Guided Land Leveler (Unit price, Rs, lakh = 4.50)

Laser guided land leveler has been adapted for precision land leveling and grading. Besides savings in irrigation water (20-25 per cent) and avoiding water logging the laser leveled field well supported the crops for uniform growth and yield. The benefits to the farmers is Rs 10/- lakh per unit per year (Annual use = 200h, Area = 80-100 ha).

Tillage and Seed Bed Preparation

For major seed spices the seedbed is prepared by 2-3 harrowing and the field is made trash and clod free with leveled/smooth surface. The field is bunded for making beds along with laying of channels for irrigation.

Animal Drawn Improved Tillage Implements

Animal drawn improved blade harrow (Unit price, Rs=3000/- and field capacity, ha/h = 0.07), patella harrow (Unit price Rs. 2500/- and field capacity = 0.30), wedge plough (Unit price Rs.700/-), Bund former (Unit price RS. 1000/-), and multipurpose tool carrier with matching attachments (Unit price Rs. 10000/- and field capacity, ha/h=0.12) were developed for shallow tillage with insitu retention

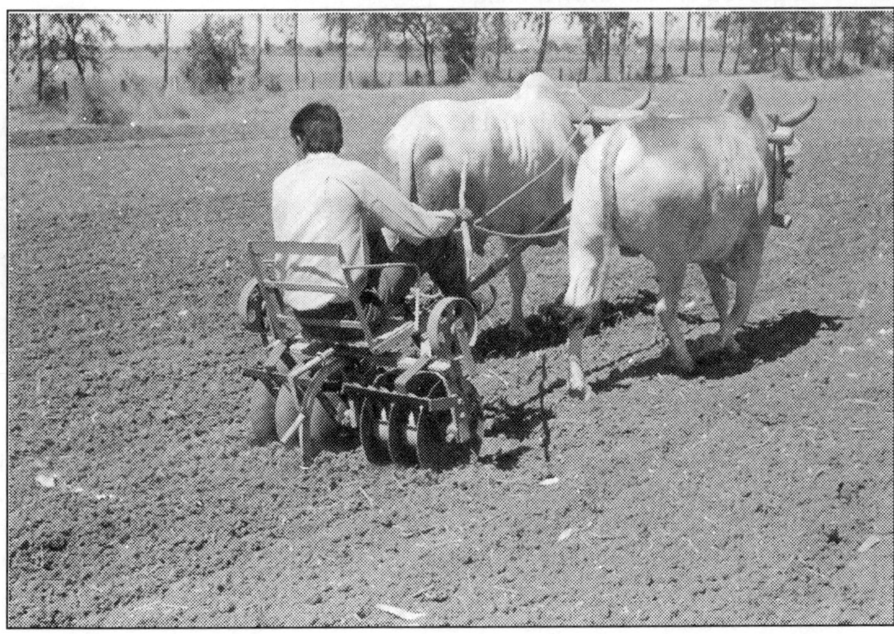

of soil moisture, and weed control. Improved package of animal drawn equipment showed 30-40 per cent higher field capacity with 30 per cent higher command area over the traditional package of farmers (Field capacity, ha/h = 0.04 and command area per season, ha = 1.5). The monetary benefits by adoption of improved animal drawn tillage equipment and yoke is Rs. 500/- per ha. Considering command area of 2 ha per season and 20000 no. of improved animal drawn tillage equipment the accrued benefit is Rs. 200/- lakh per annum.

Light Weight Power Tiller

- Weight, kg = 140
- Unit price, Rs = 75000/-
- Field capacity, ha/h = 0.08
- Command area, ha = 10
- Annual benefit per unit, Rs = 20000/-

Tractor mounted rotavator (Unit price, 70000/- and field capacity, ha/h = 0.25) being a single pass shallow tillage equipment may be adopted. Using rotavator the cost of seedbed preparation was Rs 1500/- per ha than the conventional method (Rs. 2700/- per ha). The equipment is commercially available. Considering its annual command area of 50 ha annual savings in cost of operation per rotavator is Rs 60000/-.

Tractor Operated Manure Spreader (2t Capacity)

10-15 tonne of FYM are generally applied for major seed spices prior to/during tillage. Tractor operated manure spreader was developed at CIAE, Bhopal (Unit price, Rs = 1.3 lakh and field capacity, ha/h = 0.40). Using tractor operated manure spreader the cost of manure application per ha was Rs 100/- compared to the traditional practice (Rs 1800/- per ha). Considering the annual command area of 50 ha annual savings in cost of FYM application per unit is Rs 35000/-.

Tractor Mounted Bund Former (Unit price, Rs = 10000/-)

For making plots/beds and laying of field channels bund former may be used. It's cost of operation is Rs 360/- per hour. For making beds of 4x3 m as required for cumin along with the channel for irrigation the savings in bund forming per ha by use of bund former may be 50 per cent compared to making beds/channels manually by spade (Rs/ha = 7000/-). Considering annual command area of 50 ha the annual saving per unit is Rs 1.75 lakh.

Seeding and Fertilizer Application

Several designs of manual, animal drawn and tractor drawn drills/planters have been developed and those can be adapted with suitable modification of the seed metering unit and furrow openers suiting to the requirements of major seed spices (seed rate, size/shape, test weight, and depth of placement). The recommended seeding practices is given in Table 17.2.

Table 17.2: Recommended Seeding Practices of Major Seed Spices

Sl.No.	*Major Seed Spice*	*Varieties*	*Seed Rate, kg/ha*	*Fertilizer Rate, kg/ha, $N:P_2O_5:K_2O$*	*Ave. Depth of Seeding/ Placement, cm*	*Row Spacing, cm*	*Plant Spacing, cm*
1.	Cumin	RZ-19, RZ 209, RZ 223, GC-1, GC-2, GC-3, GC-4	12-16	30:20:20 Basal: 15:20:20	1.5	25	–
2.	Fennel	NRCSS-AF-1, RF-101, 125, 143, GF-1, 2, 11 and others	8-10 (Dry seeding) and 2.5: transplanting	90:40:30	–	45–60	20
3.	Coriander	NRCSS-Acr-1, Gujarat coriander-1, 2, others	10-12 (irrigated) 15-20 (Rain fed)	60:30:20 (irrigated 40:30:20 (rainfed)	2.0–2.5	25–30	10–15
4.	Fennugreek	NRCSS AM-2 Rmt 1-143, 303, 305 (RAU) CO-1 (TNAU), PUSA Kasuri and others	20-25 (8-10: Kasuri)	25:20:20 No top dressing	2.5 (For kasuri: 1.0–1.5)	25–30	10–15

Manual and animal drawn drills/dibblers and planters have been developed for different crops. The field capacity of animal drawn drills/planters varied from 0.08 to 0.15 ha/h with savings in seeds and fertilizer (10-15 per cent) and cost of seeding/planting by 40 per cent compared to the traditional method of seeding behind country plough (Cost of seeding, Rs./ha = 650/-). The unit price of the animal drawn 2-row drill is 3000/-. The annual command area per drill is 5 ha. Considering 5000 no. of drills known to be used by the farmers the annual benefit accured is Rs. 65/- lakh by adoption of the drill/planter.

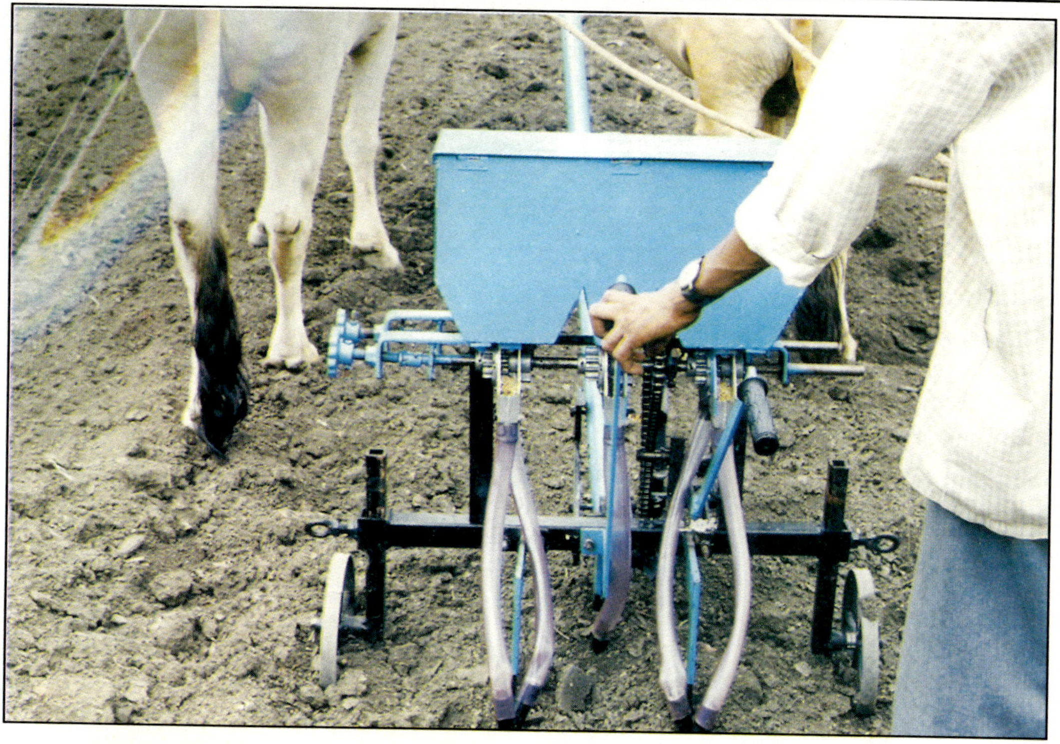

Tractor Operated Inclined Plate Planter

Tractor operated inclined plate planter 6 row (Unit price, Rs 25000/-) was developed for planting different crops by selecting metering plates with appropriate size of cells. The unit showed field capacity of 0.4 ha/h. The saving in cost of operation compared to the conventional method (cost of operation, Rs/ha = 1550/-) was Rs 600/-. Considering the annual command area of 30 ha per machine the annual benefit per machine is Rs 18000/-. Around 50 no. of machines are in use by the farmers giving total annual benefits of Rs 9/- lakh.

As the cultural practices for almost all the major seed spices indicated to protect the sown seeds from direct sunlight and maintain adequate moisture in the soil during the period of germination/ initial establishment, the drills may be modified for desired seed rates and shallow depth placement of seeds with proper soil coverage. The possible modifications may involve cup fed/inclined plate/ vertical roller cells metering, narrow tapered shoe type furrow opener for placement of seeds without excessive exposure of top soil, low height of drop of seed from the seed box to the furrow opener and suitable agitator to prevent entanglement of seeds in seed box and to avoid seed drops in bunches to drill rows.

Intercultural Operations

Manual weeding by hand hoe is time and labour consuming and often delay the weeding operation affecting the performance of crops when laboures are not available in the season. For energy efficient and cost effective weeding, improved weeders were developed for weeding and inter-cultivation in row seeded crops.

Manually operated wheel hoe and grubber (Unit price, Rs = 50-250/- and capacity, ha/h = 0.01) are useful for weeding in dray lands. Compared to the manual hand hoe weeding (Cost of weeding, Rs/ha = 3000/-) weeding by improved manual weeder saved average of Rs. 1600/- per ha. The total

economic benefits by adoption of manual weeder (10000 no.) is estimated to be Rs 160/- lakh per year through saving in labour and timely weeding leading to increased yield.

Animal drawn improved blade hoe and tine cultivator (Unit price, Rs = 1500/- and field capacity, ha/h = 0.8) were developed for weeding in row seeded crops. The command area of these weeders is 05 ha per season and around 5000 units are in use by farmers. Compared the traditional hand hoe weeding (Cost of operation, Rs/ha = 3000/-) the animal drawn weeders saved Rs = 2500/- per ha. The total annual benefits accrued is Rs = 625/- lakh.

Self propelled power weeders (Unit price, Rs 70000/- and field capacity, ha/h = 0.12) was developed for weeding in wider row crops. Compared to the manual hand hoe weeding (Cost of weeding, Rs/ha = 3000/-) it saved Rs 1600/- per ha. The command area per season is 15 ha and 300 units are available in the country. The total annual beefits accrued is Rs 72/- lakh.

Tractor Drawn Weeder (Unit price, Rs = 15000/- and field capacity ha/h = 0.35)

Narrow tire (Size:8.3x32.0-6) was adapted by replacing the rear wheel tires (Size: 12.4x28.0-6 play rating) of tractor and it was used for inter-cultivation in soybean successfully at work rate of 0.35 ha/h. The cost of the narrow wheel with tires/tube was Rs 20000/- per pair. The cost of weeding and inter-cultivation was Rs 1000/- per ha. The annual command area of narrow wheeled tractor for weeding was 30 ha. The annual/seasonal savings due to tractor drawn weeder compared to manual hand hoe weeding is Rs 60000/- per unit.

Plant Protection

Animal drawn sprayer (Unit price, Rs = 25000/- field capacity, ha/h = 0.4 and tank capacity, 1 = 200) has been developed and modified for higher performance. The sprayer with 6 no. of nozzles on 5.4 m boom operates at ground clearance of 01 m. The conventional practice is spraying of pesticides by use of knapsack sprayer or foot sprayer (Rs/ha = 700/-). The benefit over the conventional method is Rs 500/- per ha. The annual command area of the equipment is 40 ha. Presently 25 units are in use by farmers. The net benefit per year is Rs 05/- lakh.

Tractor operated aero blast sprayer (Uint price, Rs 70000/- and field capacity, ha/h = 1.7) has been developed for spraying on field/horticultural crops. The rate of application could be varied from 100 to 400 l/ha. The cost of operation is Rs 500/ha as against Rs. 700/ha by conventional methods. One machine has the command area of 100 ha/year and there are 100 units in operation. The total monetary benefit of adopting this technology is Rs 20 lakh/annum.

Harvesting

Improved sickle (Unit price, Rs = 60/- and field capacity, sq. m/h = 150). Improved designs of sickle with serrated blades are commercially available. Compared to the traditional local sickles (output, sq. m/h = 120 and cost of harvesting, Rs/ha = 1040) the saving per unit of improved sickle while harvesting one ha crop was Rs 170/-per season. Considering command area of 2 ha (135h) per season per unit of improved sickle the total saving per season is Rs 340/-.

Tractor mounted reaper (Unit price, Rs 50000/- and field capacity ha/h = 0.25). Commercial model of tractor side mounted reaper is available for harvesting of soybean and may be adapted with desired modifications for harvesting of few selected major seed spices. The possible modifications may include improvements on cutting mechanism, collocation in trays, intermittent gathering in heaps to reduce losses and to make harvesting faster and cost effective.

Threshing

Multicrop thresher powered by 5 hp electric motor is suitable for threshing wheat, sorghum, chickpea, soybean, rice, rape-seed mustard, oil seed and others. The capacity varies from 200-400 kg/h depending on the type of crop. It saved 25-30 per cent on cost of threshing operation compared to threshing by conventional practice (Cost of threshing, Rs/kg = 0.60 for rice). The unit may be adapted to thresh major seed spices with modifications of threshing mechanism (rubberized rasp bar type), optimizing cylinder speed and concave clearance, changing sieves to size and with oscillating mechanism and optimizing blower/aspirator speed and its positioning.

Production Economics of Major Seed Spices

(1) Production Economics of Seed Cumin

Cost of Cultivation (Rs/ha)

Sl.No.	*Unit Operations*	*Improved Practices*		*Traditional Practices*	
		Equipment	*Cost*	*Equipment*	*Cost*
1.	Seed bed preparation including bunding and field channel laying	Rotavator Bund former Manual spade	2300	Harrow/Cultivator Manual spade	5100
2.	FYM application (10t) Fertilizer top dressings (02)	Manure spreader Broadcaster	1100 110	Manual Manual	1800 220
3.	Sowing and basal application of fertilizer	Seed cum fertilizer drill	700	Hand broadcasting	220

Contd...

Contd...

Sl.No.	Unit Operations	Improved Practices		Traditional Practices	
		Equipment	Cost	Equipment	Cost
4.	Weeding/interculture	Self propelled weeder	1400	Manual hand hoe	3300
5.	Plant protection spraying	Aeroblast sprayer	500	Manual knapsack sprayer	700
6.	Irrigation (4 no.)	Electric pump set	3900	Electric pump set	3900
7.	Harvesting and collection	Manual: improved sickle	2420	Manual: local sickle	2750
8.	Threshing (yield, t/ha = 1)	Multicrop thresher	1500	Manual	2970
	Sub total:	–	13930	–	20960
9.	Cost of Agro inputs				
	(i) Seeds: 16 kg		3200		3200
	(ii) Fertilizer		900		900
	$N:P_2O_5:K_2O$(30:20:20):DAP Urea and MOP)				
	(iii) FYM = 10t		3000		3000
	P.P. Chemicals		1500		1500
	Sub total:		8600		8600
	Grand total:		2,2530		2,9560
10.	Yield Kg/ha		700		700
	Cost of production Rs/kg		32.00		42.00
	Market Price, Rs/kg		150		150
	Gross income, Rs/ha		105000		105000
	Net income, Rs/ha		82470		75440

The benefit due to mechanization over the traditional practice, Rs/ha = 7030/-. Considering 10 per cent mechanization of the present total area under cumin (3.49 lakh ha) in the country, the net annual benefit due to mechanization is Rs. 2453/- lakh.

(2) Production Economics of Seed Fennel (Transplanted)

Cost of Cultivation (Rs/ha)

Sl.No.	Unit Operations	Improved Practices		Traditional Practices	
		Equipment	Cost	Equipment	Cost
1.	Seed bed preparation including bunding and field channel laying	Rotavator Bund former Manual spade	2300	Harrow/Cultivator Manual spade	5100
2.	FYM application (10t)	Manure spreader	1100	Manual	1800
3.	a) Seedling preparation: bed, sowing and maintenance-6 weeks.	Manual	2200	Manual	2200
	b) Transplanting	Manual	1650	Manual	1650

Contd...

Contd...

Sl.No.	Unit Operations	Improved Practices		Traditional Practices	
		Equipment	Cost	Equipment	Cost
4.	Weeding/interculture	Self propelled weeder	1400	Animal weeder and manual	2000
5.	Plant protection spraying	Aeroblast sprayer	500	Manual knapsack sprayer	700
6.	Top dressing of fertilizer	Broadcaster	250	Manual	480
7.	Irrigation (7 no.)	Electric pump set	7000	Electric pump set	7000
8.	Harvesting and collection	Manual improved sickle	1800	Manual local sickle	2500
9	Threshing (yield, t/ha = 1.5)	Multicrop thresher	2000	Manual	4500
	Sub total:	–	20200	–	27930
10.	Cost of Agro inputs				
	(i) Seeds: 2.5 kg		375		3200
	(ii) Fertilizer		2000		2000
	$N:P_2O_5:K_2O$(90:40:30):DAP Urea and MOP)				
	(iii) FYM = 10t		3000		3000
	P.P. Chemicals		1000		1000
	Sub total:		6375		6375
	Grand total:		26575		34305
11.	Yield Kg/ha		1500		1500
	Cost of production Rs/kg		17.00		22.00
	Market Price, Rs/kg		100		100
	Gross income, Rs/ha		150000		150000
	Net income, Rs/ha		123425		115695

The benefit due to mechanization over the traditional practice, Rs/ha = 7730/-. Considering 10 per cent mechanization of the present total area under fennel (0.44 lakh ha) in the country, the net annual benefit due to mechanization is Rs. 340/- lakh.

(3) Production Economics of Seed Coriander

Cost of Cultivation (Rs/ha)

Sl.No.	Unit Operations	Improved Practices		Traditional Practices	
		Equipment.	Cost	Equipment	Cost
1.	Seed bed preparation including bunding and field channel laying	Rotavator Bund former Manual spade	2300	Harrow/Cultivator Manual spade	5100
2.	FYM application (10t)	Manure spreader	1100	Manual	1800
	Fertilizer top dressings (02)	Broadcaster	110	Manual	220
3.	Sowing and basal application of fertilizer	Seed cum fertilizer drill	700	Hand broadcasting	220

Contd...

Contd...

Sl.No.	Unit Operations	Improved Practices		Traditional Practices	
		Equipment	Cost	Equipment	Cost
4.	Weeding/interculture	Self propelled weeder	1400	Manual hand hoe	3300
5.	Plant protection spraying	Aeroblast sprayer	500	Manual knapsack sprayer	700
6.	Irrigation (5 no.)	Electric pump set	4500	Electric pump set	4500
7.	Harvesting and collection	Manual: improved sickle	2200	Manual: local sickle	2500
8.	Threshing (yield, t/ha = 1.2)	Multicrop thresher	1500	Manual	2200
	Sub total:	–	14310	–	20540
9.	Cost of Agro inputs				
	(i) Seeds: 20 kg		2000		2000
	(ii) Fertilizer $N:P_2O_5:K_2O$(40:30:20):DAP Urea and MOP)		900		900
	(iii) FYM = 10t		3000		3000
	P.P. Chemicals		250		250
	Sub total:		6250		6250
	Grand total:		20560		26790
10.	Yield Kg/ha		1200		1200
	Cost of production Rs/kg		12.00		17.00
	Market Price, Rs/kg		80.00		80.00
	Gross income, Rs/ha		96000		96000
	Net income, Rs/ha		75440		69210

The benefit due to mechanization over the traditional practice, Rs/ha = 6230/-. Considering 10 per cent mechanization of the present total area under coriander (3.40 lakh ha) in the country, the net annual benefit due to mechanization is Rs. 2118/-.

(4) Production Economics of Seed Fenugreek

Cost of Cultivation (Rs/ha)

Sl.No.	Unit Operations	Improved Practices		Traditional Practices	
		Equipment	Cost	Equipment	Cost
1.	Seed bed preparation including bunding and field channel laying	Rotavator Bund former Manual spade	2300	Harrow/Cultivator Manual spade	5100
2.	FYM application (10t)	Manure spreader	1100	Manual	1800
3.	Sowing and basal application of fertilizer	Seed cum fertilizer drill	700	Hand broadcasting	220
4.	Weeding/interculture	Self propelled weeder	1400	Manual hand hoe	3300

Contd...

Contd...

Sl.No.	Unit Operations	Improved Practices		Traditional Practices	
		Equipment	Cost	Equipment	Cost
5.	Plant protection spraying	Aeroblast sprayer	500	Manual knapsack sprayer	700
6.	Irrigation (5 no.)	Electric pump set	4500	Electric pump set	4500
7.	Harvesting and collection	Manual: improved sickle	2200	Manual: local sickle	2500
8.	Threshing (yield, t/ha = 1)	Multicrop thresher	1500	Manual	2500
	Sub total:	–	14200	–	20620
9.	Cost of Agro inputs				
	(i) Seeds: 20 kg		2000		2000
	(ii) Fertilizer		800		800
	$N:P_2O_5:K_2O$(25:20:20):DAP Urea and MOP)				
	(iii) FYM = 10t		3000		3000
	P.P. Chemicals		750		750
	Sub total:		6550		6550
	Grand total:		20750		20750
10.	Yield Kg/ha		1500		1500
	Cost of production Rs/kg		14.00		18.00
	Market Price, Rs/kg		50.00		50.00
	Gross income, Rs/ha		75000		75000
	Net income, Rs/ha		54250		47830

The benefit due to mechanization over the traditional practice, Rs/ha = 6420/-. Considering 10 per cent mechanization of the present total area under Fenugreek (0.33 lakh ha) in the country, the net annual benefit due to mechanization is Rs 212/- lakh.

Conclusions

- Low mechanization of major seed spices in India is considered to be one of the factors contributing for low yield and higher cost of cultivation. By adopting few selected equipment with suitable modifications as per the requirements placed by the major seed spices the unit operations can be mechanized for more time, energy and cost-effectiveness.
- By 10 per cent mechanization level of selected unit operations the impact in-term of economic benefits may be visible as estimated to be, Rs (lakh) 2453/-, 340/-, 2118/- and 212/- per year respectively for cumin, fennel, coriander and fenugreek in the country.

References

CIAE, Bhopal, 2007. *Achievements of CIAE, Bhopal and Impact of the Technologies Developed* (Draft report) CIAE, Bhopal.

CRSS: A profile, CRSS, SDAU, Jagudan, 2007.

Malhotra, S.K. and Vashistha, B.B., 2008. *Package of Practices for Production of Seed Spices.* NRCSS, Ajmer.

Recent Advances in Seed Spices (2011) *Pages* **186–199**
Editors: **Y. Ravindrababu, R.K. Jaiman, K.D. Patel, N.R. Patel and S.B.S. Tikka**
Published by: **DAYA PUBLISHING HOUSE, NEW DELHI**

18

Resource Conservation Technologies in Seed Spices: Prospects and Possibilities

S.K. Rautaray, B.K. Garg and Tarun Kapur
Central Institute of Agricultural Engineeering, Nabibagh, Berasia Road, Bhopal – 462 038

Introduction

The concepts of conservation agriculture has emerged as a way for transition to the sustainability of conventionally managed intensive agricultural production system. The issues of conservation have assumed importance in view of widespread resource degradation and the need to reduce production costs, increase profitability and make agriculture more competitive. The conventional mode of agriculture through intensive agricultural practices although achieved the goal of production but led to degradation of natural resources. The growing concerns for sustainable agriculture is being viewed as a positive response to limits of both low-input traditional agriculture and intensive modern agriculture relying on high levels of inputs.

Conservation agriculture universally refers to the system of raising crops without or minimum tilling the soil while maintaining crop residues in-situ on the soil surface with appropriate crop combinations while Resource Conservation Technologies (RCTs) refer to those practices that enhance resource or input use efficiency such as the zero or reduced tillage practices that save fuel and improve plot level water productivity, land leveling practices that help save water, FIRB practices that save water and improve crop productivity and many more. But some Resource Conservation Technologies although look attractive showing immediate advantage may be unsustainable in the long term. A zero tillage practice without residue retention and without suitable crop rotation at a particular site may be more harmful than the conventional practice. Therefore the RCTs being the specific component of the conservation agricultural system (farm equipment selection, crop residue and mulch management, crop establishment methods etc.) tend to be different across the environment and therefore need to be developed through adaptive research suiting to the local conditions.

Resource Conservation Technologies

Over the past two decades nationally rapid strides have been made to evolve and spread RCTs like zero and reduced tillage system, effective management of crop residues and planting systems which enhance water and nutrient conservation. Direct seeding involves growing crops without mechanical seedbed preparation and with minimal soil disturbance since the harvest of the previous crop. The term direct seeding is understood in conservation agriculture systems as synonymous with no-till farming. Planting refers to the precise placing of large seeds (maize/beans); whereas seeding usually refers to a continuous flow of seed as in the case of small cereals (wheat/barley). The equipment penetrates the soil cover, opens a seeding slit and places the seed into that slit. The size of the seed slit and the associated movement of soil are to be kept at the minimum possible. Ideally the seed slit is completely covered by mulch again after seeding and no loose soil should be visible on the surface.

Land preparation for seeding or planting under no-tillage involves slashing the weeds, previous crop residues or cover crops; or spraying herbicides for weed control, and seedling directly through the mulch. Crop residues are retained either completely or to a suitable amount to guarantee the complete soil cover, and fertilizer and amendments are either broadcasted on the soil surface or applied during seeding.

With the development of direct drilling machines almost all research work was based on the criteria of firstly attempting to define the responses of direct drilled seeds in relation to soil micro-environments. Thereafter machine design specifications especially of critical components were finalized to bring about the more desirable micro-environmental conditions. The performance of equipment was concentrated both on biological and mechanical parameters.

Selection of most appropriate equipment (Combinations) for specific situation is essential in respect of field requirements (Alleviate soil compaction, soil loosening only in crop rows or surface soil loosening with or without straw mulching); soil working condition (High moisture-plastic, moist friable or dry-friable soils); type of equipment (Discs or inverted 'T' openers, narrow rotary strips or wide rotary gangs) and cultivation practices. Besides, the chosen equipment (Design: implement geometry) should match to the power unit. Skilled operation is important to maintain work rate and avoid fuel losses.

Concerns Leading to Interest in Conservation Agricultural Machinery and Possible Resource Conservation Technologies

Water Regime	*Concerns*	*Possible RCTs*
☆ High rainfall	☆ Rapid erosion and land degradation	☆ Zero tillage
	☆ Nutrient lose	☆ Residue cover
		☆ Sensible crop rotations
☆ Low rainfall	☆ Late sowing	☆ Direct drilling
	☆ Drought stress	§ Residue cover
	☆ Low soil fetility	§ Residue+crop rotation+manune+ biopesticide
	☆ High weeds	

Contd...

Contd...

Water Regime	*Concerns*	*Possible RCTs*
✰ Dryland	✰ Drought stress	✰ Straw cover
	✰ Soil erosion	✰ Sub-soiling for in-situ moisture conservation
✰ Irrigated	✰ Ground water depletion	✰ Water Management
	✰ High cost of pumping	✰ Efficient use of input resources
	✰ High cost of production	✰ Profitable crop rotations
	✰ Scarcity of labour	✰ Controlled traffic cultivation
	✰ Compaction	✰ Permanent bed system

Machinery for Zero Tillage and Minimum Tillage Cultivation

Zero till drilling, strip till drilling, roto till drilling and slit till drilling of wheat after harvest of rice were compared to the conventional tillage sowing as practiced by farmers. Brief specifications of the direct drilling machines are given in Table 18.1.

Table 18.1: Specifications of Direct Drilling Machines

Sl.No.	*Particulars*	*Zero Till Drill*	*Strip Till Drill*	*Roto Till Drill*	*Slit Till Drill*
1.	Source of power	35/45 hp tractor	45 hp tractor	45 hp tractor	45 hp tractor
2.	Type/no. of furrow openers	Inverted 'T' type/09-11	Shoe type/09	Shoe type/11	Rotary blades and narrow shoe type/07
3.	Row spacings, mm	180 (Adjustable)	200 (Fixed)	180 (Adjustable)	230
4.	Working width, mm	1600-2000	1800	2000	1600
5.	Drive wheel	Angle lug-front mounted	Angle lug-side mounted	Star lug-rear hinged	Peg type-side mounted
6.	Weight, kg	210	250	280	300
7.	Unit price, Rs	30000	60000	70000	55000

The sowings at shallow depth (50-60 mm) under residual moisture condition (21.8-23.6 per cent) with 100 kg/ha seed rate and fertilizer dose of N:P:K::120:50:30 kg/ha were compared to the conventional practice of 03 tillage operations by duck-foot sweep cultivator (Size = 1800 mm) and sowing by seed cum fertilizer drill (09 row/row spacing = 180 mm). The time required as per actual field capacity of the machines, fuel used and cost of operations are given in Table 18.2.

The results showed that no tillage drilling was the most time, energy and cost-effective for 70.1, 67.2 and 67.3 per cent respectively over the conventional practice. The roto tillage seeding combined with full width shallow tillage in single pass operation was 60.3 per cent energy efficient and 64.4 per cent cost-effective compared to the conventional practice. The strip tillage seeding was advantageous over conventional tillage seeding but for the intermittent strip tillage the operational energy and cost requirements were higher compared to the roto tillage and zero tillage seeding. The slit till drilling showed higher performance in terms of time, fuel and cost effectiveness than the other test drills.

Table 18.2: Minimum Tillage Seeding Compared to Conventional Tillage-Sowing of Wheat

Sl.No.	*Particular(s)*	*Zero Till Seeding*	*Strip Till Seeding*	*Roto Till Seeding (Figure 18.3)*	*Conv. Tillage (3 passes)-Sowing*	*Slit Till Seeding*
1.	Time, h/ha	3.23 (70.1)	4.17 (61.2)	3.45 (68.1)	10.80	2.50 (76.8)
2.	Fuel used, 1/ha	11.50 (66.8)	17.50 (49.4)	13.80 (60.1)	34.60	10.00 (71.1)
3.	Operational energy, MJ/ha	648.96 (67.2)	1001.76 (49.3)	783.60 (60.3)	1976.11	565.00 (71.4)
4.	Cost of operation, Rs./ha	1100 (67.6)	1460 (57.1)	1210 (64.4)	3400	1000 (70.6)

(.....) values show percent savings over conventional practice.

Figure 18.1: Zero Till Seeding

Figure 18.2: Strip Till Seeding

Figure 18.3: Roto Till Seeding

Figure 18.4: Slit Till Seeding

Cultural practices specific to these conservation systems were developed in terms of frequency of irrigation and fertilizer applications. First irrigation of 40-50 mm was critical for all the direct seeding systems for initial establishment especially in zero tillage seeding. Performance of direct drilled wheat (Table 18.3) showed that in direct drilling systems although the grain yields were at par, the benefit-cost ratio were higher for 15.2-23.4 per cent with savings in operational energy for 8.4-14.7 per cent compared to the conventional practice.

Table 18.3: Production Economics of Direct Drilled Wheat (HI 8498) in Vertisols

Particulars	*Zero Till Drilled*	*Strip Till Drilled*	*Roto Till Drilled*	*Conservationally Sown*	*Slit Till Drilled*
Grain yield,t/ha	3.84	3.62	3.78	3.60	4.00
Cost of production, Rs/ha	9000	10000	9500	12000	8500
Benefit-cost ratio	3.20	2.72	2.98	2.25	3.50

Machines for Residue Incorporated Tillage-Seeding

The straw incorporated tillage (Figures 18.5 and 18.6) showed that (Table 18.4) direct rotavation under chopped straw condition followed by drilling was 28.5 per cent energy efficient and 24.9 per cent cost-effective compared to the mould board plough + rotavator + drill operations. The mould board plough and rotavator operations gave almost straw free surface (Straw incorporation = 89.7 per

Figure 18.5: Chopping of Straw by Stubble Shaver

Figure 18.6: Rotavator Tillage with Straw Incorporation

cent at 50-125 mm depths) for unimpaired drillings. With direct rotavation the average incorporations of straw was 60.4 per cent and it affected the performance of drill due to chockings. However the value of benefit-cost ratio were at par in straw/non-straw fields indicating no immediate advantage of straw incorporated cultivation although it may prove to be beneficial in long run being of organic base.

Controlled Traffic Rotary Slit Drill

It may be inferred as observed that the effectiveness of straw incorporation should not necessarily be a guide to overall benefit-cost. Therefore direct rotavation under chopped straw condition was considered to be more appropriate being shallower working system (Depth = 50-100 mm) with higher work rate and substantially less energy demanding than the mould board plough based cultivation system.

Direct drilling following controlled traffic measures (Figure 18.7) was compared to full plot sowing under zero/minimum tillage condition and conventional tillage seeding. In straw fields the field condition was represented by straw density of 0.34 kg/sq.m (Average length of loose straw > 150 mm = 272 mm: 55 per cent) and average bulk density of soil within 100 mm depth profile was 1.14 g/cc at average soil moisture content of 23.83 per cent db. The cone index of soil varied from 86 to 274 kPa from surface 25mm to 150 mm depth at corresponding soil moisture variation of 22.42 and 24.85 per cent db respectively. Seed rate (wheat) of 90 kg/ha was maintained with rotary slit drill operated in controlled

Table 18.4: Energy and Cost of Residue Incorporation for Subsequent Sowings

Type of Straw Field	*Treatment*	*Implement Used*	*Time* Required h/ha*	*Direct Energy Used, MJ/ha*	*Cost of Operation, Rs/ha*	*Amount of Straw Incorporated, Per cent*	*Total Operational Energy, MJ/ha*	*Total Cost of Operation, Rs/ha*
Combine harvested	T1	Stubble shaver (1)	2.75	508.06	511.99	–		
rice and wheat		MB plough(1)	5.13	1151.46	1041.84	76.70		
straw field		Rotavator (1)	3.00	644.58	617.84	13.00		
		Seed-Fetilizer drill (1)	3.71	696.63	678.93	–	3000.73	2895.60
Combine harvested	T2	Stubble Shaver (1)	2.75	508.06	511.99	–		
rice and wheat		Rotavator (1)	3.58	721.30	768.71	60.43		
straw fields		Seed-fetilizer Drill (1)	4.88	916.39	893.04	–	2145.75	2173.54
Non-Straw rice	T3	Duck foot sweeps (3)	7.71	1446.87	1486.90	–		
and wheat fields		Seed fertilizer Drill (1)	2.99	560.53	546.26	–	2007.10	3400.00

( ) Figures show number of passes.

Figure 18.7: Rotary Zero Till Slit Drill

paths whereas in full plot sowings, seed rate of 100 kg/ha was maintained. In full plot sowings fertilizer was applied at the rate of N:P:K::120:60:40 kg/ha and in controlled traffic fields N:P:K was applied @ 100:50:30: kg/ha respectively. Field observations on soil, straw, weeds and crops were recorded at 15,35,60,90 and 110 days after sowing (DAS). The cultural operations including irrigations were maintained as per requirements.

The machine performance of the rotary zero till slit drill compared to other test drills is given in Table 18.5.

Comparative Performance of the Controlled Traffic Rotary Zero Till Slit Drill

The rotary zero till slit drill was evaluated in rice straw and nonstraw fields for controlled path sowings. Zero till drill (Inverted T type furrow openers) and roto till drill were evaluated in straw and non straw fields for full plot sowings. Conventional sowing by use of seed cum fertilizer drill after preparatory tillage (3 passes) was evaluated for comparison.

The results showed that controlled path sowings in straw and nonstraw field was advantageous in terms of higher field capacity (0.42 ha/h) and lower fuel consumption (9.31/ha) compared to the other test drill (Table 18.5). Due to fixed path the tractor could be operated faster at low throttle sowing the crop in non-wheeled areas. In straw fields the rotary cutters in combination with the press wheels gave uniform seeding in slit with in-situ retention of cut straw on the soil surface. With linear zero till

Table 18.5: Comparative Performance of Rotary Zero Till Slit Drill for Wheat in Rice Straw and Non Straw Fields*

Sl.No	Particulars	Straw Fields				Non Straw Fields			
		Control Traffic Sowing	Full Plot Sowing			Control Traffic Sowing	Full Plot Sowing		
		Rotary Slit Drill (ST-1)	Linear Zero Till Drill (ST-2)	Roto Till Drill (ST-3)	Conv. Drill After 3-Tillage (ST-4)	Rotary Slit Drill (NST-1)	Linear Zero Till Drill (NST-2)	Roto Till Drill (NST-3)	Conv. Drill After 3-Tillage (NST-4)
1.	Eff Effective width of coverage, mm	1600	2000	2000	2000	1600	2000	2000	2000
2.	Effective speed of operation km/h	3.0	1.2	1.4	2.4	3.2	2.0	2.1	2.6
3.	Effective field capacity, ha/h	0.40	0.15	0.18	0.38	0.44	0.32	0.34	0.44
4.	Effective depth of operation, mm	98	114	88	134	101	126	105	148
5.	Field efficiency, per cent	83.3	62.5	64.3	79.2	86.3	80.0	81.0	84.6
6.	Fuel consumption, l/h	3.7 (9.3)	3.8 (25.3)	308 (21.1)	3.7** (40.1)	3.7 (11.9)	3.8 (11.9)	3.8 (11.2)	3.7** (38.8)
7.	Wheel slip, per cent	9.1	21.7	18.1	10.7	8.6	19.8	17.0	10.6

*: Field condition: Combine/Reaper harvested rice fields.

☆ Straw density, kg/sq.m = 0.34

☆ Average length of loose straw>150 mm = 272 mm

☆ Average height of stubbles, mm = 285 (combine harvested) and 168 (Reaper harvested)

☆ per cent of loose straw left on the surface = 55

☆ Average bulk density of soil in 100 mm depth profile, g/cc = 1.14 (Ave.soil moisture = 22.80 per cent db)

**: Include 30.4 l/ha@3.8 l/h for 3 tillage operations.

drill having inverted 'T' type furrow openers the drill was required to be lifted frequently to clear the chocking of drill tines due to loose straw being gathered. The result was low field capacity and non-uniform drilling of seeds. With roto till drill the loose straw got wrapped around the rotary and the broken/bent straw being incorporated at shallow depths often chocked the drill tines demanding frequent stop-go operations and causing nonuniform drilling.

In nonstraw fields the performance of linear zero till drill and roto till drill although found lower to the slit drill but may be treated as satisfactory. In conventional sowing with preparatory tillage(s) the work rate of seed cum fertilizer drill in non-straw field was higher by 15.8 per cent than the straw field (work rate, ha/h = 0.38) as in the straw fields the buried straw many a time got entangled to the furrow openers causing uneven sowing.

Raised Bed Planting on Fresh and Permanent Beds

Raised bed planting of wheat (HI-8498) by bed planter on fresh and permanent beds were compared to flat sown and no tillage condition (Table 18.6).

Table 18.6: Energy and Economic of Raised Bed Planting of Wheat Compared to Flat Sown

Sl.No.	*Particulars*	*Planting on Fresh Beds with Preparatory Tillage*	*Planting on Permanent Beds*	*Flat Sowing Zero Tillage*	*Conv. Flat Sowing*
1.	Time required, h/ha	13.04	4.80 (55.6) [63.2]	3.23	10.82
2.	Operational energy, MJ/ha	2605.36	1154.03 (41.6) [55.7]	648.96	1976.11
3.	Cost of operation, Rs/ha	2479.84	1060.80 (44.3) [57.2]	639.54	1903.04

( ) Per cent savings over conv. practice, [ ] Per cent savings over fresh bed planting.

It was found that planting on permanent beds was energy efficient and cost effective by 41.6 and 44.3 per cent respectively over the conventional flat sowing of wheat (03 tillage + 01 sowing operation). Similar advantages of time, energy and cost-effectiveness for 63.2, 55.7 and 57.2 per cent respectively were found for planting on permanent beds than on fresh beds (03 tillage + 01 planting operation).

Table 18.7: Production Economics and Operational Energy in Raised Bed Wheat

Particular	*Raised Bed Wheat*		*Zero-Tillage Wheat–Flat Sown*	*Conv. Flat Sown Wheat*
	Fresh Bed	*Permanent Bed*		
Grain yield, t/ha	5.03	5.08	4.84	4.60
Cost of production, Rs/ha	10030	8540	8635	10710
Benefit-cpst ratio	3.26	3.87	3.64	2.79
Operational energy, MJ/ha	8750	7684	8444	9516
Sp/operational energy, MJ/kg	1.74	1.51	1.74	2.07
Sp.cost of production, Rs/kg	1.99	1.68	1.78	2.33

The production economics (Table 18.7) showed that although the grain yields on fresh and permanent beds were at par the benefit-cost ratio in respect of the permanent beds was higher by 15.8 per cent over the fresh beds mainly because of the reduced cost of cultivation (14.9 per cent) due to the elimination of preparatory tillage.

Further on permanent beds than in fresh beds the water retention was higher in furrows due to compaction effect by tractor wheels and it helped to provide catchment effect for longer period by slow infiltration of water to the root zone of crops on beds. In fresh beds wheat although there found 30 per cent savings of water compared to the conventional flat sown (Water requirement = 600 mm/ha) further savings of 10 per cent of irrigation water was recorded in permanent beds over the fresh beds.

Compared to zero tillage flat sown the cultivation on permanent beds being virtually of zero tillage on beds showed 4.7 per cent higher grain yield with higher benefit-cost ratio of 5.9 per cent may be due to the combined zero tillage and bed effects in terms of better stand establishment and growth. Compared to flat sown the performance of crops on fresh and permanent beds were found higher in terms of higher grain yield (8.5-9.4 per cent) and higher benefit-cost ratio (14.4-27.9 per cent). Cultivation on permanent beds was found to be energy efficient for 9.0, 12.2 and 19.3 per cent over zero tillage flat sown, fresh beds and conventional practice respectively.

Traffic and Compaction

Once-over tillage-seeding practice, although depend on soil conditions which remain for only limited periods on many soils, has imposed higher loads on the soil due to higher weight of the tractor (1.5-2.0 tonne) and soil reactions including compaction. High throttle operation for faster travel in single pass operation especially with zero till drill in wet soil gave wheel slippage (30-40 per cent) and hampered operational control of the drill besides causing non uniform dropping/bunching of seeds in drill row. Due to compaction of soil under the wheels (Bulk density = 2.26 g/cc at soil moisture = 25.62 per cent db in vertisols) the drill tines just behind the wheels remained above the ground level without contact of soil and dropped the seeds virtually on the surface of the soil affecting germination and making the seeds more vulnerable to birds.

It was found that in zero tillage wheat the performance of crops on wheeled area was low due to erratic and uneven establishment of crop at the initial stage. During first flood irrigation which is critical to the initial establishment the seeds remained partially/fully submerged due to more accumulation of water in wheeled area for longer time creating temporary logging condition. Penetrometer observations to 150 mm depth in wheeled area showed signicant increase in soil strength compared to the adjacent non-wheeled zone (Soil resistance = 72.4-208.6 KPa in profile of 25-150 mm) in vertisols.

Sample observations form wheeled areas showed lower wheat yields (3.02 t/ha) than the non-wheeled areas (4.28 t/ha) and it was more pronounced at the headlands due to repeated passes of the tractor compounding the effect of compaction (Soil resistance = 168.6 – 280.4 KPa from surface to 150 mm depth).

Need for Controlled Traffic Research

Following the adoption of minimum and zero tillage techniques by many farmers the main focus of researchers currently drawing attention is towards the controlled traffic cultivation systems. The production operations therefore from tillage-seeding to harvesting may have to be planned to grow crops in between the wheels by following the fixed path operations in the field. To this, the controlled traffic lining and bed systems (Flat-bed / Broad-bed) may be the immediate need to go someway towards avoiding the ill effects of wheels. During combine harvesting the controlled traffic lines may only be breached but in longer run one may plan for controlled traffic harvesting or tracks on the combine to reduce compaction.

Adoption of RCTs in Seed Spices

Development and adoption of RCT practices in seed spices the common factors learned to address the issues of productivity and sustainability are summarized as follows:

- ☆ Emphasis on development and adaptation of implements may be using prototypes from elsewhere. Without locally adapted implements RCTs rarely can make success.
- ☆ Participation of private sector in implement development, adaptation, manufacture and marketing.
- ☆ Technical advise of lead Institutions may be recognized to help guide and bring the development and adaptation of RCT practices to local circumstances.
- ☆ Strong local organizations and innovative farmers in the development of RCT practices.
- ☆ Crisis (Soil erosion / land degradation, herbicide resistant weeds, soil compaction, prohibition of crop residues burning etc) that brings with it willingness to consider departure from conventional practices. Creative information exchange and needs-drive research by knowledge sharing to important crises at hand are truly the key factors.

Researchable Issues of RCTs in Seed Spices

- ☆ Precise land leveling
- ☆ Subsoiling
- ☆ Zero / minimum tillage drill with loose residues
- ☆ Controlled traffic cultivation machinery
- ☆ Permanent bed cultivation machinery
- ☆ Machinery for residue handling
- ☆ Public-private partnership for machinery development, adaptation, manufacturer and marketing.
- ☆ Farmers participatory on-farm–research to adapt RCTs than tradional / modern practices.
- ☆ Linking RCTs to appropriate agribusiness strategies (custom hire) to increase employment in areas where it is adopted and to aid promotion of resource conservation machineries.

Conclusions

Adaptation of resource conservation technologies in seed spices may show multiplicity of benefits as in other crops.

- ☆ *Reduction in the cost of cultivation*: A key factor promoting adoption of zero tillage technology that the cost of wheat cultivation has been reduced by 1500-2000 Rs / ha mainly on account of savings in fuel, labour and material inputs.
- ☆ *Reduced incidence of weeds*: The studies have indicated reduced incidence of phalaris minor in wheat in long run resulting into reduced use of weedicides.

- ☆ *Savings in water and nutrients*: Experimental results and farmers perception have indicated 25-35 per cent savings in irrigation water more particularly in laser leveled and bed planted fields.
- ☆ *Increased yield*: Timely planting and properly managed wheat fields have shown higher yields by 05 per cent compared to the traditional practice.
- ☆ *Environmental benefits*: RCTs involving zero/minimum tillage and surface managed crop residue system has offered excellent opportunity to eliminate in-field burning of crop residues which contribute to large amount of green house gasses (CO_2, CO, N_2O, SO_2). Burning of crop residues also cause loss of plant nutrients, which could be recycled when properly managed).
- ☆ *Agri business development:* Custom hiring for on-farm employment and promotion of resource conservation machinery.

References

Rautaray, S.K., 2005. Machinery for conservation agriculture: Progress and needs. In: *Conservation Agriculture: Status and Prospects,* (Eds.) I.P. Abrol *et al.*, CASA, New Delhi, pp. 43-49.

Rautaray, S.K. and Garg, B.K , 2006. Study of straw covered zero tillage seeding and controlled traffic effect in rice-wheat cultivation for increasing productivity. In: *Annual Progress Report (RPF-II)*, Agricultural Mechanization Division, CIAE, Bhopal, April 2006.

Recent Advances in Seed Spices (2011) *Pages* **200–209**
Editors: **Y. Ravindrababu, R.K. Jaiman, K.D. Patel, N.R. Patel and S.B.S. Tikka**
Published by: **DAYA PUBLISHING HOUSE, NEW DELHI**

19

Quality Issues in Seed Spices for Export

S. Siddaramappa
Deputy Director (Development), Spices Board, Ahmedabad, Gujarat

Introduction

India is the largest producer,consumer and exporter of spices in the world. The world looks upon India as kingdom of spices. No country in the world produces as many kinds of spices as India. Because of varying climates from tropical to subtropical and temperate, almost all spices are grown in the country with a wide range from highly pungent and hot black pepper and chillies to the mild and spicy seed spices like coriander and cumin. India's annual production of spices is estimated at over 4.00 million tons from an area of 2.53 million hectares. In the international market, we face stiff competition from low cost economies of the far east such as Vietnam, Indonesia, China, Thailand, etc. However, India remains to be the largest exporter of spices in spite of the fact that our exports are only 8 to 10 per cent of the production.

Importance of Export

Export form the most important source of foreign exchange for developing countries. All developing countries, including, of course, India strive to increase their exports as a means for mobilizing resources for their development plans. Indian economy is basically agrarian and hence exports of food and agricultural products assume crucial significance in our export efforts. Growth in agro-exports not only brings in additional foreign exchange for the country but benefits a large number of people involved in the production, processing and exports of such products.

Export Performance of Spices

The estimated world trade in spices is 8,50,000 tons valued at US$ 2200 million. Out of which, India still commands a significant share of 52 per cent in volume and 50 per cent in value. During the year 2007-08, we have exported 444250 MT of spices valued at Rs.4,435.50 crores (US$1101.80 million) in which seed spices, *Viz.*, coriander, cumin, fennel, fenugreek and other seeds accounts for 82100 MT for a value of Rs. 507.75 crores. The value added forms of spices like spice oils and oleoresins, curry powder/paste and mint products have shown significant growth in this period. The export figures of individual seed spices are given below:

Coriander

During 2007-08 we have exported 26000 tons of coriander valued Rs.110.25 crores as against 20500 tons valued Rs.74.62 crores in 2006-07. It is reported that the east European countries like Romania and Bulgaria, where coriander is produced, has suffered drought condition and the supply from these origins are on a lower side.

Cumin

In case of cumin, during 2007-08 we have exported 28000 tons valued Rs.291.50 crores as against 26000 tonns valued Rs. 201.50 crores in 2006-07. The export unit value realization has gone up from Rs.77.50 per kg in 2006-07 to Rs.104.11 per kg in 2007-08.

Celery

During 2007-08, we have exported 2900 tons of celery valued Rs.13.25 crores as against 3550 tons valued Rs.13.20 crores in 2006-07.

Fennel

In 2007-08 we have exported 5250 tons of fennel valued Rs. 28.50 crores as against 3575 tons valued Rs. 23.80 crores in 2006-07.

Fenugreek

Export of fenugreek in 2007-08 has increased to 11,100 tones valued Rs. 33 crores as against 8500 tons valued Rs. 26.98 crores in 2006-07.

Other Seeds

In 2007-08 we have exported seed spices like Mustard seed, Aniseed, Dill seed, Poppy seed etc. together 8850 tons valued Rs. 31.25 crores as against 8000 tons valued Rs. 22.40 crores in 2006-07.

What is Quality ?

Quality of any product is assessed by means of its intrinsic as well as extrinsic parameters. In case of spices, the quality retales to size, appearance, colour, flavour etc., moisture, volatile oil, major chemical constituents present in the spices determines the intrinsic quality. Physical contaminants, generally termed as macro contaminants constitute the extrinsic quality of the produce which will determines whether the product is consumable or not. The occurrence of the physical contaminants is mainly arising from unhygienic and unscientific practices followed in the industry. Among these are pests, micro organisms, foreign matter, poisonous matter or impurities. The harmful effects of various contaminants is summarized in Table 19.1.

Quality Requirements of Foods for Export

Developed countries are the major markets for our food exports, USA, Japan, Canada, Australia, and the European countries have their own stringent food laws and regulations. The main objective of the laws is to protect the health and safety of their citizens. They allow import of food materials only when they conform to the provision of their food laws and regulations.

Table 19.1: Harmful Effect of Various Contaminants

Contaminants	*Harmful Effects*
☆ Excreta of animals	Carrier of disease and add filth to spice.
☆ Insects/pests	Destroy spices, leave excreta and carrier of the disease.
☆ Mites, Psocides and spiders.	Leaves insect fragments harmful to health and leave excreta.
☆ Molds	Produce toxins harmfull to health and deteriorates the spices.
☆ Extreneous matter	Harmful to health and may cause injury.
☆ Chemicals, pesticides	Extremely harmful and poisonous may cause cancer.
☆ Rodent hair and human hair	Stomach disorder, harmful to health, add filth to spices.

Why Food Quality

The developed countries give top priority to the health of their citizens. The laws with respect to items of food are meant to protect the consumers from food of inferior quality, or those which are likely to be contaminated by impurities or poisonous substances. Therefore any food item that we export, be it marine products, cashews, pepper, cardamom or ginger, it is important that the product conforms to the quality standards demanded by the importing country. In the context of thousands of people getting infected with food borne diseases or even dying of food poisoning, it is only just and reasonable that countries which rotten, spoiled, infected with micro-organisms or contaminated by other impurities are either destroyed by the import inspection authorities or sent back to the exporting country. This not only results in loss of market but also damages the exporting country's reputation.

Table 19.2: Cleanliness Specification for Spices in Major Importing Countires

Name of Country/ Spice	*Extraneous Matter Per cent/wt.*	*Moisture Per cent/wt.*	*Total Ash Per cent/wt.*	*Acid Insoluble Ash Per cent/wt.*
		Germany		
Chilli	–	10.0	8.0	1.0
Celery	–	10.0	12.0	2.5
Coriander	–	–	8.0	2.5
Dill	–	12.5	7.0	1.0
Fennel	–	10.0	10.0	1.0
Fenugreek	–	10.0	5.0	1.0
Ginger	–	12.5	7.0	1.0
Pepper	–	12.0	7.0	1.0
Turmeric	–	10.0	7.0	1.5
		Netherlands		
Chilli	–	12.0	7.2	---
Celery	–	12.0	10.0	2.5
Coriander	1.5	10.0	8.0	2.5

Contd...

Table 19.2–Contd...

Name of Country/ Spice	*Extraneous Matter Per cent/wt.*	*Moisture Per cent/wt.*	*Total Ash Per cent/wt.*	*Acid Insoluble Ash Per cent/wt.*
Dill	–	10.0	12.5	0.5
Fennel	–	12.0	12.0	--
Fenugreek	2	6.0	–	--
Ginger	–	10.0	8.0	3.0
Pepper	–	12.0	7.0	2.0
Turmeric	–	12.0	7.0	3.0
		U.K.		
Chilli	–	7.5	–	--
Celery	1.0	14.0	11.0	2.0
Coriander	2.0	10.0	8.0	1.5
Dill	1.0	–	–	–
Fennel	1.0	7.5	9.0	1.0
Fenugreek	2	6.0	–	–
Ginger	1	12.0	6.0	1.0
Pepper	2	12.0	7.0	1.0
Turmeric	–	–	–	–

Table 19.3: Maximum Pesticides Residues Limits in Netherlands and United Kingdom

Active Substance	*Limiting Values in PPM*	
	Netherlands	*United Kingdom*
HCH without Lindane	0.02	0.02
Lindane	0.02	--
Hexachlorobenzene	–	0.01
Aldrin and Dieldrin	0.03	0.01
Sum of DDT	0.15	0.05
Malathion	0.05	8.00
Dicofol	0.05	0.50
Chlorpyrifos	0.01	--
Ethion	0.01	--
Chlordan	0.01	0.02
Parathion	0.10	1.00
Parathion methyl	0.10	0.20
Mevinphos	0.05	--
Sum of Endosulfan	0.02	0.10
Phosalon	1.00	0.10

Contd...

Table 19.3–Contd...

Active Substance	Limiting Values in PPM	
	Netherlands	United Kingdom
Vinclozolin	–	0.10
Dimethoat	0.01	0.05
Quintozen	–	1.00
Metacriphos	–	---
Heptachlor and epoxid	0.21	0.01
Methidathion	–	---
Diazinon	0.05	0.05
Fenitrothion	0.05	0.05
Bromophos	–	---
Mecarbam	–	---
Methoxychlor	0.05	---
Omethoat	–	0.20
Dichlorvos	0.05	---
Phosmet	0.01	---
Methyl bromide	–	0.10
Tetradifon	–	

Table 19.4: Guidelines on Quality Improvement of Cumin and Fennel ASTA Cleanliness Specifications (Effective April 28, 1999)

Name of Spice Whole Insect Dead	By Count/lb Excreta Mammalian	By Mg/lb Excreta Other	By Mg/lb Mold	Per cent by Wgt Insect Defiled/ Infested	Per cent by Wgt Extraneous Foreign Matter	Per cent by Wgt
Cumin	4	3.0	5.00	1.00	1.00	0.50
Fennel	*	*	*	1.00	1.00	0.5·0

*: 20 per cent or more of the subsamples contain any rodent other excreta, or whole insects, or an average of 3 mg lib or more mammalian excreta, the lot must be reconditioned.

In addition to ASTA cleanliness specifications, spices imported to USA has to comply with the Defect Action Levels (DAL) of FDA. At present FDA Defect Action Levels for Cumin and Fennel

FDA Defect Action Levels for Cumin and Fennel comprise of:

Defect Action Level

Cumin

Sand and grit average of 9.5 per cent or more ash and I or 1.5 per cent or more acid insoluble ash

Fennel

1. Insect 20 per cent or more of subsamples contain insect
2. Mammalian excreta

20 per cent or more subsamples contain mammalian excreta or average of 3 mg. or more of mammalian excreta per pound.

Besides the above DAL, FDA has also surveillance programmes to monitor for pesticide residues in spices.

Pesticide Residue

The tolerance limit for various pesticide residues in cumin and fennel as per CFR is as follows:

- ☆ Tolerance Limits for certain pesticides
- ☆ Pesticide Tolerance Limit
- ☆ BHC: 0.05 ppm
- ☆ Dichlorvos: 0.1 ppm

With the formation of European Union (EU) this market has adopted harmonised standards for all foods imported including spices. Germany is at present monitoring for the presence of aflatoxin and pesticide residues in spices. The unified European market is likely to adopt the requirements enforced by Germany.

Artificial Colours in Chilli and Turmeric

Chilli and Turmeric are spices of commercial importance contributing fairly well in export trade. It is reported that nearly 92 per cent of the produce is consumed domestically and only the remaining 8 per cent is exported.

Problem Areas

1. Pesticide Residues

Pesticide residues more than the prescribed limits are seen in chilli and turmeric. Top priority is given to control this problem. Farmer's co-operation and good thinking in this regard is praise worthy.

2. Sudan

Sudan is a type of chemical which is found mixed with chilli products to yield extra red colour for products. The chemical is carcinogenic. Because of this practice, few of our export consignments in 2003 were sent back by some European countries.

The Spices Board has taken up this issue as a serious one and started introducing stringent controls. As per the present condition the export of chilli will be permitted only if the consignments are certified Sudan free after testing of the pre-shipment sample by the Spices Board. In this problem, role of farmers is nil, but at the end the suffers are farmers because of the setback in the export of chillies.

3. Rhodamine B

This could be controlled by farmers only. While storing the chilli and turmeric in gunny bags/ plastic bags, names, lot numbers, date etc., are marked on the bags with an ink (mixing ink powder in water).

All these colours are artificial and most of them are carcinogenic having a chemical called RHODAMINE B. It is an established fact that RHODAMINE B is carcinogenic. Even a minute quantity of 2 PPB (2mg/Kg) of this Rhodamine B is not permissible in the European countries and the consignment will be rejected. The following control measures are recommended for the safety of consumers and to steamline exports.

1. Change of artificial colours: Should be replaced with the permissible food grade colours like Carmosine, Erythrosine, indigo, Caramine, brilliant Blue FCF etc.
2. Bags once used should not be used again. Lack of proper processing has resulted in a series of rejection of Indian consignments abroad. Europe and USA have issued Rapid Alert for Indian Chilly and its all products containing chilli and also Turmeric powder which has resulted in introducing compulsory pre shipment inspection of the products for the presence of Sudan dye and also for the high level of Aflatoxin in the products.

Steps to be Taken at Farmers Level

Pre-Harvest and Postharvest Operations

The quality of any product depends upon the quality of raw materials and the practices adopted in processing, packing, storing and transportation. In the case of agricultural products, therefore, constant care right from pre-harvest operations till the product reach the consumers becomes imperative.

Pre-harvest Operation: Some Critical Aspects

The planting material may be treated with suitable insecticidal fungicide, but only at the recommendation and supervision of experts. Pesticides which are banned in the importing countries should not be used in cumin and fennel.

In case the crop is affected by diseases or infested by pests, apply fungicides or insecticides, only after consulting experts and at the dosage and schedule recommended by them. It is to be borne in mind that cumin and fennel are being checked for the presence of pesticide residues by the importing countries.

Harvesting: Important Points

Harvest the crop only when it is fully matured. Maturity is indicated by the drying up of the plant including the base of the stem. While harvesting care should be taken not to cause any damage to the seeds.

Processing: Things to Remember

Processing of cumin and fennel consists of drying and cleaning. Sun drying is done on clean cemented yards or other suitable clean surfaces. The material is occasionally turned over to ensure uniform drying. The material should be heaped and covered during night time to ensure protection from rain. No colouring material should be used to improve the appearance of the product as chemicals and artificial colours are highly objected to by the importing countries.

Storage Precautions

The material should be stored ensuring protection from dampness. Dunnage should be provided to stack the packed bags to prevent moisture ingress from the floor. Care should be taken to stack the bags 50 to 60 cms. away from the wall.

No insecticide should under any circumstances, be used directly on the dried material. Stored material should be subjected to periodic fumigation for which only authorised persons should be engaged.

Insects, rodents and other animals should be effectively prevented from getting access to the premises where the material is stored. Stored product should be periodically exposed to the sun.

If care is taken in all stages of cultivation, harvesting, postharvest handling. processing. packing storage and transportation by following sound methods and practices, we will be able to prevent contamination and deterioration of quality in any farm produce and ensure consumer satisfaction.

Conformity to the quality requirements of the buying countries and price competitiveness are the key factors which determine our survival in the international market today. Improved productivity and quality will enhance the farmers' income and increase the country's foreign exchange earnings much needed for our developmental activities.

The maximum produce process and prosper' holds true in the case of exports of spices from India. The message is eloquent–we have to produce more cumin and fennel through higher production and improved productivity to meet the demands of the domestic and export markets and should process at adopting good processing practices creating value addition and of course have to export the product meeting the requirements of the importing countries which in turn would lead to significant growth in exports.

Initiatives of Spices Board for Postharvest Improvement of Spices

In order to educate the growers about various practices that are to be followed for improving the quality of product at farm level, Spices Board is implementing the following schemes and programmes for promotion of export of spices.

Construction of Drying Yards

In order to enable growers to dry seed spices under hygienic conditions, the Board will assist growers in constructing drying Yards. The subsidy offered is 90 per cent of the cost subject to a maximum of Rs. 540/- per sq.mtr for scheduled tribe growers. 50 per cent of the cost subject to a maximum of Rs. 300/- per sq.mtr for growers belonging to scheduled caste and 25 per cent of the cost subject to a maximum of Rs. 150/- per sq.mtr for general category growers. The estimated cost of construction of concrete drying yard is Rs. 600/- per Sq.mtr.

In addition, construction of community drying yards by NGOs, Farmers Groups (SHG) and Associations of Farmers and Agriculture Market Committees will also be supported by extending 50 per cent of the cost of construction as subsidy.

Supply of Polythene Sheets

Polythene sheets (of size 8X6m. 250 GSM, 5 layered) will be supplied at 50 per cent subsidy to tribal growers and 33.33 per cent to other growers for drying spices under hygienic conditions. Small and marginal growers owning upto 1ha. and 1-2 ha. are eligible for 2 Nos. and 3 Nos. of sheets, respectively. The Board will arrange for centralized purchase of polythene sheets and supply it to growers after collecting the non-subsidy portion from them.

Scheme for Vermicompost Units

In order to enable the growers to produce organic farm inputs, particularly vermi compost, setting up of vermi compost units in the growers field will be assisted by providing Rs. 2000/- as grant-in-aid per unit towards 20 per cent of the cost of establishing one unit with a production capacity of one tonne vermi compost per cycle.

Quality Improvement Training Programmes

This programme aims at educating the farmers, traders, exporters, officers of state Horticultural/ Agricultural Departments and representatives of NGOs to improve the quality of spices at harvest, postharvest and storage level to meet the requirements of quality specifications of the importing countries as well as to create a general awareness on hygienic handling of spices.

The training programmes are conducted by Board through its various offices. However, in areas where the Board's Offices are not functioning, implementation will be arranged through the State Agricultural/Horticultural departments and identified NGOs.

Organic Farm Certification

The scheme aims to help growers/processors of spices in acquiring organic certification which is a pre-requisite for marketing organic spices. Group of farmers, NGOs and farmers co-operative Societies/Associations will be assisted in acquiring certification for their farms/processing units by meeting 50 per cent cost of the certification, subject to a maximum of Rs.75,000/-. Individual farmers and processors will be given assistance of 50 per cent of the cost of certification with a maximum of Rs. 25,000/- per certification. The assistance will be available to the beneficiaries not more than twice during the plan period.

Promoting Production of Organic Spices

Spice like ginger, turmeric, chilli and kokam are suitable for organic cultivation in Maharashtra, Orissa, Chattisgarh, Uttaranchal, Bihar, etc. There is large scope for promoting organic farming for cumin, fennel, fenugreek, coriander, dill, etc. in the states of Gujarat and Rajasthan.

The scheme will support small and marginal growers owning up to four hectares to take up organic cultivation by providing a subsidy of 12.5 per cent of the cost subject to a maximum of Rs.4,250/- per ha. as part of the production cost. The subsidies will be disbursed in two instalments of Rs.2,250 and 2,000/-. The first instalment will be given immediately after sowing/planting and second instalment will be given immediately after harvest in the case of annual crops and during the second year in the case of perennial crops. The programme is implemented through selected NGOs by providing them NGO Retention Fee @ Rs. 250/- per hectare and @ Rs. 250/- per ha. to meet the cost of certification.

Market Development and Export Promotion Schemes

The Spices Board is implementing the following schemes and programmes for promotion of export of various spices and spices products from the country.

Indian Spice Logo

It is awarded to those who export spices in branded consumer packs. The award will be given after necessary inspections and prior certification under HACCP is a pre-condition for awarding Indian Spice Logo.

19 Spice Exporters are awarded with Indian Spice Logo and the Board has registered with Trade Registry Authorities in 18 countries.

Spice House Certificate

It is awarded to exporters of spice who possess the specified infrastructure facilities for cleaning, processing, grading, packing and warehousing as the case may be.

So far 58 units have been awarded Spice House Certificates.

Registration of Brand Name

Registration of brand is given to those retail packs which satisfy stipulated standards of packaging/ labelling. This is to assist the exporters in the export in branded consumer packs and protect the individual brand name.

Adoption of Hitech and Technology Upgradation

Trade Promotion

1. Packaging development and bar coding.
2. Sending business samples abroad.
3. Trade promotion tours.
4. Printing of Promotion Brochures.
5. Participation in Trade Fairs/Exhibitions abroad.

Recent Advances in Seed Spices (2011) *Pages* **210–227**
Editors: **Y. Ravindrababu, R.K. Jaiman, K.D. Patel, N.R. Patel and S.B.S. Tikka**
Published by: **DAYA PUBLISHING HOUSE, NEW DELHI**

20

Role of AICRPS in Seed Spices Research

M. Anandaraj, J. Rema and Johny A. Kallupurackkal
All India Coordinated Research Project on Spices,
Indian Institute of Spices Research, Calicut, Kerala

All India Coordinated Research Project on Spices (AICRPS) is located in Kerala with its head quarters at Indian Institute of Spices Research, Calicut. The research programmes of AICRPS are coordinated through a network of its 19 Coordinating, 8 Co-opting and 7 Voluntary centers which are in operation in different agro-ecological regions in twenty two states of the country. The AICRPS is the largest network in spices research in the country which came into existence in 1986.

The All India Coordinated Research Project on Spices conducts and coordinates research on 12 major spices *viz*, black pepper (*Piper nigrum*); cardamom (*Elettaria cardamomum*); large cardamom (*Amomum subulatum*); ginger (*Zingiber officinale*), turmeric (*Curcuma longa*), cinnamon (*Cinnamomum verum*); clove (*Syzygium aromaticum*); nutmeg (*Myristica fragrans*); coriander (*Coriandrum sativum*); cumin (*Cuminum cyminum*); fennel (*Foeniculum vulgare*); fenugreek (*Trigonella foenum-graecum*) with specific mandates and a budget of Rs.1400 lakh for the XI Plan and a sanctioned staff strength of 83 in various categories. The major mandates and objectives of AICRP on spices are to evolve high yielding, high quality varieties suitable for various agro-ecological situations, and that are tolerant/resistant to pests and diseases; standardize agro-techniques for different agro-climatic conditions; evolve measures for management of major pests and diseases; Work as an interface between State Agricultural Universities (SAUs) and Indian Council of Agricultural Research (ICAR).

Among the 109 spices listed by ISO, 20 are seed spices. Seed spices are mostly annuals and the dried fruits or seeds are used as spices. Coriander, cumin, fennel and fenugreek are considered as major seed spices on the basis of area, production and importance in national economy and ajawain, aniseed, celery, caraway, dill, and nigella are considered as minor seed spices.

Seed spices form as one of the important groups of spices crops in India. Almost all the states in India grow one or more seed spices. But major growing belt spreads from arid to semi arid regions, covering large area in Rajasthan and Gujarat. Most of the seed spices are either grown under rain fed production system or in limited water conditions. To overcome the production constraints in seed spices, various research and extension efforts are in progress.

Seed Spices Research Centre Under AICRPS

The research on seed spices are mainly carried out by the AICRPS Centres based in SAU's (Table 20.1) in 9 States namely, Rajasthan, Gujarat, Uttar Pradesh, Bihar, Haryana, Chattisgarh, Madhya Pradesh, Andhra Pradesh and Tamil Nadu so as to cover all the major seed spice growing area in India.

Table 20.1: Seed Spices Centres Under AICRPS

Seed spice	*State/Centre*
Coriander (8)	Jobner, Udaipur (Rajasthan), Jagudan (Gujarat), Coimbatore (Tamil Nadu), Guntur (Andhra Pradesh), Hissar (Haryana), Dholi (Bihar), Raigarh (Chattisgarh), Kumarganj (Uttar Pradesh), Jabalpur (Madhya Pradesh).
Cumin (4)	Jagudan (Gujarat), Jobner, Udaipur (Rajasthan), Hissar (Haryana), Kumarganj (Uttar Pradesh), Jabalpur (Madhya Pradesh)
Fennel (5)	Jagudan (Gujarat), Jobner, Udaipur (Rajasthan), Hissar (Haryana), Dholi (Bihar), Kumarganj (Uttar Pradesh), Jabalpur (Madhya Pradesh)
Fenugreek (8)	Jagudan (Gujarat), Jobner, Udaipur (Rajasthan), Guntur (Andhra Pradesh), Coimbatore (Tamil Nadu), Hissar (Haryana), Dholi (Bihar), Raigarh (Chattisgarh), Kumarganj (Uttar Pradesh), Jabalpur (Madhya Pradesh)

The seed spices are mainly cultivated in Rajasthan and Gujarat with a sizeable acreage in Madhya Pradesh, Haryana, Punjab, Uttar Pradesh, Andhra Pradesh and Bihar. India is the largest producer, consumer and exporter of seed spices in the world. Seed spices account for about 37 per cent area and 18 per cent of the production of the total spices in the country. On an average India produces about 5.0 to 6.0 lakh tonnes of seed spices annually and about 10 per cent of the production is exported. Seed spices contribute around 20 per cent of export and 13 per cent in the export earnings. India earned more than Rs. 200 crores by exporting seed spices in the form of raw as well as value-added products. Rajasthan and Gujarat together contribute more than 80 per cent of the total seed spices produced in the country. The area, production and productivity of major seed spices are given in Tables 20.2–20.6.

Table 20.2: Area, Production, Productivity (2004-05) and Export of Seed Spices in India (2005–06)

Seed Spices	*Area '000 ha*	*Production '000mt*	*Productivity kg/ha*	*Export*	
				Qty.MT	*Value Rs Lakhs*
Coriander	374.7	267.6	714	22750	6410.00
Cumin	367.7	176.1	479	12000	8800.00
Fennel	39.9	51.1	1279	4100	1960.00
Fenugreek	50.4	62.8	1245	12250	2400.00
Ajwain	15.0	5.3	450	935	400
Dill(sowa)	16.5	13.1	750	925	276
Celery	0.7	1.0	1303	3400	1265

Table 20.3: State Wise Area and Production of Coriander 2005–2006

State	*Area (Hect)*	*Production (Tons)*	*Productivity (kg)*
Andhra Pradesh	28110	12730	453
Assam	20210	18970	938
Bihar	2302	1667	724
Chhattisgarh	4090	1010	247
Gujarat	8670	18420	2125
Karnataka	7800	1190	153
Madhya Pradesh	101350	38130	376
Orissa	19060	9040	474
Rajasthan	136740	142350	1041
Tamil Nadu	21062	6424	305
Uttar Pradesh	6116	3450	564
Uttaranchal	439	830	1890

Table 20.4: State Wise Area and Production of Cumin 2005–2006

State	*Area (Hect)*	*Production (Tons)*	*Productivity (kg)*
Gujarat	267920	147615	551
Rajasthan	135113	52239	387

Table 20.5: State Wise Area and Production of Fennal 2005–2006

State	*Area (Hect)*	*Production (Tons)*	*Productivity (kg)*
Gujarat	36251	57236	1579
Rajasthan	3895	3657	939

Table 20.6: State Wise Area and Production of Fenugreek 2005–2006

State	*Area (Hect)*	*Production (Tons)*	*Productivity (kg)*
Gujarat	4851	5934	1223
Rajasthan	27292	29173	1069
Uttar Pradesh	257	186	724
Uttaranchal	261	415	1728

Major Limitations in Seed Spices Cultivation

1. Seed spices are traditionally cultivated in marginal lands with low fertility.
2. Lack of suitable varieties for specific agro-climatic zones.
3. Limited variability in germplasm.
4. Inherent nature of slow germination
5. Non-availability of quality seed of recommended

6. Low and erratic rainfall in area and limited surface water resources
7. High incidence of diseases in seed spice crops
8. Lack of adoption of improved technologies.
9. Lack of organized marketing system.
10. Problems during postharvest storage

Specific Problems in Major Seed Spices

Cumin

- ☆ Poor germination
- ☆ Cultivation under rain fed conditions
- ☆ Lack of bold seeded varieties with high volatile oil
- ☆ Umbel blight appearing at flowering stage
- ☆ Powdery mildew &Wilt disease

Coriander

- ☆ Early death of plants due to damping off
- ☆ Lack of dual purpose and bold seeded varieties
- ☆ Lack varieties having high oil content
- ☆ Powdery mildew resulting in no seed formation or small and shriveled seed

Fennel

- ☆ Lack of early maturing varieties
- ☆ Lack of short stature and synchronous flowering varieties
- ☆ Lack of varieties with high green *saunf* yield and volatile oil
- ☆ Lodging of plants and brittleness of branches causing loss during windy months
- ☆ Damping off during early stage of growth.
- ☆ Occurrence of umbel and leaf blight
- ☆ Sugary disease reducing seed setting

Fenugreek

- ☆ Damping off at germination stage
- ☆ Powdery mildew
- ☆ Aphids

Role of AICRPS in Overcoming these Limitations

To increase the productivity of spices from the present level emphasis has to be given for popularizing high yielding varieties/hybrids together with adoption of an improved production technology and expanding areas to non traditional areas. AICRPS has standardized several production technologies for increasing productivity, including management of major diseases and pests. Production technologies developed at various Research Centres have been demonstrated, but adoption level of such technologies by growers is still very low.

Research in Seed Spices: Conservation of Genetic Resources

Collection and conservation of genetic resources of seed spices is a major mandate of the project. Systematic surveys of the various seed spices growing areas of the different states were conducted to collect the variability of seed spices. Germplasm of major seed spices is also introduced and these exotic germplasm is used for widening the genetic base. The germplasm collected are conserved at various AICRPS centres. At present 1911 accessions of coriander, 526 cumin, 625 fennel and 978 fenugreek are conserved at various centres (Table 20.7). These genetic resources are utilized to develop need based varieties suitable for various agro climatic zones.

Table 20.7: Germplasm Holding of Major Seed Spices at AICRPS Centres

Crop/Center	*Indigenous*		*Exotic*	*Total*
	Cultivated	*Wild and Related sp.*		
Coriander				
Coimbatore	274	–	–	274
Dholi	97	3	–	100
Guntur	248	–	–	248
Hissar	251	–	–	251
Jagudan	86	22	–	108
Jobner	753	102	–	855
Kumarganj	75	–	–	75
Total	1784	127	–	1911
Cumin				
Jagudan	218	–	7	225
Jobner	281	–	20	301
Total	499	–	27	526
Fennel				
Dholi	39	–	–	39
Hissar	122	–	–	122
Jagudan	135	–	4	139
Jobner	261	–	20	281
Kumarganj	44	–	–	44
Total	601	–	24	625
Fenugreek				
Dholi	107	–	–	107
Guntur	125	–	–	125
Hissar	240	–	–	240
Jagudan	64	–	–	64
Jobner	353	–	12	365
Kumarganj	77	–	–	77
Total	966		12	978

Improved Varities

Crop improvement programmes utilizing the indigenous and exotic germplasm conserved were in progress for the last three decades and has resulted in the development of over 50 improved varieties in major seed spices. This includes 20 coriander, 8 cumin, 13 fenugreek and 9 fennel varieties. The improved varieties are identified for high yield, high quality resistant/tolerant to major pest and diseases. Dual purpose varieties in coriander and fenugreek suitable for greens and grains, varieties suitable for saline, alkaline and drought prone areas has also been released.

The introduced germplasm is used in breeding programmes and two varieties one each in fennel (RF-125) and cumin (Guj Cum-3) has been released from the exotic germplasm for cultivation in the country.

Technologies for Increasing Production

Seed spices are cultivated traditionally in marginal lands with low fertility. Lack of information to the growers about the standard production technologies is the one of the main hurdles for successful commercial venture in seed spices crops. AICRPS has taken necessary steps to standardize agro-techniques for commercial production of seed spices under semi-arid to arid region for rainfed and irrigated production systems following an integrated approach.

Coriander

Among the seed spices, coriander occupies the top place in terms of area, production and export (Table 20.8). Several technologies for maximizing production were developed at AICRPS centres.

- ☆ For an early crop, sowing is done in the mid-May which will be ready for harvest by July-August and for late crop, sowing in the mid-October and harvesting in January-February are optimum (Table 20.9).
- ☆ Spacing was standardized for different states (Table 20.10).
- ☆ A pre-sowing seed hardening treatment with 1 per cent Potassium di-hydrogen phosphate is needed for rainfed crop.
- ☆ Seeds are to be split in to two halves by crushing and then sown at a depth not more than 2-3 cm. Soaking the seeds in water for 8-12 hours is essential to enhance germination.
- ☆ First irrigation should be given immediately after sowing and subsequent 4 irrigations at 30-35, 60-70, 90-100 and 115-125 days after sowing.
- ☆ Weeding and hoeing are to be done at 30-35 days after sowing under rainfed condition and under irrigated condition at 40-45 days after sowing.
- ☆ Integrated weed management with pre-emergence application of pendimethalin @ 1.0 kg/ha, supplemented with hand weeding once in 50 days after sowing for Rajasthan and application of fluchloralin @ 1.5 kg/ha as pre-sowing or pendimethalin @ 1.5 kg/ha as pre-emergence for Haryana are recommended.
- ☆ Foliar application of micronutrients *viz.* Copper, Zinc, and Molybdenum resulted in increase in the volatile oil content.
- ☆ Coriander leaf plucking to the extent of 50 per cent at 75 days after sowing has been recommended to increase the returns.
- ☆ Harvesting of coriander crop at 50 per cent grains turning yellow has been recommended to give a good luster to the grains.

☆ Schedules for fertilizers and manures (Table 20.11).

☆ In case of frost, one light irrigation can save the crop in irrigated areas. In rainfed areas, spray 0.1 per cent H_2SO_4 and smoking around the border of the field.

Table 20.8: Major Coriander Poducing States with Zones/Districts

State	High Productivity Zones/Districts	Agro-climatic Region
Rajasthan	Baran, Thalawar Kota	Northern dry region
Madhya Pradesh and Chattisgarh	Guna, Raigarh, Mandsour	Central plateau and hill region
Andhra Pradesh	Kurnool, Cuddaph Ananthapur	Southern plateau and hill region
Tamil Nadu	Trichy	East cost plains and hill region
Uttar Pradesh	Varanasi, Mirzapur, Gajipur, Jaunpur, Somabhadra, Hamirpur, Jhansi, Devaria, Kanpur, Barabanki, Badun	Upper gangetic plains
Gujarat	Dantiwada, Junagadh	

Table 20.9: Sowing Time Recommended for Different States in Coriander

State	Sowing time
Rajasthan	First fortnight of October or first week of November/Mid of Oct to mid of Nov.
Gujarat	Last week of October to first fortnight of November
Tamil Nadu	June–July (kharif) September–October (rabi)
Andhra Pradesh	First fortnight of November
Bihar	First week of October
Uttar Pradesh	Middle of October to first week of November

Table 20.10: Spacing Recommended for Different States in Coriander

State	Spacing
Gujarat	–40 x 15 cm or broadcasting
Tamil Nadu	–15 x 15 cm
Andhra Pradesh	–25 x 15 cm
Uttar Pradesh	–30 x 10 cm
Rajasthan	–30 x 10 cm

Cumin

Cumin occupies maximum area and production, next to coriander, among the seed spices cultivated in India. In production, Gujarat and Rajasthan hold the almost entire share of the country. The high productivity zones/districts of these states are presented in Table 20.12. The technologies developed and disseminated for adoption by different centres under AICRPS are enumerated below:

☆ Minimum two weedings and one hoeing are needed. Herbicide tebutryn 0.5 kg a.i/ha or oxidiazone 0.5 kg/ha or pendimethalin 1.0 kg/ha as pre-emergence treatment can be used for weed control.

☆ Plants must be thinned out to a spacing of 12–15 cm in broadcast crop and 5 cm in line-sown crop during first weeding.

Table 20.11: Manure and Fertilizer Recommendation for Coriander

State	*Recommendation*
Rajasthan	10-20 T of well decomposed FYM + 60: 30: 40 Kg NPK/ha + neem cake. The FYM is applied at the time of field preparation, out of which 20 kg N, 30 kg P_2O_5 and 20 kg K_2O/ha is applied at the time of sowing and the rest in two equal splits at time of irrigation and second at flowering stage.
Gujarat	20 T FYM + 60: 60: 30 kg NPK/ha. N applied in two split doses as basal and at flower inititiation stage snf P and K as single basal dose in last ploughing.
Tamil Nadu	10-15 T FYM + 20: 40: 40 NPK/ha.
Andhra Pradesh	10-15 FYM + 20: 30: 40: 20 kg NPK/ha in single dose in last ploughing.
Bihar	60: 40: 20 kg NPK/ha under irrigation and apply N in two equal splits at sowing and 60 DAS.
Uttar Pradesh	60: 30: 30 NPK/ha.
Haryana	60 kg N and 37.5 kg P_2O/ha.

☆ Light irrigation to be given after sowing and second irrigation 8-10 days after when seeds start germinating, thereafter 3-5 times and at 7-8 and 15 days after sowing (at 15-25 days interval).

☆ Sowing on 15th October is the most appropriate time for reducing disease incidence.

☆ Before sowing, soak the seeds in water for 24 hours and treat with fungicide *viz.* captaf/agrosan/difoltan/bavistin/thiram @ 3g/kg of seed to reduce the wilt.

☆ Various recommendations/schedules for sowing (Table 20.13), seed rate and spacing (Table 20.14), fertilizers and manures (Table 20.15) were developed.

Table 20.12: Major Cumin Producing States with Zones/Districts

State	*High Productivity Zones/Districts*	*Agro-climatic Region*
Rajasthan	Jaipur, Jodhpur, Jalore, Barmer, Nagaur, Ajmer, Pali, Tonk, Bhilwara	Western dry region
Gujarat	Banaskantha, Mehasana	Gujarat plain and hill Region

Table 20.13: Sowing Time Recommended for Cumin

State	*Sowing Time*
Gujarat	First week of November
Rajasthan	15th–30th November
Tamil Nadu	First fortnight of November

Table 20.14: Seed Rate and Spacing Recommended for Cumin

State	*Seed Rate*	*Spacing* *Line Sowing–Row Spacing*
Gujarat	16 – 20 kg/ha	22.5 to 30 cm or 15 cm x 30 cm
Rajasthan	12 – 15 kg/ha	30 x 30 cm or 20 x 30 cm
Tamil Nadu	8 – 10 kg/ha	30 x 15 cm

Table 20.15: Fertilizer and Manure Recommendation for Cumin

State	*Recommendation*
Gujarat	30kg N and 15kg P/ha. Half N and full P+10 t compost/FYM as basal dose and the balance N given are monthly after sowing as top dressing. 25:20:20 kg/ha is also recommended.
Rajasthan	30 kg N and 20 kg P/ha. The N is given in two splits at 30 and 60 DAS and full P as basal +10t compost/ FYM at sowing.
Tamil Nadu	70:60:40kg NPK/ha 30kg N and P and K are applied as basal and the remaining N in two equal splits at 30 and 60 DAS.

* 30:30:20 kg NPK/ha *i.e.* ½ N and 30 kg P_2O_5 and 20 kg K may be added at the time of land preparation and 50 kg N broadcasted with irrigation.

Fennel

Fennel is cultivated mainly in Gujarat and Rajasthan with 98 per cent share in production (Table 20.16). Various technologies were developed for accelerating the production of fennel in the country.

- ☆ Sow in the nursery 45 to 60 days prior to the date of transplanting.
- ☆ Nursery sowing: June – July for transplanting in August – September.
- ☆ September–October for transplanting in November.
- ☆ Direct sowing by broadcasting or line sowing (45 cm apart) to a depth of 2 cm.
- ☆ For better germination, soak the seeds in water and treat with bavistin/thiram/agrosan GN @2g/kg seed.
- ☆ One or two weedings and hoeing are recommended (first weeding 30 DAS while plants are about 5 cm tall; excess plants can be removed keeping plant to plant distance of 20 cm within rows in line sown crop.
- ☆ Integrated weed management strategies may be adopted with pre-emergent weedicide pendimethalin @1.0 kg/ha, supplemented with one hand weeding at 50 days after sowing for Rajasthan and Haryana.
- ☆ For producing chewing type of fennel, umbels are to be harvested at 30-40 days after pollination, when seed size is just half of fully developed seeds.
- ☆ Recommendations on seed rate and spacing (Table 20.17), sowing time (Table 20.18), fertilizers and manures (Table 20.19) were made by different centres.
- ☆ In frost prone areas, spray with 0.1 per cent solution of sulphuric acid, irrigation of the crop prior to incidence of frost, creating smoke cover in early morning before sunshine.

Table 20.16: Major Fennel Producing States with Zones/Districts

State	*High Productivity Zones/Districts*	*Agro-climatic Region*
Gujarat	Sabarkantha, Mehasana	Gujarat plains and hill regions
Rajasthan	Sirohi, Tonk, Jodhpur, Ajmer, Bhilwara	Western dry region

Table 20.17: Seed Rate and Spacing Recommended for Fennel

State	*Direct Sowing*	*Transplanting*	*Spacing*
Rajasthan	10-12 kg/ha	3-4 kg/ha	60 x 30 cm between the rows and 20 cm with in rows
Gujarat	9-12 kg/ha	3-5 kg/ha	*Rabi*: 45 x 10 cm and *kharif*: 60 x 40 cm
Haryana	9-12 kg/ha	3-5 kg/ha	30 x 20 cm, 90 x 60 cm/100 x 60 cm is recommended for seedling transplanting

Table 20.18: Sowing Time for Fennel

State	*Sowing Time*
Rajasthan	From 1st fortnight of October to mid November for direct sowing. For transplanted crop, nursery in June and transplanting in August/September.
Gujarat	For transplanted crop, first week of September in nursery and transplanting in 2nd half of October and first fortnight of October for direct sowing.
Bihar	*Rabi*:-November1st week of September in nursery and transplanting in first week of October.
Haryana	First fortnight of October.

Table 20.19: Fertilizer Recommendations for Different States in Fennel

State	*Recommendation*
Gujarat	N 45 kg/ha, P_2O_5 3 kg/ha. For rabi crop 90 kg N, 45 kg P_2O_5, K_2O 20 kg/ha and 25 T FYM (as basal dressing). N is given in 3 splits ie. 50 per cent N and P as basal 25 per cent N at 30 DAS and balance 25 per cent at 60 DAS and for Rabi crop 90:30:00 NPK/ha.
Rajasthan	For soils of poor fertility status: 90 kg N, 40 kg P_2O_5/ha as basal and 30 kg N after 30 DAS and 30 G after 60 DA transplanting in the irrigation.
Haryana	50 kg N and 30 kg/P_2O_5. N is applied in 3 splits, 50 per cent as basal and at 30 and 60 DAS with irrigation.
Bihar	60 kg N, 40 kg P_2O_5 and 20 kg K_2O.
Uttar Pradesh	60: 50: 10 kg NPK/ha.
	15 kg N, 50 kg P/ha as basal dose followed by 15-30 kg N/ha as top dressing.

Fenugreek

Among the seed spices, fenugreek occupies the third position, next to coriander and cumin in terms of area and production. It is mainly grown in central and western India in certain specific zones (Table 20.20). Technologies developed for increasing the production and productivity of fenugreek are as follows:

- ☆ Soak the seeds in water for 6-8 hours and treat with bavistin or agrosan GN or thiram @ 3 g/kg of seed against root rot before sowing. Seeds are drilled or dibbled in line at a depth less than 5.0 cm or broadcasted.
- ☆ Treating the seeds with *Rhizobium meliloti* culture is also recommended.
- ☆ Sowing should be done 30 cm apart in rows with a plant to plant spacing of 10 cm at a depth less than 5 cm (line sowing).
- ☆ The field should be irrigated immediately after sowing. The crop requires 6-7 irrigations in light soils and 4-5 irrigation in heavy soil.
- ☆ Two weedings and hoeing, first at 10-20 days after sowing, along with thinning of the plants and second weeding at 50 days after sowing. Integrated weed management using pre-sowing application of pendimethalin @ 1kg/ha (Haryana), pre-plant application of fluchlorin @ 0.75 kg/ha in 500-600 litres of water, supplemented with hand weeding once in 50 DAS.
- ☆ For different states, different sowing time (Table 20.21), seed rate and spacing (Table 20.22) and fertilizer schedules (Table 20.23) are recommended depending on the soil and climatic conditions.
- ☆ Seeds are harvested at 45th day after anthesis, when seeds attain physiological maturity (Tamil Nadu).
- ☆ For safer storage of seeds, the moisture level should be at 7.73 per cent.

Table 20.20: Major Fenugreek Producing States with Zones/Districts

State	*High Productivity Zones*	*Agro-climatic Region*
Gujarat	Mehasana, Banaskantha	Gujarat plain and hill region
Rajasthan	Sikar, Chittorgarh, Jaipur, Pali, Jhalwar	Western dry region
Madhya Pradesh	Ratlam, Mandsaur, Neemuch, Ujjain, Rajgarh, Shajapur, Guna, Dhar, Sebore, Betul, Khargone	Central plateau and hill region
Chattisgarh	Jabalpur, Chattarpur, Indore	Central plateau and hill region
Haryana		West plateau and hill region
Maharashtra		Central plateau and hill region

Table 20.21: Sowing Time Recommended for Different States in Fenugreek

State	*Sowing Time*
Rajasthan	1st week of October to last week of November
Gujarat	Last week of September to 1st week of October
Bihar	Middle of October
Uttar Pradesh	October – November
Tamil Nadu	First week of October
Haryana	Middle of October up to November (1st week of November is ideal)
Andhra Pradesh	First week of October

* 10-15 t FYM is to be applied at the time of field preparation.

Table 20.22: Seed Rate and Spacing Recommended for Fenugreek

State	*Seed Rate*	*Spacing*
Rajasthan	20-25 kg/ha	25 cm x 10 cm row spacing
Gujarat	20-25 kg/ha	15 x 10 and 30 x 10 cm
Bihar	20-25 kg/ha	20 x 10 cm
Uttar Pradesh	15-20 kg/ha	30 x 10 cm
Tamil Nadu	–	15 x 10 cm or 22.5 x 10 cm
Haryana	–	30 x 10 cm
Andhra Pradesh	30 kg/ha	–

Table 20.23: Fertilizer Recommendation for Fenugreek

State	*Recommendation*
Rajasthan	10 FYM at the time of field preparation. 40 kg N, 40 kg P_2O_5 (20 kg N and 40 kg P_2O_5ha at the time of sowing)
Gujarat	40 kg P_2O_5/ha. 40 kg N, 25 kg P and in two equal splits as basal and top
Bihar	40 kg N, 60 kg P_2O, 20 kg K_2O
Uttar Pradesh	60: 50: 10 kg NPK/ha
Tamil Nadu	N 50 kg/ha, P 25 kg/ha and 40 kg K/ha is recommended
Haryana	40 kg/ha P

Organic Cultivation of Seed Spices

Research has been initiated to standardize agrotechniques utilizing organic products for increasing the nutrient status of soil (Table 20.24), for growth promotion (Table 20.25) and for controlling diseases in seed spices.

Table 20.24: Biofertilizer Recommendation for Seed Spices

State	*Recommendation*
	Corriander
Tamil Nadu	Inorganic nitrogen 50 per cent + *Azospirillum* (1.5 kg/ha seed treatment) + 5 t/ha FYM
Bihar	Inorganic nitrogen 100 per cent + *Azospirillum* (1.5 kg/ha seed treatment) + 5 t/ha FYM
Uttar Pradesh	*Azospirillum* (1.5 kg/ha seed treatment) + 10 t/ha FYM
Haryana	Inorganic nitrogen 100 per cent + *Azospirillum* (1.5 kg/ha seed treatment) + 5 t/ha FYM
	Fennel
Bihar	Inorganic nitrogen (100 per cent) + FYM 5t/ha + *Azospirillum*(1.5 kg/ha seed treatment)
Uttar Pradesh	*Azospirillum* (1.5 kg/ha seed treatment) + 10 t/ha FYM
	Fenugreek
Tamil Nadu	Inorganic nitrogen 100 per cent + *Azospirillum* (1.5 kg/ha seed treatment) + 5 t/ha FYM
Bihar	Inorganic nitrogen 100 per cent + *Azospirillum* (1.5 kg/ha seed treatment)+ 5 t/ha FYM

Table 20.25: Recommended Bioregulators for Seed Spices

State	Recommendation
	Corriander
Rajasthan	Two foliar sprays of 50 ppm NAA or 1.0 ml/l triacontanol at 40 and 60 DAS
Tamil Nadu	Spraying triacontanol 0.5ml/l after 40 days of sowing
	Fenugreek
Bihar	Three sprays (25,45 and 70 DAS) of GA @ 50 ppm
Rajasthan	2 sprays (40 and 60 DAS) of NAA 50 ppm
Uttar Pradesh	Spraying tricantanol 1 ml/l of water after 30, 40 and 60 DAS.

Utilization of Rhizobacteria for Growth Promotion in Seed Spices

Studies in AICRPS centres in seed spices indicate that treatment of coriander with rhizobacterial strain has enhanced growth and increased the yield. Treating rhizobacteria helps in increasing germination in seed spices. The study is continued in major seed spices growing centres.

Identification of Source of Drought Tolerance

Occurrence of abiotic stresses *viz.* salinity, drought and frost are the major hurdles for successful production.Screening of germplasm for identification of drought tolerant accessions is in progress in seed spices at various centres. Varieties suitable for drought conditions has been identified in seed spices.

Quality Evalation in Seed Spices

Seed spices with high quality is the need of the hour to meet the international demand. Several high quality varieties have been released (Table 20.8). The germplasm accessions are also evaluated for identifying high oil yielding accessions in seed spices.

Production of Leafy Type Coriander during Off-Season

Efforts are on to produce leafy type coriander at Coimbatore, Guntur and Hissar centres. The highest leaf yield of 3.18 t/ha was registered by the cultivar DH-202 followed by the check Pant Haritma (3.15 t/ha) and LCC- 234 (3.10 t/ha). At Guntur highest greens yield was recorded in LCC-244 (18.9 t/ha) followed by LCC-233 (18.5 t/ha) and LCC-234 (18.0 t/ha), which were significantly superior to check Local (9.07 t/ha).

Integrated Pest and Diseases Management

Diseases account for a major crop loss at various developmental stage of the crop in all the major seed spices. Systematic work on integrated pest and disease management to ensure quality products is being undertaken at various research centres. Occurrence of high diseases and insect-pests pressure viz, wilts, blights, rots, powdery mildews, aphids and store grain pests has resulted in poor yield and productivity. Consequently, the quality of spices is highly affected. The management strategies developed for coriander, cumin, fennel and fenugreek are given below.

Coriander

Disease

Powdery Mildew

Two sprays (250 l each) with Karathane 48EC/Dinocap/Bavistin (0.01 per cent) at disease initiation and 10 days after with carbendazim (0.1 per cent). Two sprays of wettable sulphur (@ 2.5 g/l) or dusting of sulphur @ 25 kg/ha thrice is also recommended. Ecofriendly techniques of using 5 per cent onion leaf extract as foliar spray three times can also protect the plants.

Root Rot

Drenching Brassicol or Bavistin 0.1 per cent twice at 20 days intervals.

Wilt Complex

Sow in the first and second week of November for reducing disease incidence. Seed treatment with *Trichoderma viride* (CFU 10^6 @ 4 g/kg seed and neem cake application @ 150 kg/ha following foliar spray of hexaconazone (0.05 per cent). Biocontrol agent *T. viride* in combination with foliar spray of either Hexaconazole or Thiophenate methyl on 25th, 40th and 55th DAS and soil application of neem cake @ 150 kg/ha. Use resistant varieties–Co-3, Sadhana, Surabhi. Adopt crop rotation, give deep summer ploughing, use disease free seeds. Seed dressing as well as soil drenching with carbendazim 0.2 per cent. Seed pelleting with *Trichoderma viride* (with CFU 10^6 @ 4g/kg of seed) plus neem cake application (150 kg/ha). Seed dressing with 1.5 g Bavistin + 1.5 g Thiram/kg seed and spraying twice with carbendazim 0.1 per cent.

Grain Mould

Spray 0.1 per cent carbendazim (0.1 per cent) at 20 days after flowering or captan (0.2 per cent) twice at grain set and at 15 days interval.

Stem Gall

Selection of disease free seeds, collection and destruction of affected plants, growing coriander variety RCr-41, seed treatment with Agrosan GN/Thiram (2.5g/kg) or Bavistin (1g/kg) of seed. Spraying of Bayleten or Systhane (0.4 ml/l of water) repeat after 20 days.

Pest

Aphids and White Fly

Endosulphan (0.05 per cent) twice just at the time of floral initiation and after 15 days there after with monocrotophos or quinalphos (0.1 per cent) or dimethoate (1.5 per cent) or methyl demethon (2.01 per cent) or phosphomiden (0.5 per cent) thrice @ 400-500 l/ha at 10 days interval.

Cumin

Disease

Wilt

Use healthy and disease free seeds, seed treatment with a mixture of carbendazim + captan (1:1) @ 4g/kg of seed, soil solarisation during summer months and dry summer ploughing. Organic amendments with neem cake @ 10q/ha. Integrated management by *T. harzianum* application as seed treatment as well as soil application along with spraying of topsin (0.07 per cent) or indofil (0.3 per cent) and neem oil. Three year crop rotation sequence to accommodate cumin once in three years such as cumin-clusterbean-wheat clusterbean-mustard is recommended to reduce wilt menace of cumin in Rajasthan.

Blight

The disease appears at flowering and spread rapidly in cloudy and cool weather. As a prophylatic measure seed dressing with carbendazim or captan (0.1 per cent) as well as foliar application of any fungicides such as Indofil M-45 (0.2 per cent)/blitox-50, cupramar or fytolan or blue copper or bavistin (0.1 per cent each) or copper oxychloride (0.1 per cent) at the time of flowering followed by fortnightly intervals if there is rain or heavy dew. Seed dressing with captan orthiram @ 2.5 g/kg seed accompanied by spraying the crop with mancozeb @ 0.2 per cent thrice at an interval of 15 days from 45 DAS can also be adopted.

Powdery Mildew

Adopt control measures at the initial stage of appearance of disease symptoms. Dusting of sulphur @ 25 kg/ha or spraying 400-500 Lits/ha of wettable sulphur or dinocap (0.25 per cent) or karathane LC or calaxin (0.05 to 0.1 per cent) and repeat after 15-20 days commencing from 45 days after germination.

Pest

Aphids

Endosulfan has been considered relatively safer which can be sprayed @ 0.07 per cent concentration. However careful spray of 0.03 per cent phosphamidon (100 EC) or dimethoate (0.03 per cent) or monocrotophos (0.036 per cent) @ 400-500 lits/ha also is recommended. Repeat the spray after 10 days if required

Fennel

Disease

Powdery Mildew

Spray Karathane (0.1 per cent) or wettable sulphur (0.25 per cent) once at flower initiation stage and another after 15 days. (500-800 Lits). Dusting with sulphur 300 mesh (20-25 kg/ha) also controls the disease.

Sugary Disease

Vector control by spraying with Phosphomidon (0.2 per cent) or Dimethoate (Rogar) (0.33 per cent) or Quinalphos/DDVP @ 0.05 per cent at 15 days interval (3 sprays) commencing from 50 per cent emergence of flowering to check aphids that spread the disease. Stop application of fertilizer as well as irrigation.

Pest

Gall Midge

Five sprays of the insecticide (Methylode moton 0.025 per cent, Monocrotophos –0.036 per cent, Thiometon 0.03 per cent and DDVP 0.05 per cent) at 15 days interval commencing from emergence of flowers.

Fenugreek

Disease

Root rot

Seed treatment with thiram or captan @ 2 to 3 g/kg. Application of neem cake @ 150 kg/ha combined with seed pelleting with antagonistics like *T.viride, T.harzianum* (Take formulation 4 g/kg of seed) followed by soil application of neem cake 150 kg/ha. Drenching carbendazim or brassicol (0.1 per cent) once at initial appearance and after one month. Deep summer ploughing of field and adoption of crop rotation.

Powdery Mildew

Dusting with 300 mesh sulphur (25 kg/ha) or by spraying wettable sulphur or Dinocap (0.25 per cent) or kerathan L.C. (0.1 per cent) twice at flowering stage and 15 days interval commencing from 40 DAS or spraying dinocap or karathane (0.1 per cent) @ 400-500 Lits. solution/ha.

Downy Mildew

Spray 400-500 Lits solution of 0.2 per cent Blitox/Difolatan/fitolen. Repeat the spray after 10 days if needed spray of zineb (3g/Lit) 15 days interval is also recommended.

Damping Off

Spray carbendazim (1g/Lit) cultural operation like crop rotation and removal of diseased plants. To minimise the source of infection seed treatment with benomyl (3g/kg seed) or carbendazim (3 g/kg seed) is recommended in reducing the seedling mortality.

Pest

Aphids

Spray endosulfan (0.07 per cent) or dimethoate (0.3 per cent) or 0.33 per cent phosphomidon C or 1.0 per cent malathion or monocrotophos 400-500 Lits per ha.

Planting Material Production

Production of elite/nucleus planting material/foundation seeds of improved varieties are taken up by the AICRPS centres. The centres are also executing the Integrated Programme for Development of Spices (IPDS) of Govt. of India, Dept. of Agric. And Co-operation as well as Spices Board sponsored programme for production of foundation seed of improved varieties of seed spices.

Technology Refinement and Dissimination

AICRPS has on going programmes for refining technologies evolved from various research programmes and field demonstrations so that the technology dissiminates to the end users.

Jagudan

1. Demonstration of wilt resistant variety of cumin (GC-4)
2. Demonstrations on high yielding variety of fennel(GF-11)

Jobner

1. Demonstration of the improved high yielding varieties of coriander
2. Demonstration of improved high yielding varieities of cumin and
3. Demonstration of improved high yielding varieties of fennel

Hissar

1. Demonstration of the high yielding varieties of fenugreek
2. Demonstration of the high yielding varieties of coriander

Coimbatore

1. Biocontrol of root rot of fenugreek
2. Demonstration of released varieties of coriander

Guntur

1. Demonstration of improved varieties of coriander

Impact of the Technology/Variety

Guntur

Due to the collaborative efforts of the AICRP on Spices and Department of Horticulture, A.P. Government, the state of Andhra Pradesh surpassed the national average productivity in Coriander (national average productivity 609 kg/ha) by reaching the productivity of 700 kg/ha (2002-03), which is completely from rainfed production systems. This is a great achievement on behalf of the A.P. state as Coriander is grown in other states of our country as an Irrigated-Dry crop having high productivity.

Success Stories

Coriander

Andhra Pradesh

Variety Sudha

The farmers of Kurnool district and Prakasam districts, reported that the performance of Sudha was very good in comparison to the varieties grown traditionally there. The entire area was now covered by high yielding variety Sudha (LCC-128).

Varieties Sindhu and Sadhana

Area covered by the high yielding varieties like sindhu and sadhana is around 60 per cent.

Bihar

Pant Haritima and Rajendra Swati

Since Corriander var Pant Haritima and Rajendra Swati are resistant to stem gall disease of coriander and as these varieties are of dual purpose (leafy and seed purpose 70-75 per cent coriander growing area of Bihar is with these varieties.

Gujarat

Variety GC-2 released in 1992 with high yielding capacity, better grain quality and GC-3 released in 1998 with wilt resistance have become popular among the farmers. The productivity gains at the farmers field increased from 315 kg/ha during 1992-93 to 422 kg/ha in 2001-02. Also during the same period, the area under cultivation has increased from 1.16 lakh ha to 1.45 lakh ha. Recently released variety GC-4 have high yield potential wilt resistance and better grain quality, compared to GC-2 and GC-3; which will make cumin crop still more profitable.

Fennel

Bihar

Rajendra Kanti, H. M. 444

Farmers in Bihar are replacing traditional fenugreek variety by high yielding and insect pest resistant fenugreek cultivar of Rajendra Kanti, H. M. 444. As a result about 50-52 per cent fenugreek growing area of Bihar are under cultivation with these varieties.

Gujarat

Kharif fennel variety GF-2 released in 1997, being instrumental for increasing the productivity from 1316 kg/ha (1992-93) to 1503 kg/ha (1998-99). This variety is also grown under rabi season. During 2003, variety Guj. Fennel – 11 released for rabi sowing has 12 per cent grain increase over GF-2. This will be further helped in gaining more productivity on farmers field.

Fenugreek

Gujarat

Variety G.Methi – 1 released in 1999 have helped in increasing the productivity from 1034 kg/ha to 1138 kg/ha as well as area under cultivation from 5800 ha to 7280 ha during 1992-93 to 2001-02. Thereby making the crop more remunerative.

The proposed variety GM-2 released in 2006 with its superiority in yield (1920 kg/ha), grain size and colour and tolerance to powdery mildew as well as better resistance against root rot and downy mildew diseases will provide best alternative to GM-1 and Hissar Sonali.

Future Thrust

- ☆ Breeding and selection of drought resistant/tolerant genotypes of seed spices offer the best long term solution to minimize the risk and adverse impact of limited water availability especially under rain fed production system.
- ☆ Breeding varieties of internationally acceptable quality
- ☆ Organic cultivation of seed spices
- ☆ The postharvest management of seed spices especially during storage.
- ☆ Value added products for Indian and International cuisine for increasing demand for spices and its value added products globally.
- ☆ Exploring the nutraceutical properties of seed spices and developing varieties for this trait.

Recent Advances in Seed Spices (2011) *Pages* **228–234**
Editors: **Y. Ravindrababu, R.K. Jaiman, K.D. Patel, N.R. Patel and S.B.S. Tikka**
Published by: **DAYA PUBLISHING HOUSE, NEW DELHI**

21

Development of Seed Spices in India

M. Tamil Selvan
Director, Directorate of Arecanut and Spices Development,
(Ministry of Agriculture, Government of India), Calicut – 673 005, Kerala

Spices comprise fragrant products of plant origin used for flavouring, foods and beverages. India which is often referred to as land of spices grows about 63 spices and many of them occupy sizeable area under cultivation. Among different group of spices, seed spices comprise the single largest group of spices with over 17 items coming under it.

India is the largest producer of seed spices with a production of about 5.45 lakh tonnes of seed spices annually from an estimated area of about 9.51 lakh ha. Seed spices contribute about 37 per cent and 14 per cent of total area and production of spices in the country, respectively. This group of spices has a prominent place in our national economy because of its large domestic consumption and growing demand for export. Being annual crops these are grown extensively in rotation with food crops and also as inter/mixed crops under rainfed/irrigated conditions. Seed spices are mainly cultivated in the states of Rajasthan and Gujarat with a sizeable area in the states of Madhya Pradesh, Haryana, Punjab, Uttar Pradesh, Andhra Pradesh and Bihar. The states of Rajasthan and Gujarat hold very important positions in terms of area and production of seed spices.

Though 17 seed spices are grown in the country, coriander, cumin, fennel and fenugreek are pillars of economics importance. Ajwain seed, dill seed, celery and poppy seed too contribute a minor share.

Area and production of seed spices in the country has shown significant increase in the last three decades with seasonal fluctuations. Area and production of seed spices in the country during the last two years are given in Table 21.1.

Table 21.1: Area and Production of Seed Spices in the Country during the Last Two Years

Crops	2006-07			2007-08		
	Area ('000 ha)	*Production ('000 tonnes)*	*Productivity (kg/ha)*	*Area ('000 ha)*	*Production ('000 tonnes)*	*Productivity (kg/ha)*
Coriander	321.34	232.84	725	397.51	242.84	611
Cumin	363.72	129.77	357	429.38	172.47	402
Fennel	48.47	65.51	1352	48.53	62.25	1283
Fenugreek	45.86	53.76	1172	55.17	55.45	1005
Ajawain	16.49	9.87	599	3.24	1.63	503
Celery	1.15	1.49	1,293	1.15	1.49	1,293
Dill/Poppy Seed	16.55	9.22	557	16.55	9.22	557
Total	**813.59**	**502.46**	**618**	**951.54**	**545.35**	**573**

The trend in area and production of seed spices in the country for the last twenty years shows a highly fluctuating trend. Though area and production of seed spices registered an increase during the period, but the productivity was stagnant during the same period. These fluctuations in area, production and productivity were mainly due to disease epidemics, periodic occurrence of drought and price of these commodities.

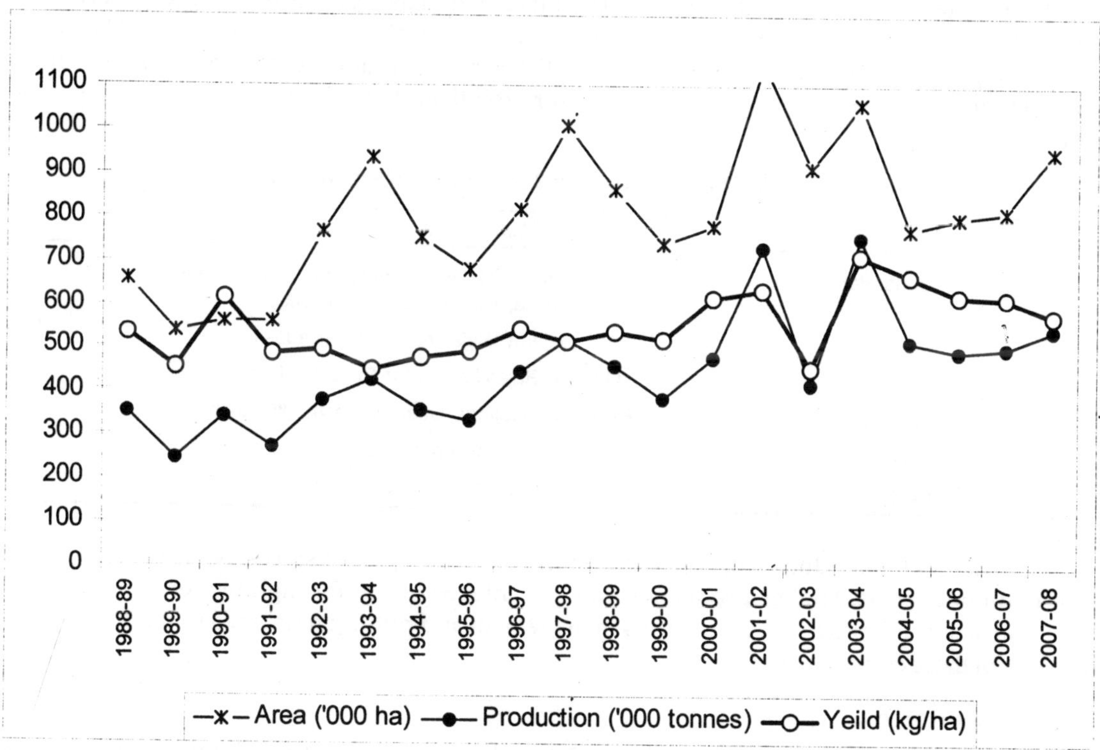

Figure 21.1: Trend in Area, Production and Productivity of Seed Spices in India

The major reasons for low productivity in seed spices are:

1. Cultivation of these crops traditionally on marginal lands with low fertility,
2. Lack of adequate scientific support for production and quality improvement,
3. Non-availability of sufficient number of high yielding varieties to suit different agro-climatic regions,
4. Lack of adoption of recommended package of practices,
5. High incidence of insects, pest and diseases,
6. Inadequate response to fertilizers,
7. Limited availability in the germplasm of seed spices,
8. Slow germination and poor stand of the crop,
9. High dependency on nature,
10. Lack of proper adoption of pre-and postharvest technology and inadequate marketing facilities.

The above reasons have been by and large static for a long time and knowledge in the case of application of fertilizers and other agronomic practices, specific techniques have been advocated and are under transfer to the farmers.

Since our country is blessed with varied agro-climatic conditions *i.e.*, tropics, sub-tropics and nearly temperate zone, India is in an advantageous position as a producer of large variety of seed spices. No other country in the world has such a broad supply base of seed spices. The climatic conditions prevailing in Rajasthan, Gujarat, Uttar Pradesh and some other adjoining states in the arid and semi-arid region are very much conducive for growth and development of wide range of seed spices.

Table 21.2: Major Production Zones of Seed Spices in India

Seed Spice	*Major States of Production*
Coriander	Rajasthan, Madhya Pradesh, Andhra Pradesh, Tamil Nadu, Orissa, Karnataka, Uttar Pradesh
Cumin	Gujarat, Rajasthan, Uttar Pradesh
Fennel	Gujarat, Rajasthan, Uttar Pradesh, Haryana
Fenugreek	Rajasthan, Gujarat, Uttar Pradesh, Uttaranchal
Ajawain	Rajasthan, Gujarat
Dil/Poppy/Celery	Rajasthan, Gujarat, Punjab

Still there is tremendous scope for increasing production of seed spices by introducing them in non-traditional areas. Higher production can be achieved easily through higher yields by better adoption of cultural practices, biotic stress management and putting more area under high yielding varieties of these crops.

Export Scenario

India exported 0.44 million tonnes of spices valued at US $1101.8 million (Rs 4435.5 crores) during the year 2007-08, to more than 150 countries around the world. Among the spices exported, seed spices accounts for 18.5 per cent of total volume and 11.4 per cent of total foreign exchange earnings by spices. During 2007-08, India exported 82100 tonnes of seed spices valued at Rs 507.75 crores. Export of various spices and spice products from India during the last three years are given in Table 21.3.

Table 21.3: Item-Wise Export of Spices from India during the Last Three Years

Spices	*2005-06*		*2006-07*		*2007-08*	
	Quantity (tonnes)	*Value (Rs in lakhs)*	*Quantity (tonnes)*	*Value (Rs in lakhs)*	*Quantity (tonnes)*	*Value (Rs in lakhs)*
Pepper	17,363	15,095	28,750	30,620	35,000	51,950
Cardamom (S)	863	2,682	650	2,236	500	2,475
Cardamom (L)	1,046	1,155	1,500	1,695	1,325	1,500
Chilli	113,174	40,301	148,500	80,775	209,000	109,750
Ginger	9,411	4,296	7,500	3,975	6,700	2,800
Turmeric	46,405	15,286	51,500	16,480	49,250	15,700
Coriander	23,756	6,771	20,500	7,462	26,000	11,025
Cumin	12,879	9,819	26,000	20,150	28,000	29,150
Celery	4,165	1,501	3,550	1,321	2,900	1,325
Fennel	5,725	2,782	3,575	2,380	5,250	2,850
Fenugreek	15,525	3,403	8,500	2,699	11,100	3,300
Other seeds (1)	12,670	3,322	8,000	2,240	8,850	3,125
Garlic	34,688	4,798	11,500	2,128	675	400
Nutmeg and Mace	1,530	3,117	2,100	4,274	1,300	2,875
Vanilla	72	1,227	125	1,996	200	1,775
Other spices (2)	21,134	7,493	19,500	7,280	19,000	8,100
Curry powder/Paste	9,340	7,838	9,500	8,693	11,500	11,100
Mint Products (3)	14,544	81,321	16,250	110,095	21,100	128,050
Oils and Oleoresins	6,074	50,557	6,250	51,079	6,600	56,300
Total	**350,363**	**262,762**	**373,750**	**357,575**	**444,250**	**443,550**

(1) Include Ajawain seed, Dill seed, Poppy seed, Aniseed, Mustard etc. (2) Include Asafoetida, Cinnamon, Cassia, Cambodege, Saffron, Spices (NES) etc (3) Include menthol, menthol crystals and mint oils.

Seed spices export shown substantial increase in both quantity and value in the recent past. Among seed spices, cumin is the major export earner followed by coriander, fenugreek, fennel, celery etc.

Production Constraints

It does not appear to have any formidable impediment in increasing the production of seed spices in the country. What actually matters is that the Central and State Governments have not given encouragement to the cultivation of seed spices as given to cash crops and food crops. Seed spices cultivation is mostly undertaken by small and marginal farmers. The cultivation practices followed by them are traditional and hence the yields obtained are low.

Development Programmes

Though the development programmes for spices have started from the beginning of the Five Year Plans, seed spices got significant place in the development front only from the VIII Five Year Plan, where Integrated Programme for Development of Spices (IPDS) was introduced with a financial outlay of Rs 102.82 million (about 10 per cent of the total spices allocation) was incurred for seed spices. The development programmes contemplated are production of nucleus seeds, establishment of demonstration cum seed multiplication plots, distribution of minikits. In the case of seed spices, even though a good number of improved varieties have been released from Research Stations, a major share of area under these crops are still with low yielding traditional varieties with poor quality attributes. Achievements under seed production porgramme assigned to different agencies were not satisfactory due to lack of adequate infrastructure and poor quality attributes for seed production, processing storage etc. However, a network of demonstration cum seed multiplication plots established in farmer's field throughout the growing tracts and minikit distribution programme have contributed to a certain extent for more area coverage under high yielding varieties and also in productivity improvement. As a result area and production of coriander, cumin, fennel and fenugreek significantly increased. However productivity of all these crops could not be increased to an appreciable level.

The nucleus seed production programme has been continued under the National Horticulture Mission (NHM), which was introduced during 2005-06, providing 100 per cent assistance to State Agricultural Universities (SAUs) for nucleus seed production programme. During the first three years of NHM (2005-06 to 2007-08) 228.75 tonnes of nucleus seeds of various seed spices were produced at various SAU centres and ICAR institutes. In 2008-09, it is expected to produce 95 tonnes of nucleus seeds of various seed spices. Assistance was also provided for establishing seed infrastructure including seed storage and processing units at various SAUs located in seed spices producing areas.

The Directorate also extended assistance for establishing frontline demonstration plots in SAUs/ Farmers Fields to disseminate the latest technologies developed in cultivation of seed spices and 14 such demonstration plots are established during the current year.

The State Department of Horticulture initiates appropriate measures for the development of seed spices and for the welfare of seed spice producing farmers. Since Agriculture is the state subject, the Centrally Sponsored Schemes under National Horticulture Mission (NHM) is directly implemented by the State Horticulture Mission (SHM). The NHM funds are being made available to the SHM for development of seed spices under the following components.

1.. Production of planting material.
   (*a*) Model Nurseries – Public and private.
   (*b*) Small Nurseries – Public and Private.
   (*c*) Upgradation of tissue culture laboratories.
2. Establishment of New Gardens – Area Expansion.
3. Integrated pest management – Integrated Nutrient management.
4. Establishment of disease forecasting unit.

5. Establishment of Bio Control Lab.
6. Establishment of leaf analysis laboratory.
7. Organic farming.
8. Establishment of vermicompost unit.
9. Organic certification.
10. Training programmes for farmers/extension workers.
11. Creation of water resources.
12. Demonstration in Farmers fields.
13. Frontline demonstration in public sector.
14. Postharvest management.

    Establishment of cold storage.

    (*a*) Establishment of Pack houses.

    (*b*) Construction of drying platforms.
15. Development of market Yards.
16. Farm mechanization and postharvest tools.
17. Transfer of Technology programmes

Eventhough the research carried out at the Agricultural Universities has resulted in a number of high yielding varieties of seed spices, they are yet to become popular among the majority of farmers as their cultivation is still confined to the neighbouring areas of the research stations. Cultural practices to be adopted under varied agro-climatic conditions have also been worked out to some extent and these recommendations are yet to reach the farming community due to poor extension support.

The crops being grown in dry areas, prolonged drought situations and erratic monsoons limit their timely sowing and adoption of other cultural practices which have profound influence on yield and quality of the produce.

Important Measures to Increase the Productivity of Seed Spices

1. The most important measure to increase the production of seed spices is stepping up of productivity through popularization of high yielding varieties and adoption of scientific cultivation practices. The Agricultural Universities in the State of Gujarat, Rajasthan, Andhra Pradesh, Tamil Nadu, Madhya Pradesh etc. have evolved a number of high yielding varieties of seed spices. Large scale multiplication of seeds of these varieties has to be taken up in the Government Farms and the seeds made available to the farmers at reasonable cost, in order to replace the existing low yielding local varieties within the shortest period.
2. Various inputs like organic manures, fertilizers, plant protection chemicals etc. required to the scientific cultivation practices should be made available to the farmers in time at reasonable cost. Adequate financial assistance backed by attractive subsidy may also be given to the needy farmers.
3. The extension machinery attached to the State Agriculture/Horticulture departments need to be strengthened for effective transfer of technology. Extension personnel have to be periodically trained to keep the abreast with the latest technology developed on crop management.

4. There had been a severe setback in the production of seed spices in the recent years, owing to continuous drought situation prevailed in the major growing states like Rajasthan and Gujarat. Water sources have to be developed to mitigate the problems created by such recurring drought and erratic monsoon. Long term loans have to be extended to create irrigation facilities.
5. Research has to be intensified so as to evolve more high yielding varieties with pests and disease tolerance/resistance and better quality attribute, and also to find out and suggest effective control measures against pests and diseases and agronomic practices to be adopted under varied agro climatic situations.

Recent Advances in Seed Spices (2011) *Pages* **235–238**
Editors: **Y. Ravindrababu, R.K. Jaiman, K.D. Patel, N.R. Patel and S.B.S. Tikka**
Published by: **DAYA PUBLISHING HOUSE, NEW DELHI**

22

Postharvest Handling of Seed Spices

G. Lal and M.M. Anwar
National Research Centre on Seed Spices, Tabiji, Ajmer, Rajasthan

ABSTRACT

India "The Home of Spices" is the leading country in production, consumption and export of seed spices and more than 80 per cent of total production is contributed by Rajasthan and Gujarat. Coriander, cumin, fenugreek and fennel are considered to be major seed spices. Improper postharvest handling leads to sever loss in quantity as well as quality of valuable seed spices. Contaminations at various stages during harvesting, threshing, packaging, transportation, storage etc. are the main causes of quality deterioration. Since the seed spices are vital adjuncts to food and drugs, the quality and purity have to be of the highest level. The aspects related to quality begin with the harvesting of raw spices and reached to the consumption at consumers end. During this the produce is to be travelled through various stages of handling mentioned above. For high quality spices, their products and for market acceptance, it is mandatory to encounter the various factors *viz.*, proper variety, crop management practices (especially nutrient, weed and water management), harvesting stage, best drying, cleaning, grading, packaging and storage practices. Efficient management of raw material at different stages is very necessary for up gradation of quality standards. Use of modern instrumentation will lead to preparation of various products like spice powder, spice paste, liquid spices, spice blends, concentrations, essential oils and oleoresins.

Introduction

Indian is known for its spices since ancient times. India is the world's largest producer, consumer and exporter of seed spices. About 8.3 lakh hectare area is grown in seed spices in the country of which the major share of about 5 lakh hectare is covered in Rajasthan alone. Total production of seed spices in the country 6-7 lakh tones annually. The climatic condition prevailing in Rajasthan, Gujarat and other adjoining states in the arid and semi-arid region are very much conducive for growth and development of wide range of seed spices. The major production area of seed spices are Rajasthan, Gujarat, Haryana, Maharashtra, Punjab, Tamil Nadu, M.P., U.P. etc. Major seed spices of India are coriander, cumin, fenugreek and fennel which occupy large area. The minor seed spices consist of dill, ajawain, celery, nigella and aniseed. These are used as test enhancer, flavouring agents, aromatics, appetizer, digestive, antiflatulents and other beneficial effects to human health.

Seed spices are important export oriented commodities and about 12 per cent of the total spices production (75,100 tone/year) is exported in raw as well as value added products realizing foreign exchange worth Rs. 208.36 crores. In the export markets besides quality whole seed spices, the demand for essential oil and oleoresin is also increasing.

Quality standards of seed spices as per the requirement of importing countries are to be maintained. Our growers, traders and exporters are still not so concerned about the different quality parameters and how to achieve them in a proper fashion? To obtain the quality produce from the field and to maintain it till the produce reach to the consumer, there is a need to harvest the spices at right time, right maturity and by a right method and careful handling is necessary during different postharvest operations.

Harvesting

- ☆ Harvesting is one of the major important factors that determine the quality of the produce.
- ☆ The stage of harvest varies from crop to crop.
- ☆ Improved method of harvesting at proper stage offers quality produce as well as overcomes the problem of contamination.
- ☆ Harvesting should take place between half to two third of the grains are ripe.
- ☆ To minimize the breakage, the plants should be cut during the early morning or in the late evening.
- ☆ The whole plant is removed from the ground and dried in the field.
- ☆ After harvest, the harvested crop put in open yard under sun to attain the optimum moisture level for threshing.
- ☆ Different types of harvest equipments for seed spices-
  - (*a*) Self propelled reaper.
  - (*b*) Power tiller operated reaper

Threshing

- ☆ Adoption of mechanized threshing can improve the quality and reduced the postharvest losses.
- ☆ The seeds are obtained by threshing/beating the dried plants with sticks.
- ☆ Some threshing equipments suggested by Stressors (1995) can be used for proper threshing of seed spices.
  - (*a*) Paddle Thresher
  - (*b*) Axial flow thresher
  - (*c*) Groundnut Stripper

Drying

- The moisture content of seed spices is normally greater than 20 per cent at the harvest, which is reduced to 14 per cent for threshing.
- Drying under direct sunlight affects the volatile oil content.
- Mechanized drying could enhance the quality and reduce the postharvest losses.
- Green house drying may also be a good option for drying the seed spices.
- Some drying systems for seed spices crops (Roberts, 1998) are:
  - (*a*) IGSI Grain dryer
  - (*b*) Portable batch dryer
  - (*c*) Tray dryer
  - (*d*) Solar cabinet dryer

Cleaning/Grading

- The types, shapes, characteristic of all impurities varied in nature cannot be separated by one process or by one machine (Pruthi, 1987)
- The cleaning/grading operation is done at Mandi level and at the level of processors.
- The cost of cleaning at Mandi level is born directly or indirectly by farmers. The uniform size of seed spices is necessary for processing.
- The whole dried seeds are usually packed into sacks and stored in a cool dry place.
- The seed spices can be stored in jute or poly bags.
- The fennel can be stored for one year but for coriander, it can be advocated at farmers level to avoid distress sale.
- The efficient grinding of spices prevents changes with respect to flavor and pungency.
- During conventional grinding *i.e.*, grinding without cooling the spices, the temperature increases to over 90° C can occur in fast rotating mills which leads to substantial losses in volatile oils which leads to loss in quality, an indirect water cooling system has been developed to debussed the loss (Singh, and Goswami, 1999)

Storage/Packing

- Farmers generally sell seed spices immediately in the same season.
- Traders and processors do the storage for further processing for value addition. The whole dried seeds are usually packed into sacks and stored in a cool dry place.
- The seed spices can be stored in jute or poly bags. The poly and cotton bags help to retain volatile compounds compared to jute bags.
- The fennel can be stored for one year but coriander is recommended to store for six months only. Since storage of seed spices is simple, it can be advocated at farmer's level to avoid sale during harvesting or glut period.

Value Addition at Processors Level

The value added products from seed spices at the level of processors are:

1. Ground spices
2. Spices extraction
3. Curry power

Ground Spices

- ✰ The efficient grinding of spices prevents changes with respect to flavor and pungency.
- ✰ During conventional grinding that is grinding without cooling the spices, the temperature increase to over 90° C, can occur in fast rotating mills which leads to substantial losses in volatile oils consequently it reduces the quality (Schaub, 1995).
- ✰ In an effort to decrease the loss in volatile oils by using cooling agents, an indirect water-cooling system has been developed.

Spices Extraction

- ✰ Spices extracted compound serve as the alternate to whole and ground spices.
- ✰ The Essential oil and oleoresin are principal extracted compounds for value added modules.
- ✰ The essential oil is the aromatic, volatile compounds present in seed spices.This oil content varies from crop to crop.
- ✰ In coriander seed it ranges from 0.1 to 4 per cent (Agrawal *et al.*, 2002) in cumin seeds 2.3 to 4.8 per cent, in fennel 1.5 to 1.8 per cent.

Curry Powder

- ✰ Curry powder is a ready- made mix of various spices in ground form. The number of spices varies from 5 to 20 depending upon the producer and its end use (Rumpf, 1990).
- ✰ The ingredients of curry change with the need for different foods. Curry powder is used to flavor foods and coriander is as a major ingredient in curry powder where it can be as much as 24 per cent. Coriander is also used to a large extent in the chemical industry in Russia.
- ✰ The export trade in curry powder at present is dominated by India.

References

Agrawal, S., Divakara Sastry, E.V. and Sharma, R.K., 2002. *Seed Spices: Production, Quality and Export*. Pointer Pub., Jaipur, p. 1–315.

Pruthi, J.S., 1987. Cryo milling: The latest technology of cold grinding of spices condiments. *Proceedings of Symposium, AFST(I)*, New Delhi, p. 38.

Roberts, A.W., 1998. Particle technology: Reflections and horizons: An engineering perspective. *Trans. Ichems E*, 76A: 775–796.

Rumpf, H., 1990. *Particle Technology* (English edition). Powder Technology Series, London, Chapman and Hall.

Schaub, 1995. Feinmahlen mi dem cryogen kaltmahlverfahren. *Gasaktuel*, 45: 19–24.

Singh, K.K. and Goswami, T.K., 1999. Design of a cryogenic grinding systems for spices. *Journal of Food Engineering*, 39(44): 359–368.

Stressors, 1995. Aktueller Stand der Mahltechnik. *KHD Symposium*, pp. 15–28.

Recent Advances in Seed Spices (2011) *Pages* **239–243**
Editors: **Y. Ravindrababu, R.K. Jaiman, K.D. Patel, N.R. Patel and S.B.S. Tikka**
Published by: **DAYA PUBLISHING HOUSE, NEW DELHI**

23

Characterization and Evaluation of Indigenous Dill Seed (*Anethum sowa* Kurtz) Germplasm under North Gujarat Conditions

D.B. Prajapati, K.P. Patel, M.M. Patel and Y. Ravindrababu
Centre for Research on Seed Spices, Sardarkrushinagar Dantiwada Agricultural University, Jagudan – 382 710, Dist. Mehsana, Gujarat

ABSTRACT

Indian Dill (*Anethum sowa* Kurtz) is one of the important minor seed spice crops grown under irrigated as well as unirrigated condition in Gujarat, Rajasthan, M.P., J and K, Orissa and Punjab. Another species of Dill is European Dill (*Anethum graveolens* L.) is spread in limited extend in Gujarat. Indian Dill a native India has huge genetic variability for various traits. The germplasm consisting of 48 genotypes (Released varieties and local entries) collected from different districts of Gujarat were studied based on eight quantitative and two qualitative traits. Wide variability existed among the genotypes for all the morphological and economic traits.

Days to 50 per cent flowering (72–87), Days to Maturity (146–163), Plant height (106–153 cm), Number of umbellate (16.2–24.9), Number of seeds (13.8–28.7) and yield/net plot (349–449 gm) are highly variables based on the variance. The correlation coefficients of the grains yield and some yield attributing characters *viz.*, Number of branches, Number of umbels, Number of umbellate and 1000 seed weight revealed that high positive correlation. Number of seeds showed high positive correlation with Number of umbels and Number of umbellate. Promising lines identified performing better than the Gujarat Dill-1, check variety.

Introduction

Indian Dill (*Anethum sowa* Kurtz) is one of the important minor seed spice crops grown under irrigated as well as unirrigated condition in Gujarat. Dill seed is the native of India and Europe and till today it occurs in natural from at many places. The major dill seed producing countries are India, Germany, Hungary, Netherlands, Pakistan and United States. In India, Gujarat, Rajasthan, M.P., J&K, Orissa and Punjab are the leading states in respect of area and production. The districts of Mehsana, Banaskantha, Patan, Surendranagar and Ahmedabad are leading in Gujarat, where it is cultivated in scattered pockets. The dill somewhere raised by farmers on bunds of the main crops, hence lacking the real statistics. The area, production and productivity of dill seed in Gujarat during 2005-06 were 4.83 hectares, 3.82 tonnes and 790 kg/ha, respectively (DAO, 2005). The share of Gujarat in area and production is 52 per cent and 68 per cent, respectively towards total area and production of country (2005-06). Seed of dill are largely consumed within the country and still a small quantity is exported to the western countries. On average India exports dill seed worth of rupees 20 millions.

Dill seed works as mild stomach medicine for children. Seed of dill is rich in carvone, which is popularly used as carminative, aromatic, stimulant and diuretic in ayurvedic and unani medicines. Decoction prepared from dill seed is given to women after delivery as milk stimulating agent. Dill seed is also fed to animals after calving. Seed of dill seed yield 2.5 to 3.5 per cent essential oil rich in carvone, which imparts to oil its characteristics flavor. Thus, both seed and oil used in medicine. Preparation of dill water and similar products are used in flatulent, abdominal and colic pain.

Since last three years there is drastically decrease in area and production in India. There are some causes for declining the crop status. Obviously, having shattering habit of the grains, non-synchronous maturity of crop, as delaying in harvesting of early umbels automatically and shattered on plot, next year the seed grows as weeds deteriorates the remaining crop.

Indian dill has huge genetic variability for existed among the genotypes for all the morphological and economic traits. So attempts were made to evaluate the germplasms with characterization for all characters which is prerequisite for the utilization of the available crop diversity in the crop improvement programme. Hence, these collections were characterized to assess the variability and identify the promising accessions for different traits and may useful for solving the above burning problems.

Materials and Methods

The preliminary characterized and evaluation of 48 genotypes (Released varieties and local entries) of dill germplasm collections were carried out at the Centre for Research on Seed Spices, Sardarkrushinagar Dantiwada Agricultural University, (23052' N, 72043' E, 70 m MSL), Jagudan, Gujarat during 2005-06 and 2006-07.The accessions originated from different corner of state mainly from Saurashtra, Ahmedabad, Patan and Mehsana. These are tall, dwarf, shattering habit, late, early, solid stem, high yielding and local land race are the type of materials.

The accessions were grown in an augmented Block design (ABD) with check varieties (Guj.Dill-1, Guj.Dill-2, Guj. Dill-3) in each block with two rows of 4 meter length with row to row spacing of 90 cm and plant to plant spacing was 30 cm. All the recommended crop protection measures were followed timely to control insect pest for successful raising the crop. A uniform fertilizer dose of 20 kg N and 40 kg P_2O_5 ha^{-1} was applied and normal agronomic practices were followed. Observations were taken on five randomly selected plants from each of the replications for quantitative and qualitative characters as per minimal descriptor developed by NRCs, Malhotra and Vashishtha, (2008). Mean data from each of the replications werc used to analyzed statistically.

Results and Discussion

The preliminary characterized and evaluation of 48 genotypes (Released varieties and local entries) of dill germplasm collections were carried out for 14 agro-morphological and quality characters. Wide range of variability was recorded in those characters. Similarly, Joshi *et al.* (1963) reported the germplasm of European dill had marked variability for yield potential. Majority of accessions showed shattering habit, non lustrous seeds or brown colors and presence of stalk of the fruit. However, variation was observed in height, No. of umbles, umbellate, seeds wt., 1000 seeds wt. and yield are highly variables.

The quantitative characters also showed wide variation in the evaluated dill germplasm.The range of variability and frequency observed in qualitative traits are given in Table 23.1; those are determining grain quality and economic traits of the crops. The results of descriptive statistical analysis are presented in Table 23.2. The characters of Days to 50 per cent flowering (72-87), Days to Maturity (146-163), Plant height (106-153cm), Number of umbellate (16.2-24.9), Number of seeds (13.8-28.7) and yield/net plot (349–449 gm) are highly variables based on the variance.

Table 23.1: Characterization and Primary Evaluation of Dillseed

Characters	*Variation*	*No. of Accessions*
Seed size	Bold	8
	Medium	22
	Small	18
Seed colour	Brown	20
	Dark Brown	12
	Light Brown	10
	Dull Brown	06
Presence of stalk of fruit after threshing	Absence	09
	Presence	39
Type of stem	Hollow.	40
	Solid	08
Shattering habit	Absence	12
	Presence	36
Fruit type	Spitting type	38
	Non-Spitting type	06
Fruit shape	Ellipsoidal	21
	Ovate	18
	Cylindrical	09

Table 23.2: Statistical Analysis of Quantitative Characters

Sl.No.	Characters	Range	Min	Max	Mean	SD	Variance
1.	Days to 50 per cent flowering	15	72	87	79	3.81	14.52
2.	Days to Maturity (Early)	17	146	163	157	4.22	17.81
3.	Plant height(cm) (Dwarf)	47	106	153	132	10.91	119.03
4.	No. of Branches/Plant (More branches)	2.3	4.7	7.0	5.7	0.48	0.23
5.	No. of Umbels/Plant (More umbels)	12.1	5.9	18.0	10.5	3.49	12.18
6.	No.of Umbellates/Umbel (More umbellates)	8.7	16.2	24.9	20.8	1.76	3.10
7.	No. of Seeds/Umbel (More seeds)	14.7	13.8	28.7	24.7	2.45	6.00
8.	1000 seed weight (gm)	1.1	3.5	4.6	4.2	0.27	0.07
9.	Grain Yield (gm/plot) (High yield)	199	250	449	349	53.46	2857.97

Table 23.3: Correlation Coefficient of Yield Attributing Characters

Sr No	Entry	50% Flowering	Maturity	Plant ht. (cm)	Br/Plant	Umb/ Plant	Umbellate/ Umbel	Seeds/ Umbellate	Test wt. (gm)	Yield/ Plot (g)
1.	50% Flowering	1								
2.	Maturity	0.6251**	1							
3.	Plant ht	0.3874**	0.608**	1						
4.	Br/Plant	–0.025	0.341	0.417**	1					
5.	Umb/Plant	–0.368**	–0.379**	–0.455**	–0.783	1				
6.	Umbellate/ Umbel	0.0192	–0.117	–0.158	0.058	0.078	1			
7.	Seeds/ Umbellate	0.1971	0.482	–0.042	0.031	0.302**	0.360**	1		
8.	Test wt. (gm)	–0.230	–0.192	–0.262*	–0.010	0.192	0.173	–0.148	1	
9.	Yield/Plot (g)	–0.022	–0.408	–0.106	0.306**	0.058	0.115	–0.084	0.141	1

*: Singnificant at 1 per cent level

**: Singnificant at 5 per cent level.

The correlation coefficients of the grains yield and some yield attributing characters *viz.*, number of branches, number of umbels, number of umbellate and 1000 seed weight revealed that high positive correlation. Number of seeds showed high positive correlation with number of umbels and number of umbellate. Promising lines identified performing better than the Gujarat Dill-1, check variety are listed in Table 23.4.

Table 23.4: Promising Accessions of Dill Identified from Evaluation

Sl.No.	*Character*	*GD-1 (Ch)*	*Details*	*No. of Entries*	*Promising Entries Identified*
1.	Days to 50 per cent flowering	83	< 73	3	JD-95-73, JD-02-18, Jamanagar Local, JD-95-161
2.	Days to Maturity (Early)	156	<151	3	JD-95-79, Jamanagar Local, JD-95-161
3.	Plant height(cm) (Dwarf)	126	< 117	4	JD-95-29, JD-95-38, JD-02-09, JD-95-161
4.	No. of Branches/Plant (More branches)	5.7	> 6	3	JD-02-06, JD-02-09, JD-95-161
5.	No. of Umbels/Plant (More umbels)	22.8	>16	4	JD-95-49, JD-02-93, JD-95-43, JD-95-161
6.	No.of Umbellates/Umbel (More umbellates)	23.0	>23	3	JD-95-29, JD-95-40, JD-95-161
7.	No. of Seeds/Umbe (More seeds)	29.5	>28	3	JD-95-40, JD-95-41, JD-95-122
8.	1000 seed weight (gm)	3.7	>4.5	2	JD-95-1, JD-95-15
9.	Grain Yield (gm/plot) (High yield)	1062 Kg/ha	>450	5	JD-95-38, JD-02-09, JD-02-19, JD-95-15, JD-95-161

Conclusion

The accession JD-95-161 found better in respect to grain yield, grain quality and non-shattering and non-spitting habit of grain. The plant type is early; synchronize in maturity, dwarf plant nature and solid stem. For strengthening genetic resources of dill the collections from the indigenous sources may not help creation of adequate variability. Hence, it is suggested that systematic efforts should be initiated immediate to procure European dill or exotic materials from the centre of origin or from other area of world to enrich our indigenous germplasms for proper population and further hybridization programme.

References

DOA, 2009. *District-Wise Area and Production of Seed Ṡpices in Gujarat*. Directorate of Agriculture, Govt. of Gujarat, Gandhinagar.

Malhotra, S.K. and Vashishta, B.B., 2006. *A Manual on Minimal Descriptors of Seed Spices*. National Research Centre on Seed Spices, Ajmer, Rajasthan.

Joshi, B.S., Ramanujam, S. and Saxena, M.B.L., 1963. Improvement of some essential oil bearing spices. *Plants. Bull. Res. Lab., Jammu*, 1: 94–100.

Recent Advances in Seed Spices (2011) *Pages* **244–249**
Editors: **Y. Ravindrababu, R.K. Jaiman, K.D. Patel, N.R. Patel and S.B.S. Tikka**
Published by: **DAYA PUBLISHING HOUSE, NEW DELHI**

24

Association of Characters and their Direct and Indirect Contribution for Fresh Yield in Fenugreek (*Trigonella foenum-graecum* L.)

Yudhvir Singh, Sunil Dutt and Madhu Sharma
Department of Vegetable Science and Floriculture,
CSK HPKV, Palampur, Himachal Pradesh

ABSTRACT

Thirty-eight genotypes of fenugreek (*Trigonella foenum-graecum* L.) on evaluation at experimental farm of Department of Vegetable Science and Floriculture, CSKHPKV Palampur, revealed significant differences for different characters *viz.*, plant height, number of branches per plant, number of leaves per plant, days to green harvesting, Vitamin C, chlorophyll, calcium, moisture content and fresh yield. Heritability in broad sense was high for plant height, number of leaves per plant, days to green harvesting, moisture content and fresh yield and moderate for number of branches per plant. Plant height, number of leaves per plant, number of branches per plant exhibited significant positive correlation with fresh yield while days to green harvesting exhibited negative significant correlation with fresh yield. Path coefficient analysis revealed that plant height, number of leaves per plant, number of branches per plant, moisture content and chlorophyll content were the most important traits as they exhibited direct positive effects on fresh yield and thus would serve as good selection indices to attain maximum fresh yield.

Introduction

Fenugreek (*Trigonella foenum-graecum* L.) belonging to the family Fabaceae, is an important leafy vegetable as well as an important spice. It is also called as Bird's Foot and Greek hay. It is the third largest seed spice in India after coriander and cumin. Fenugreek leaves are rich in minerals, protein, vitamin A and vitamin C. Its seeds substantially contain 'Diosgenin' which is used as a starting material in the synthesis of sex hormones. Indian women use the seeds of fenugreek for its power to promote lactation.

For effective improvement in its herbage and seed yield potential, the knowledge of the magnitude of variation, heritability and genetic advance is a pre-requisite for planning a sound breeding programme based on selection techniques. The presence of variability in any crop is the basis for all crop improvement expected by selection. Improvement in any population depends on the amount of genetic variability present in the population; wider the genetic variability greater the scope for selection. Further, the association analysis of yield with its component traits and partitioning of components of yield into direct and indirect effects is of paramount importance in planning of an effective breeding programme.

The phenotypic correlation indicates the extent of the observed relationship between two characters while genotypic correlation provides an estimate of inherent association between genes controlling any two characters. Genotypic correlations in particular are thus helpful in the construction of selection indices and permit the prediction of correlated response. They further permit the evaluation of the relative influence of various characters on yield. However, merely relying on association between various characters may not always lead to fruitful results, as unfavourable associations between desired characters under selection may result in genetic slippage (Dickerson, 1955). Path analysis provides more reliable information on favourable and unfavourable associations among the characters.

In view of the lack of adequate information on fresh yield components in fenugreek (*Trigonella foenum-graecum* L.) an investigation was undertaken to analyze the variability, heritability, genetic advance and correlation coefficients among different characters.

Materials and Methods

The material used in the present study consisted of 34 lines and 4 released standard varieties *viz.*, Pusa Early Bunching, Hissar Sonali, Rajindra Kranti and Rmt-1. All the 38 genotypes were grown at research farm of department of Vegetable Science and Floriculture, CSK HPKV, Palampur, Himachal Pradesh, during 2002-03, in a completely randomized block design with 3 replications. The seeds were sown at row to row and plant to plant spacing of 30 and 4-5 cm, respectively. Ten random competitive plants were used to record observation on nine characters *viz.*, plant height, number of branches per plant, number of leaves per plant, days to green harvesting, moisture content, fresh yield, vitamin-C, calcium and chlorophyll content. The mean values were used to determine the range mean, phenotypic and genotypic coefficient of variation heritability and genetic advance. The analysis of variance was calculated according to Panse and Sukhatme (5). The phenotypic and genotypic coefficients of variation were calculated according to Burton and Devane (2). The broad sense heritability was computed according to the method of Johnson *et al.* (4). The extent of expected genetic advance as percentage of mean was calculated by the method described by Burton and Devane (2).

The genotypic and phenotypic correlations were calculated by the method described by Al-Jibouri *et al.* (1) and path coefficient analysis as per Dewey and Lu (3).

Results and Discussion

The analysis of variance indicated that there was significant variation for all the characters under investigation. To understand the nature of observed variability in the germplasm the partitioning was done into genotype and environmental components. The results clearly indicated that genotypic coefficients of variation were of higher magnitude than the phenotypic coefficients of variations for all the characters under study indicating higher genetic variability and little influence of environment on these characters. The estimates of PCV and GCV were moderate for plant height, number of leaves per plant, fresh yield, calcium content and Vitamin-C, however these were high for chlorophyll content which indicated that there was substantial variability ensuring ample scope for improvement through selection.

Table 24.1: Analysis of Variance for Nine Characters in Fenugreek

Characters	Mean Sum of Squares		
	Replications (df=2)	Treatments (df=37)	Error (74)
Plant height (cm)	1.21	39.63*	74
Number of branches per plant	0.10	0.59*	0.18
Number of leaves per plant	0.47	66.54*	1.47
Days to green harvesting	3.07	117.10*	2.01
Moisture content (per cent)	0.87	10.26*	0.69
Fresh yield (q/ha)	1.56	432.80*	0.65
Vitamin C (mg/100 g)	0.57	143.80*	1.39
Chlorophyll (mg/g)	0.000028	0.018*	0.00032
Calcium (mg/100 g)	3.50	8640.40*	1.75

* Significant at 5 per cent level.

Table 24.2: Range, Mean±, Phenotypic and Genotypic Coefficients of Variation, Heritability and Genetic Advance for Different Characters in Fenugreek

Characters	Range	Mean±SE	Coefficient of Variation		Heritability (Per cent)	Genetic Advance as Percentage of mean
			Phenotypic	Genotypic		
Plant height (cm)	19.05-39.22	32.51±0.55	11.43	11.05	93.4	21.99
Number of branches per plant	2.93-5.03	4.02±0.24	14.10	9.09	41.6	12.19
Number of leaves per plant	29.00-56.35	42.23±0.70	11.39	11.03	93.6	21.97
Days to green harvesting	46.33-87.66	52.68±0.82	12.06	11.76	95.0	23.61
Moisture content	78.95-87.66	83.67±0.48	2.35	2.13	82.0	3.97
Fresh yield (q/ha)	36.80-85.62	64.52±0.47	18.64	18.60	99.5	38.24
Vitamin C (mg/100 g)	42.62-67.52	52.41±0.69	13.34	13.15	97.1	26.69
Chlorophyll (mg/g)	0.07-0.38	0.18±0.010	42.68	41.49	94.5	83.33
Calcium (mg/100 g)	205.33-408.25	296.44±0.77	18.11	18.10	99.9	37.28

High heritability alongwith high estimates of genetic advance were observed in case of fresh yield, chlorophyll and calcium content. While, high heritability alongwith moderate genetic advance was observed for plant height, number of leaves per plant, days to green harvesting and Vitamin-C which indicated the presence of additive gene effects. In the present study, the estimates of genotypic

Table 24.3: Estimates of Phenotypic (P) and Genotypic (G) Correlation Coefficients for Fresh Yield in Fenugreek

		Plant Height (cm)	No. of Branches/ Plant	No. of Leaves/ Plant	Days to Green Harvesting	Fresh Yield (q/ha)	Moisture (per cent)	Vitamin C (mg/100 g)	Chlorophyll (mg/g)	Calcium Content (mg/100 g)
Plant height (cm)	G		0.586	0.352	–0.647	0.662	0.099	0.126	–0.046	–0.085
	P		0.403**	0.328*	–0.607**	0.639**	0.082	0.122	–0.054	–0.081
No. of branches/plant	G			0.614	–0.329	0.649	0.273	0.094	0.165	–0.239
	P			0.440**	–0.192	0.420**	0.170	0.046	0.078	–0.154
No. of leaves/plant	G				–0.102	0.367	0.057	0.104	–0.007	–0.038
	P				–0.105	0.359*	0.038	0.101	–0.006	–0.038
Days to green harvesting	G					–0.365	0.319	–0.072	0.259	0.026
	P					–0.354*	0.268	–0.075	0.241	0.025
Fresh yield (q/ha)	G						0.191	0.131	0.156	–0.044
	P						0.167	0.127	0.152	–0.044
Moisture (per cent)	G							–0.081	0.157	–0.177
	P							–0.070	0.118	–0.061
Vitamin C (mg/100 g)	G								0.413	–0.061
	P								0.394*	–0.061
Chlorophyll (mg/g)	G									0.157
	P									0.151
Calcium (mg/100 g)	G									
	P									

*, **: Significant at P = 0.05 and P = 0.01, respectively.

Table 24.4: Estimate of Direct and Indirect Effects of Different Traits on Fresh Yield at Phenotypic (P) and Genotypic (g) Levels

		Plant Height (cm)	*No. of Branches/ Plant*	*No. of Leaves/ Plant*	*Days to Green Harvesting*	*Vitamin C (mg/100 g) (q/ha)*	*Chlorophyll (mg/g)*	*Calcium (mg/100 g)*	*Moisture (per cent)*	*Correlation Coefficient with Fresh Yield (q/ha)*
Plant height (cm)	G	0.465	0.247	–0.018	–0.027	0.002	–0.003	–0.008	0.003	0.662
	P	0.500	0.045	0.045	0.056	–0.003	–0.010	–0.001	0.009	0.639**
No. of branches/plant	G	0.273	0.422	–0.032	–0.014	0.002	0.012	–0.021	0.007	0.649
	P	0.201	0.111	0.060	0.018	–0.001	0.015	–0.001	0.018	0.420**
No. of leaves/plant	G	0.164	0.259	–0.051	–0.004	0.002	0.000	–0.003	0.002	0.367
	P	0.164	0.049	0.137	0.010	–0.003	–0.001	0.000	0.004	0.359*
Days to green harvesting	G	–0.301	–0.139	0.005	0.042	–0.001	0.018	0.002	0.008	–0.365
	P	–0.303	–0.021	–0.014	–0.091	0.002	0.046	0.000	0.028	–0.354*
Vitamin C (mg/100 g)	G	0.059	0.040	–0.005	–0.003	0.019	0.029	–0.005	–0.002	0.131
	P	0.061	0.005	0.014	0.007	–0.027	0.075	–0.001	–0.007	0.127
Chlorophyll (mg/g)	G	–0.021	0.069	0.000	0.011	0.008	0.071	0.014	0.004	0.156
	P	–0.027	0.009	–0.001	–0.022	–0.011	0.190	0.001	0.012	0.152
Calcium (mg/100 g)	G	–0.040	–0.101	0.002	0.001	–0.001	0.011	0.089	–0.005	–0.044
	P	–0.040	–0.017	–0.005	–0.002	0.002	0.029	0.010	–0.018	–0.043
Moisture (per cent)	G	0.046	0.115	–0.003	0.013	–0.002	0.011	–0.017	0.027	0.191
	P	0.041	0.019	0.005	–0.025	0.002	0.022	–0.002	0.104	0.167

*, **: Significant at P = 0.05 and P = 0.01, respectively. Residual effect: P = 0.050, G = 0.037; Bold values indicates direct effects.

correlations, in general are slightly higher than their phenotypic correlations. This indicated that there is a strong inherent association between characters studied and there is a little influence of environment on the phenotypic expression.

Fresh yield had significant and positive correlation with plant height, number of branches per plant and number of leaves per plant at phenotypic level. Among quality characters, Vitamin-C was found significantly and positively correlated with chlorophyll content. Days to green harvesting was significantly and negatively correlated with fresh yield. This indicates that the genotypes which took less days for green harvesting gave higher yield and thus can be exploited for fresh yield.

Path analysis is useful in partitioning the correlation coefficients into direct and indirect examination of specific forces acting to produce a given correlation and measure the relative importance of each causal factor. Hence, such studies provide a realistic basis for allocation of weightage to each attribute in deciding suitable criterion for selection (Sandhu *et al.*, 1979; Singh *et al.*, 1968). Path analysis for fresh yield at phenotypic level revealed that plant height had the maximum positive direct effect. Other major direct positively contributing traits were number of leaves per plant, number of branches per plant, chlorophyll content and moisture content. The negative direct effect of days to green harvesting was counterbalanced by its positive indirect effects via chlorophyll content and moisture content.

Thus, the present study revealed that the selection in fenugreek for fresh yield should be effectively based on plant height, number of leaves per plant number of branches per plant and days to green harvesting for maximum yield with high chlorophyll content.

References

Al-Jibouri, H.A., Millar, P.A. and Robinson, H.P., 1958. Genotypic and environmental variances and covariances in an upland cotton cross of interspecific origin. *Agronomy Journal*, 50: 633–638.

Burton, G.W. and Devane, E.H., 1953. Estimating of heritability in tall fescue (*Festuea arundinacea*) from replicated clonal material. *Agron. J.*, 45: 478–81.

Dewey, J.R. and Lu, K.H., 1959. A correlation and path analysis of components of crested wheat-grass seed production. *Agronomy Journal*, 51: 515–518.

Johnson, H.W., Robinson, H.F. and Comstock, R.E., 1955. Estimates of genetic and environmental variability in soybean. *Agron. J.*, 47: 477–83.

Panse, V.G. and Sukhatme, P.V., 1967. *Statistical Methods for Agricultural Workers.* ICAR, New Delhi.

Sandhu, T.S., Bhullar, B.S., Chemma, H.S. and Gill, A.S., 1979. Variablity and interrelationship among grain protein, yield and yield componenets in mungbean. *Indian journal of Genetics*, 39: 480–483.

Singh, S.P., Singh, B.H., Mishra, S.N. and Singh, A.B., 1968. Genotypic and phenotypic correlations among some quantitative characters in mungbean. *Madras Agricultural Journal*, 44: 233–237.

Recent Advances in Seed Spices (2011) *Pages* **250–257**
Editors: **Y. Ravindrababu, R.K. Jaiman, K.D. Patel, N.R. Patel and S.B.S. Tikka**
Published by: **DAYA PUBLISHING HOUSE, NEW DELHI**

25

Field Screening of Fenugreek Genotypes Against Powdery Mildew Disease (*Erysiphe polygoni*)

D.B. Prajapati[1], S.N. Badaya[2], N.R. Patel[1] and P.K. Patel[1]
[1]Centre for Research on Seed Spices, S.D.A.U, Jagudan, Dist. Mehsana, Gujarat
[2]Head of Department Botany, C.P. College of Agriculture, S.K.Nagar, Gujarat

ABSTRACT

Studies were undertaken for betterment of crop improvement especially through powdery mildew resistance in fenugreek (*Trigonella foenum graecum* L.). The burning problem–powdery mildew is caused by *Erysiphe polygoni*. D.C. and is one of the most destructive and major production constraints. So to overcome and also locate the source of resistance, the attempts have been made to develop powdery mildew resistance genotypes. Sixty four fenugreek genotypes were selected in 2002-03 at Main Spices Research Station, Sardarkrushinagar Dantiwada Agricultural University, Jagudan and results revealed that two genotypes *viz.*, UM-19 and UM-303 were found immune (0- 00 PDI) and eight genotypes were resistant (1-10 PDI) and fifteen genotypes were found moderately resistant (11-25 PDI) to powdery mildew. While genotypes, LFC-114 and RM-1/90 were found to be most susceptible with percent disease intensity 68.75 and 66.25, respectively.

Keywords: Fenugreek, Powdery mildew, Source of resistance, Erysiphe polygoni.

Introduction

Fenugreek (*Trigonella foenum graecum* L.) is one of the important condiments largely cultivated in Northern India during rabi season. It is grown for grain as well as vegetable purpose both. The seeds and leaves of fenugreek are widely used as a culinary spice and condiments for flavouring food preparation and also used as animal feed and green plants are used as vegetable and fodder purpose.

The important fenugreek growing countries of the world are India, Argentina, Egypt, France, Morocco and Lebanon. In India, it is mainly cultivated in the states of Rajasthan, Gujarat, Tamil Nadu, Andhra Pradesh, Uttar Pradesh, Himachal Pradesh, Madhya Pradesh and Haryana. The area, production and productivity of fenugreek in India during 2004-2005 were 50400 ha, 62800 M. tonnes and 1245 kg/ha respectively. The country has earned Rs. 38.11 crores of foreign exchange by export of 10750 tonnes of Fenugreek (D.D, Spices Board, 2009). In Gujarat, it is cultivated on an area of about 4900 hectares of land producing 5900 M. tonnes with productivity of 1204 kg/ha. (D.O.A., 2009)

Recently, in Gujarat, the GM-1 is developed as a high yielding variety of fenugreek but susceptible to powdery mildew and thus gives very low yield under infestation. This disease appears regularly and cause heavy damage. It is caused by *Erysiphe polygonii* as well as by *Levellilulla taurica*. The powdery mildew growth on plants gives a whitish appearance wherever the infection has take place. It reduces the yield as well as grain size and deteriorates the quality. So it is the most destructive and major production constraint and in Gujarat there are few reports of screening fenugreek germplasm lines for resistance to powdery mildew.

So attempts were made to study for screening against powdery mildew disease in 64 diverse genotypes collected from different corner of India. The particulars of genotypes used are listed in Table 25.1.

Table 25.1: List of Genotypes Used for the Study and Geographic Origin

Sl.No.	Genotypes	Origin/Source	Sl.No.	Genotypes	Origin/Source
1.	UM-4	Jobner, Rajasthan	33.	HM-376	Hissar, Haryana
2.	UM-18	Jobner, Rajasthan	34.	LFC-74	Lam, Andhra Pradesh
3.	UM-19	Jobner, Rajasthan	35.	LFC-82	Lam, Andhra Pradesh
4.	UM-23	Jobner, Rajasthan	36.	LFC-83	Lam, Andhra Pradesh
5.	UM-29	Jobner, Rajasthan	37.	LFC-84	Lam, Andhra Pradesh
6.	UM-70	Jobner, Rajasthan	38.	LFC-87	Lam, Andhra Pradesh
7.	UM-74	Jobner, Rajasthan	39.	LFC-92	Lam, Andhra Pradesh
8.	UM-81	Jobner, Rajasthan	40.	LFC-94	Lam, Andhra Pradesh
9.	UM-92	Jobner, Rajasthan	41.	LFC-97	Lam, Andhra Pradesh
10.	UM-97	Jobner, Rajasthan	42.	LFC-100	Lam, Andhra Pradesh
11.	UM-106	Jobner, Rajasthan	43.	LFC-102	Lam, Andhra Pradesh
12.	UM-107	Jobner, Rajasthan	44.	LFC-105	Lam, Andhra Pradesh
13.	UM-117	Jobner, Rajasthan	45.	LFC-114	Lam, Andhra Pradesh
14.	UM-121	Jobner, Rajasthan	46.	LFC-124	Lam, Andhra Pradesh
15.	UM-128	Jobner, Rajasthan	47.	RM-1/90	Dholi, Bihar

Contd...

Table 25.1–Contd...

Sl.No.	Genotypes	Origin/Source	Sl.No.	Genotypes	Origin/Source
16.	UM-144	Jobner, Rajasthan	48.	RM-9	Dholi, Bihar
17.	UM-175	Jobner, Rajasthan	49.	RM-12	Dholi, Bihar
18.	UM-205	Jobner, Rajasthan	50.	RM-13	Dholi, Bihar
19.	UM-209	Jobner, Rajasthan	51.	RM-18	Dholi, Bihar
20.	UM-210	Jobner, Rajasthan	52.	RM-27	Dholi, Bihar
21.	UM-235	Jobner, Rajasthan	53.	RM-33	Dholi, Bihar
22.	UM-294	Jobner, Rajasthan	54.	RM-44	Dholi, Bihar
23.	UM-296	Jobner, Rajasthan	55.	JFenu-182	Jagudan, Gujarat
24.	UM-303	Jobner, Rajasthan	56.	JFenu-193	Jagudan, Gujarat
25.	UM-305	Jobner, Rajasthan	57.	JFenu-228	Jagudan, Gujarat
26.	UM-323	Jobner, Rajasthan	58.	JFenu-232	Jagudan, Gujarat
27.	UM-351	Jobner, Rajasthan	59.	JFenu-239	Jagudan, Gujarat
28.	UM-352	Jobner, Rajasthan	60.	JFenu-244	Jagudan, Gujarat
29.	HM-346	Hissar, Haryana	61.	JFenu-258	Jagudan, Gujarat
30.	HM-350	Hissar, Haryana	62.	JFenu-272	Jagudan, Gujarat
31.	HM-365	Hissar, Haryana	63.	JFenu-273	Jagudan, Gujarat
32.	HM-372	Hissar, Haryana	64.	G Methi-1	Jagudan, Gujarat

Materials and Methods

The experimental material for the present investigation consisted of 64 diverse genotypes against powdery mildew during 2002-03 from geographic and genetic origin, among these 10 genotypes were included from the All India Co-ordinated Research Project on spices, Main Spices Research Station, Gujarat Agricultural University, Jagudan and 54 genotypes from different states of India. The experiment plot size has two rows of 4 meter length with row to row spacing of 30 cm and plant to plant spacing was 10 cm. All the recommended agronomics and insect pest control measures were followed timely for successful raising the crop.

Whenever the powdery mildew disease appeared in general fenugreek crop, the infected plants from that field were picked and rubbed to upper surface of the leaves of the plants to all genotypes for artificial inoculation.

The observation on powdery mildew of fenugreek was recorded by selecting twenty plants randomly from each plot. The selected plants were graded into six classes using 0-5 rating scale on the basis of percentage area of plant leaf covered with powdery growth of fungus through visual observations.

To assess the degree of resistance or susceptibility to powdery mildew under infestated field conditions all the test entries were scored from zero to five according to the intensity of disease. The description of rating scores used as given below.

Details of Disease Scoring

Rating Scale	*Description of Rating Scale*
0	Healthy
1	0-10 per cent areas of plant covered by powdery mildew
2	11-25 per cent areas of plant covered by powdery mildew
3	26-50 per cent areas of plant covered by powdery mildew
4	51-75 per cent areas of plant covered by powdery mildew
5	>75 per cent areas of plant covered by powdery mildew

The PDI was calculated by using following equation as given by McKinney (1923).

$$\text{Per cent disease index} = \frac{\text{Total numerical rating}}{\text{Total plant observed x Maximum rating}} \times 100$$

Details of Per cent Disease Intensity

On the basis of disease reaction under natural epiphytotic conditions they were grouped as follows:

Per cent Disease Intensity	*Categories*
0	Immune
1-10	Resistant (R)
11-25	Moderately resistant (MR)
26-50	Susceptible (S)
> 50	Highly susceptible (HS)

Results and Discussion

Observations revealed that two genotypes *viz.,*.UM-19 and UM-303 were found completely free from the disease (0.00 per cent PDI). Eight genotypes were found resistant to powdery mildew disease intensity ranged from 1.00-10.00 per cent. Most of the genotypes were found highly susceptible. The genotypes LFC-124 recorded highest powdery mildew severities (68.75 per cent) among all the germplasms studied. Fifteen genotypes were found moderately resistant (11-25 per cent PDI) to powdery mildew (Table 25.2). Similar, result was reported as early that, no germplasm entry was found immune but GP-75, GP-96 and PEB were that the most resistance lines to powdery mildew in fenugreek by Gupta *et al.* (1997). Same on the way based upon the mean performance, the genotypes UM-323 and UM-18 and J.Fenu-244 were superior for grain yield per plant (Table 25.3) with resistant to powdery mildew by recorded PDI is 7.50 (Table 25.2) and two genotypes *viz.,* UM-19 and UM-303 were found immune.

On contrary, no germplasm was found immune or resistant to powdery mildew in coriander reported by Desai and Patel (2002).The genotype UM-305 have resistant genes and express earliness but grain yield is low but can utilized in crossing programme. Resistant genetic resources might be utilized in breeding programme.

Table 25.2: Screening of Different Genotypes of Fenugreek Against Powdery Mildew Disease (PDI) (2002-2003)

Sl.No.	Genotypes	PDI	Powdery Mildew Reaction	Sl.No.	Genotypes	PDI	Powdery Mildew Reaction
1.	UM-4	11.25	MR	33.	HM-376	42.50	S
2.	UM-18	7.50	R	34.	LFC-74	10.00	R
3.	UM-19	0.00	I	35.	LFC-82	63.75	HS
4.	UM-23	58.75	HS	36.	LFC-83	61.25	HS
5.	UM-29	57.50	HS	37.	LFC-84	36.75	S
6.	UM-70	47.50	S	38.	LFC-87	66.25	HS
7.	UM-74	18.75	MR	39.	LFC-92	18.75	MR
8.	UM-81	25.00	MR	40.	LFC-94	62.50	HS
9.	UM-92	11.25	MR	41.	LFC-97	60.00	HS
10.	UM-97	21.25	MR	42.	LFC-100	21.25	MR
11.	UM-106	48.75	S	43.	LFC-102	20.00	MR
12.	UM-107	56.25	HS	44.	LFC-105	17.50	MR
13.	UM-117	41.25	S	45.	LFC-114	10.00	R
14.	UM-121	20.00	MR	46.	LFC-124	68.75	HS
15.	UM-128	45.00	S	47.	RM-13	48.75	S
16.	UM-144	18.75	MR	48.	RM-1/90	66.25	HS
17.	UM-175	46.25	S	49.	RM-12	18.75	MR
18.	UM-205	10.00	R	50.	RM-9	38.75	S
19.	UM-209	23.75	MR	51.	RM-18	53.75	HS
20.	UM-210	20.00	MR	52.	RM-27	53.75	HS
21.	UM-235	38.75	S	53.	RM-33	61.25	HS
22.	UM-294	37.50	S	54.	RM-44	42.50	S
23.	UM-296	37.50	S	55.	JFenu-182	27.50	S
24.	UM-303	0.00	I	56.	JFenu-193	7.50	R
25.	UM-305	8.75	R	57.	JFenu-228	33.75	S
26.	UM-323	7.50	R	58.	JFenu-232	28.75	S
27.	UM-351	28.75	S	59.	JFenu-239	41.25	S
28.	UM-352	23.75	MR	60.	JFenu-244	7.50	R
29.	HM-346	20.00	MR	61.	JFenu-258	35.00	S
30.	HM-350	36.25	S	62.	JFenu-272	8.75	R
31.	HM-365	43.75	S	63.	JFenu-273	56.25	HS
32.	HM-372	55.00	HS	64.	GM-1	31.25	S

Rank Grade

I = Immune (0.0 per cent)
R = Resistant (1-10 per cent)
MR = Moderately resistant (11-25 per cent)
S = Susceptible (26-50 per cent)
HS = Highly susceptible (> 50)

Table 25.3: Mean Performance of Different Characters of Fenugreek Genotypes

Sl.No.	Genotypes	Days to 50 per cent Flower	Days to Maturity	Plant Height (cm)	Branches per Plant	Pods per Plant	Pod Length (cm)	Seeds per Pod	1000-Seed Weight (g)	Grain Yield per Plant (g)
1.	UM-4	53.50	111.00	71.30	5.00	110.00	7.40	17.10	13.02	19.80
2.	UM-18	51.50	110.50	71.50	6.15	100.50	10.55	17.40	14.03	20.00
3.	UM-19	54.50	115.50	74.00	6.80	92.50	9.00	16.60	16.99	15.40
4.	UM-23	50.00	110.00	76.00	5.00	87.15	10.00	17.50	14.47	13.60
5.	UM-29	55.00	116.00	76.65	5.65	86.65	11.50	15.65	14.80	14.90
6.	UM-70	55.00	114.50	77.65	5.85	86.50	9.85	15.00	10.39	07.20
7.	UM-74	47.50	110.50	80.35	5.50	103.50	9.85	17.70	15.64	19.00
8.	UM-81	51.00	108.50	71.85	6.50,	96.30	9.10	15.90	13.95	17.00
9.	UM-92	51.50	110.50	69.85	6.35	113.50	8.60	15.90	13.16	17.30
10.	UM-97	47.50	109.00	81.20	5.35	84.90	9.55	17.60	8.62	05.00
11.	UM-106	50.50	109.00	89.45	7.65	92.00	8.45	15.20	14.44	15.00
12.	UM-107	51.00	112.00	76.50	4.50	103.60	9.80	15.90	11.34	10.50
13.	UM-117	50.50	109.00	68.80	5.30	106.00	8.45	15.60	15.86	19.30
14.	UM-121	48.50	109.00	87.50	6.00	76.90	9.20	16.55	14.87	13.80
15.	UM-128	49.50	111.50	85.00	5.35	99.55	9.10	16.85	13.59	15.20
16.	UM-144	48.50	110.50	94.15	6.10	86.80	8.95	14.90	11.28	14.90
17.	UM-175	50.00	111.00	84.00	4.00	61.80	9.10	15.80	14.10	13.60
18.	UM-205	50.50	110.50	84.85	5.30	90.40	8.85	17.80	15.40	19.80
19.	UM-209	50.00	110.50	83.10	5.50	79.00	8.55	15.10	13.35	09.10
20.	UM-210	49.00	108.50	72.90	5.50	77.10	8.35	14.75	12.55	13.30
21.	UM-235	49.50	110.00	79.80	6.90	77.30	8.60	14.50	16.15	09.60
22.	UM-294	51.00	112.50	76.90	7.30	86.30	8.70	12.65	14.95	15.40
23.	UM-296	50.50	113.50	95.80	6.00	101.50	7.95	14.25	12.93	19.20
24.	UM-303	50.50	112.00	70.40	6.80	96.40	8.25	14.50	15.85	13.70
25.	UM-305	43.50	096.00	29.00	6.50	29.30	7.75	12.50	13.72	11.00
26.	UM-323	50.00	111.50	68.70	5.30	125.50	8.35	16.85	14.90	21.80
27.	UM-351	49.50	109.50	72.40	5.10	104.00	8.85	15.05	12.90	14.70
28.	UM-352	50.00	109.00	73.90	6.00	87.60	9.20	16.15	13.25	13.00
29.	HM-346	49.50	110.50	72.70	6.75	81.25	9.45	15.60	12.95	12.20
30.	HM-350	53.00	112.00	74.25	6.10	97.20	8.75	16.25	14.15	18.55
31.	HM-365	62.00	123.00	86.20	3.90	40.00	8.60	13.50	12.35	07.50
32.	HM-372	61.00	119.00	76.25	5.30	76.30	7.55	16.80	14.35	14.90
33.	HM-376	48.50	110.00	75.10	5.30	78.90	8.65	16.05	13.05	10.20
34.	LFC-74	50.50	110.00	74.00	5.15	66.00	8.55	17.20	14.43	09.90

Contd...

Table 25.3–Contd...

Sl.No.	Genotypes	Days to 50 per cent Flower	Days to Maturity	Plant Height (cm)	Branches per Plant	Pods per Plant	Pod Length (cm)	Seeds per Pod	1000-Seed Weight (g)	Grain Yield per Plant (g)
35.	LFC-82	51.50	111.50	82.90	6.20	79.20	9.95	17.70	12.84	12.80
36.	LFC-83	52.00	114.00	71.00	5.90	78.00	9.90	18.85	13.25	18.60
37.	LFC-84	49.50	112.00	76.00	5.35	80.80	8.75	16.15	15.59	15.40
38.	LFC-87	48.50	110.00	80.70	4.75	84.00	9.30	17.80	14.53	17.80
39.	LFC-92	50.00	110.50	82.45	4.05	80.90	9.40	17.90	15.05	17.50
40.	LFC-94	49.00	110.50	72.30	5.30	60.50	7.90	13.90	14.44	08.20
41.	LFC-97	53.00	116.00	69.40	7.90	94.50	7.95	16.60	14.38	10.20
42.	LFC-100	52.50	114.00	73.50	5.50	71.70	8.05	11.90	13.40	11.60
43.	LFC-102	49.50	112.50	81.90	5.00	87.00	7.80	14.25	16.75	13.00
44.	LFC-105	49.00	110.50	79.80	5.30.	72.60	8.40	15.50	12.95	10.00
45.	LFC-114	48.50	108.50	72.30	5.30	79.20	8.00	15.60	14.05	12.60
46.	LFC-124	50.00	108.50	77.30	6.30	75.45	8.80	15.85	15.80	09.40
47.	RM-1/90	50.50	109.00	66.80	7.30	76.10	9.35	16.25	13.05	11.60
48.	RM-9	50.00	110.00	83.20	8.00	98.00	8.15	16.05	12.20	09.80
49.	RM-12	50.50	111.50	75.50	6.80	81.90	8.75	16.10	13.05	11.30
50.	RM-13	53.00	116.50	68.80	6.40	92.00	9.15	15.30	12.15	10.60
51.	RM-18	49.50	110.50	80.00	6.00	70.20	8.30	16.90	14.60	08.60
52.	RM-27	53.00	117.00	92.10	5.50	89.40	8.85	19.95	12.75	12.40
53.	RM-33	50.00	112.00	69.95	6.00	94.90	9.35	22.00	13.34	16.40
54.	RM-44	49.50	110.00	81.60	5.00	80.20	8.70	17.50	13.55	11.60
55.	JFenu-182	49.00	110.00	86.60	4.25	78.90	8.70	15.90	13.24	10.50
56.	JFenu-193	49.50	109.00	82.50	5.20	69.70	8.30	17.40	13.38	12.60
57.	JFenu-228	52.00	112.50	67.45	5.00	64.20	9.05	18.30	17.13	14.40
58.	JFenu-232	48.50	108.00	83.20	4.80	62.40	9.55	18.35	14.93	11.50
59.	JFenu-239	50.00	110.50	81.40	4.50	69.90	8.95	17.35	15.13	16.20
60.	JFenu-244	49.00	110.00	74.80	5.80·	121.30	8.60	16.95	15.76	20.40
61.	JFenu-258	52.00	112.50	70.50	5.60	97.60	9.25	15.95	14.03	17.20
62.	JFenu-272	53.00	111.50	71.70	5.30	90.80	8.85	16.60	13.38	14.80
63.	JFenu-273	50.00	111.00	70.40	4.90	67.10	9.30	16.30	15.11	10.40
64.	GM-1	49.00	110.00	80.10	5.30	62.00	9.60	17.30	16.16	19.80
	S.Em.±	0.84	0.95	2.75	0.45	7.05	0.31	0.68	0.32	1.37
	C.D. at 5 per cent	2.39	2.71	7.84	1.28	20.09	0.88	1.92	0.90	3.89
	C.V. per cent	2.36	1.22	5.12	11.50	12.32	4.94	5.92	3.23	14.14

Conclusion

By screening 64 genotypes, two genotypes *viz.* UM-19 and UM-303 (0.00 per cent PDI) were found completely free from disease and eight entries found resistant to disease and may be used as parents in hybridization programme for transfer of yield attributing traits and disease resistance genes in cultivated varieties.

References

Desai, V.K. and Patel, J.G., 2002. Powdery mildew of coriander caused by *Erysiphe polygoni* D.C. and its management. *M.Sc. Thesis,* Gujarat Agricultural University, Anand (Unpublished).

DOA, 2009. *District-Wise Area and Production of Seed Spices in Gujarat.* Directorate of Agriculture, Govt. of Gujarat, Gandhinagar.

Deputy Director, Spices Board, Cochin, 2009. Spices exports go up by 15 per cent in Value. *A Journal of Spices India,* 22(1): 7.

Gupta, P.P., Jhorar, B.S., Arora, R.N. and Yadav, Rajesh, 1997. Evaluation of fenugreek genetic resources against major fungal diseases, Haryana. *Plant Disease Research* 12(1): 48–51, Forage Section, Deptt. of Pl. Breeding, CCS, Haryana Agril. Univ., Hissar, India.

McKinney, H.H., 1923. Influence of soil temperature and moisture on infection of wheat seedling by *H. sativum. J. Agril. Res.,* 26: 195–217.

Recent Advances in Seed Spices (2011) *Pages* **258–262**
Editors: **Y. Ravindrababu, R.K. Jaiman, K.D. Patel, N.R. Patel and S.B.S. Tikka**
Published by: **DAYA PUBLISHING HOUSE, NEW DELHI**

26

Quality Status of Coriander in Vertisols of South East Rajasthan

Vibha Doshi, Preeti Verma, Mashiat Ali and P.K. Gupta
Agricultural Research Station, Ummedganj, Kota
Maharana Pratap University of Agriculture and Technology, Udaipur, Rajasthan

Introduction

Spices are being used for flavouring foods science ancient times. Some species have antisepetic, antibiotic antifungal (Lim *et al.*, 2007) antioxidant and preservative properties. They are also used in pharmaceuticals, cosmetics, perfumery etc. About 50 species are grown in different parts of India of which coriander is one of the major seed species of economic importance. Coriander seeds are mainly used in curries for flavouring and leaves for garnishing and making chutneys. India produces on an average about 290000 tonnes of coriander per annum from about 540500 hectares. In India coriander is cultivated in the black cotton soils of Deccan and South India under rainfed conditions and rich slit loam soils of North India. Besides, using coriander seeds for flavouring curries it is also being used in making different types of curry powders, for essential oils, olioresions, ayurvedic medicines etc. The aromatic odour and taste of coriander seeds is due to essential oil. The essential oil content depends upon the source of fruit, harvesting and postharvesting conditions along with the climatic conditions in which the crop is grown. During 2007-08, 127000 hectare area is from Humid South eastern plain zone of Rajasthan was under coriander cultivation (Department of Agriculture, Jaipur). In the present scenario of global trade market production of high quality seed spices is very important. Looking to the importance of good quality traits present study was undertaken with the objective of selection and identification of high essential oil containing variety with disease resistance and good seed yield.

Materials and Methods

A field experiment was conducted at Agricultural Research Station, Ummedganj, Kota in black clay loam soils during rabi 2006-07 and rabi 2007-08. Experiment comprises of 139 entries of coriander including 5 released varieties (CS-6, RCr-41, RC-436, RCr-480 and CS-2) collected from different parts of Rajasthan *viz.* Kota, Ramganjmandi, Jhalawar, Baran, Bundi, Chittorgarh, Ajmer, Jobner, Jaipur, Sikar and State of Madhya Pradesh, Gujarat, Haryana and Tamil Nadu. These germ plams were evaluated in a randomized block design with two replications. Each genotype was grown in 5 meter long four rows with crop geometry 30 x 15cm. All package of practices with recommended fertilizer application were adopted to raise the healthy crop. Observations were recorded on ten randomly selected plants in each of the two replications in two years. The climate condition during the cropping season is depicted in the Tables 26.1 and 26.2. The crop was harvested when 50 per cent of capsule turned yellow. The harvested crop was dried in shade to keep the roots upwards. It was dried till ten per cent moisture level was achieved. Samples for oil estimation were taken plot wise from the harvested seed.The essential oil was estimated by using clevenger's apparatus (AOAC method) suitable for essential oil lighter than water.

Table 26.1: Mean Weekly Weather Parameters and Rainfall during Crop Season (2006 to 2007)

Standard Meteorological Weeks	*Date of Week*	*Temperature (°C)*		*Rainfall (mm)*	*Relative Humidity*	
		Mean Min.	*Mean Max.*		*Min.*	*Max.*
45	Nov. 05–Nov. 11	14.3	28.2	0.20	30	89
46	Nov. 12- Nov. 18	13.8	30.9	0.00	32	86
47	Nov. 19–Nov. 25	13.7	28.1	0.20	34	84
48	Nov. 26–Dec. 02	8.7	25.0	0.00	24	81
49	Dec. 03–Dec. 09	12.7	23.1	0.00	44	85
50	Dec. 10–Dec. 16	7.1	24.2	0.40	37	87
51	Dec. 17–Dec. 23	6.5	24.2	0.60	41	89
52	Dec. 24–Dec. 31	6.7	19.9	0.80	56	90
01	Jan. 01 – Jan. 07	7.2	20.8	0.80	46	74
02	Jan. 08 – Jan. 14	4.8	21.0	1.00	39	83
03	Jan. 15 – Jan. 21	6.1	23.5	0.00	35	85
04	Jan. 22 – Jan. 28	6.6	23.4	0.00	42	88
05	Jan. 29 – Feb. 04	11.1	27.3	0.00	45	91
06	Feb. 05–Feb. 11	10.0	26.7	9.00	45	87
07	Feb. 12–Feb. 18	8.7	24.9	1.40	38	93
08	Feb. 19–Feb. 25	10.4	28.3	13.00	41	86
09	Feb. 26 – Mar. 04	10.7	27.7	3.40	37	90
10	Mar. 05–Mar. 11	8.9	30.1	1.00	36	89
11	Mar. 12–Mar. 18	12.1	31.2	0.00	38	84
12	Mar. 19–Mar. 25	14.3	33.9	0.00	39	82
13	Mar. 26–Apr. 01	13.6	37.3	0.00	37	88

Table 26.2: Mean Weekly Weather Parameters and Rainfall During Crop Season (October 2007 to April 2008)

S.W. No.*	*Period (From to)*	*Temperature (°C)*		*Relative Humidity (Per cent)*	*Rainfall (mm)*	*No. of Rainy Days*
		Max.	*Min.*			
40	Oct. 01–Oct. 07	35.14	16.88	57.43	–	
41	Oct. 08–Oct. 14	34.06	16.08	60.28	–	–
42	Oct. 15–Oct. 21	34.54	13.29	63.00	–	–
43	Oct. 22–Oct. 28	34.13	13.72	62.43	–	–
44	Oct. 29–Nov. 04	32.58	12.57	52.28	–	–
45	Nov. 05–Nov.11	32.93	13.96	65.28	–	–
46	Nov. 12–Nov.18	30.79	12.20	67.14	–	–
47	Nov. 19–Nov.25	29.59	8.99	59.14	–	–
48	Nov. 26–Dec. 02	28.87	10.23	59.71	–	–
49	Dec. 03–Dec. 09	23.91	7.24	79.14	–	–
50	Dec. 10–Dec. 16	22.57	8.15	78.85	4.60	6
51	Dec. 17–Dec. 23	24.81	5.17	74.00	–	–
52	Dec. 24–Dec. 31	24.87	8.96	73.43	–	–
01	Jan. 01–Jan 07	21.64	5.48	75.43	–	–
02	Jan. 08–Jan 14	24.64	7.19	75.86	–	–
03	Jan. 15–Jan 21	23.46	5.93	77.14	–	–
04	Jan. 22–Jan 28	19.21	2.04	79.71	–	–
05	Jan. 29–Feb. 04	21.4	4.53	74.43	–	–
06	Feb. 05–Feb. 11	19.85	3.28	71.85	–	–
07	Feb. 12–Feb. 18	24.69	2,54	71.57	–	–
08	Feb. 19–Feb. 25	26.08	7.71	76.00	–	–
09	Feb. 26–Mar. 04	29.87	7.42	67.12	–	–
10	Mar. 05–Mar. 11	31.63	11.47	67.43	–	–
11	Mar. 12–Mar. 18	31.67	12.45	68.57	4.2	3
12	Mar. 19–Mar. 25	34.96	14.88	56.28	–	–
13	Mar. 26–Apr. 01	36.67	13.68	47.00	–	–

Results and Discussion

Two years experimental data reveals that a grate variation in seed yield, yield attributes and essential oil content among the 139 germplasm lines and checks collected from different areas was found. The seed yield ranged from 9.35q/ha to 18.75.q/ha during 2006-07 and from 9.55 q/ha to 20.00 q/ha and during 2007-08. The 1000 seed weight ranged from 5.2g to 18.5g during 2006-07 and 5.8 g to 18.2 g. during 2007-08. Sharma 1982 and Gupta 1982 were also of the opinion that to achieve genetic variability collection of germpplasm should be done from different climatic zones. Joshi *et al.*, 1963 also collected and evaluated germplasm from different parts of India and abroad and identified various developed varieties *viz.* ND (D) 92, 95, 172, NP (J) 16, 24 and NP (K) 45 which out yielded 8-15 per cent more than indigenous varieties. Mathur, Anwar and Bhargava (1973) studied 63 varieties and they also reported wide variation in yield, yield attributing characters and 1000 grain weight. Similar results were also reported by Joshi *et al*. (1965 and 1967).

Essential Oil

The essential oil content ranged form 0.01 per cent to 0.4 per cent during the year 2006-07. However the essential oil content during the year 2007-08 varied from 0.1 per cent to 0.4 per cent. Sharma 1982 and Gupta 1982 reported genetic variability in Indian and European origin germplasm in essential oil, yield, yield attributing characters.

Table 26.3: Essential Oil Content of Coriander Genotypes

Sl.No.	*Line*	*Essential Oil (Per cent)*		*Mean*	*Per cent Increase Over Best Check (RCr–436)*
		2006-07	*2007-08*		
1	LGC–36	0.40	0.40	0.40	25
2	LGC–18	0.34	0.40	0.37	16
3	LGC–40	0.34	0.40	0.37	16
4	LGC–27	0.34	0.36	0.35	
5	LGC–44	0.30	0.40	0.35	
6	LGC–34	0.30	0.40	0.35	
7	LGC–39	0.30	0.40	0.35	
8	LGC–42	0.36	0.34	0.35	
	RCr–480 (C)	0.28	0.32	0.30	
	CS–2 (C)	0.30	0.34	0.32	
	RCr–436 (C)	0.30	0.34	0.33	

Table 26.4: Ancillary Data of Coriander Genotypes

Line/Variety	*Days to 50 per cent Flowering*	*Days to Maturity*	*1000–Seed wt.(g)*	*Seed Yield (kg/ha)*
LGC-18	58	108	11.8	1315
LGC-36	59	110	9.8	878
LGC 27	58	108	9.2	1412
LGC-40	59	110	13.7	1366
RCr-480 (C)	63	115	8.5	619
CS-2 (C)	63	115	11.6	216
RCr-436 (C)	61	112	12.6	1000

On the basis of two years pooled data (Table 26.3) highest essential oil content 0.4 per cent was registered by LGC 36 followed by LGC 18 (0.37) and LGC 40 (0.37) and thus an increase of 25 per cent, 16 per cent and 16 per cent in essential oil content over best check RCr 436 was obtained respectively.

Results

Table 26.3 reveals that maximum essential oil content of 0.4 per cent followed by 0.36 and 0.36 per cent was obtained from genotypes LGC 36, LGC 38 and LGC 42 respectively during the year 2006-07 however, during the year 2007-08 0.4 per cent essential oil content was obtained by many of the genotypes viz, LGC 10, 14, LGC 18, 29, LGC 34, 36, LGC 39, LGC 40, 41, 44, 45, LGC 136 and LGC 147, during the year 2007-08. The average essential oil content was more during the year 2007-08 as compared to the year 2006-07. The most possible reason for this is that during year 2007-08 winters were very cool with low temperature for prolonged period from 19th Nov. to 4th March 2008 (Table 26.1). Enhancement in yield during cool and damp weather was also reported by Bauer 1942. There is a scope for further enhancement of essential oil by the use of growth regulator as reported by (Kolev, *et al.*, 2005, Kristova *et al.*, 2005) and hencemore work in this direction may be initiated.

References

Bauer, Landw, Jahresber, 1942. In: *The Essential Oils, Vol. 3.* R.E. Kreiger Publishing Company Inc. Florida, USA.

Gupta, A., 1982. Genetic divergence I in germplasm of coriander on the basis of morphological characters and grain yield. *M.Sc. (Ag.) Thesis,* Sukhadia University, Campus, Jobner.

Joshi, B.S., 1965. Analysis of factors effecting yield and their significance for improvement of *Coriander sativum* L. *Ph.D. Thesis,* Division of Botany, IARI, New Delhi.

Joshi, B.S., Ramanujam, S. and Saxena, M.B.L., 1963. Improvement of some essential oil bearing species plants. *Bull. Regional Res. Lab. K,* 1:94–100.

Joshi, B.S., Joshi, A.B. and Ramanujam, S., 1967. Variance and co-variance in some umbelliferous species crops. I. Variability in coriander. *Ind. J. Genet.,* 27: 211–219.

Khristova, D., Ivanova, I., Nenkota, D., 2005. Study on influence of some biologically active substances upon ethereal oil crops-Coriander (*Coriandrum sativum* L.). *Crop Husbandry,* V42(3): 239–242.

Kolev, T., 2005. Effect of several plants growth regulators on the productivity and essential oil content of coriander. *Bulgarian Journal of Agricultural Science,* Bulgaria, V11(5): 571–575.

Lim, S. and Shin, S.W., 2007. Synergism in antifungal activity against coriander and Trichophyton species in combination with the essential oil of *Coriander sativum* L. and antibiotics. *Natural Product Sciences,* V13(1): 85–89.

Sharma, M.M., 1982. Genetic divergence in the germplasm of coriander. II. On the basis of oil content, seed weight, seed volume and seed yield. *M.Sc. (Ag.) Thesis,* Sukhadia University Campus, Jobner.

Recent Advances in Seed Spices (2011) *Pages* **263–272**
Editors: **Y. Ravindrababu, R.K. Jaiman, K.D. Patel, N.R. Patel and S.B.S. Tikka**
Published by: **DAYA PUBLISHING HOUSE, NEW DELHI**

27

Gujarat Dill-3: A New Dwarf, Early Dill Cultivar having Good Economic Traits in Irrigated as well as Rainfed Conditions in Gujarat

D.B. Prajapati, K.P. Patel, N.R. Patel, A.V. Agalodiya and P.K. Patel
Centre for Research on Seed Spices, S.D.A.U, Jagudan, Dist. Mehsana, Gujarat

ABSTRACT

Indian Dill (*Anethum sowa* Kurz) commonly known as sowa is one of the important minor seed spice crops grown under irrigated as well as unirrigated conditions in Gujarat. During 2004-05, India produced 13,100 MT of dill from an area of 16500 ha with productivity of 750 kg/ha. In the same year, it was grown in 10100 ha in Gujarat which produced 8600 MT with productivity of 850 kg/ha. The share of Gujarat in national production and area of dill was 61.2 and 65.7 per cent respectively. Though the productivity of dill has increased from 813 to 850 kg/ha (2001- 02 to 2004-05) after release of GD-1 and GD-2 (2001), the rate of increase is very low, because of lack of variety suitable for both rainfed as well as irrigated conditions and high susceptibility to powdery mildew. Therefore, looking to the importance of dill, there is urgent need to enhance the productivity as well as production and improve the quality of the produce to make a dill crop more profitable. It was tested in multi location and out yielded GD-1 and GD-2 by producing 23.04 and 15.04 per cent higher yield. It was also tested at Jagudan for three years (2003-04 to 2005-06) under irrigated conditions. It has produced on an average 1273 kg/ha and significantly out yielded the check varieties GD-1 (1128 kg/ha) and GD-2 (1080 kg/ha) by producing 12.85 and 17.87 per cent higher yield. On an average of 9 trials conducted, the variety GD-3 produced 1062 kg/ha, which was 18.79 and 16.19 per cent higher than the check variety GD-1 (Irrigated) and GD-2 (Rainfed), respectively, clearly indicating its high yield potential under both the varying production conditions. The yield potential of GD-3 under irrigated and rainfed conditions is 1659 and 1296 kg/ha, respectively.

The plant type is dwarf (25 cm) compared to GD -1 (126 cm) and 15 days early maturing than GD-1 (156 days).The grains are uniform in size, attractive, plumpy having dark brown colour and non splitting habit. The essential oil (2.7 per cent) is higher than both the checks. The stalk of the fruit (seed) easily detached during threshing, whereas, it remains attached in GD-1 and GD-2 and deteriorate the quality of produced.

Introduction

Indian Dill (*Anethum sowa* Kurz) commonly known as sowa is one of the important minor seed spice crops grown under irrigated as well as unirrigated conditions. The major dill seed producing countries are India, Germany, Hungary, Netherlands, Pakistan and United States. In India, Gujarat, Rajasthan, M.P., J and K, Orissa and Punjab are the leading states in respect of area and Producion. The districts of Mehsana, Banaskantha, Patan, Surendranagar and Ahmedabad are leading in Gujarat state, where it is cultivated in scattered pockets. It is well grown in saline soils. The crop was introduced as a rabi crop in the paddy-growing area of Ahmedabad and Kheda districts (Patel and Patel, 2001). The dill somewhere raised by farmers on bunds of the main crops, hence lacking the real statistics. The area, production and productivity of dill seed in Gujarat during 2005-06 were 4.83 hectares, 3.82 tonnes and 790 kg/ha, respectively (DAO, 2005). The share of Gujarat in area and production is 52 per cent and 68 per cent, respectively towards total area and production of country (2005-06). Seed of dill are largely consumed within the country and still a small quantity is exported to the western countries. On average India exports dill seed worth of rupees 20 millions.

Dill seed works as mild stomach medicine for children. Seed of dill is rich in carvone, which is popularly used as carminative, aromatic, stimulant and diuretic in ayurvedic and unani medicines. Decoction prepared from dill seed is given to women after delivery as milk stimulating agent. Dill seed is also fed to animals after calving. Seed of dill yield 2.5 to 3.5 per cent essential oil, rich in carvone, which imparts to oil its characteristics flavor. Thus, both seed and oil used in medicine. Preparation of dill water and similar products are used in flatulent, abdominal and colic pain.

During 2004-05, India produced 13,100 MT of dill from an area of 16500 ha with productivity of 750 kg/ha. In the same year, it was grown in 10100 ha in Gujarat which produced 8600 MT with productivity of 850 kg/ha. The share of Gujarat in national production and area of dill was 61.2 and 65.7 per cent respectively. Patan (4499 ha), Ahmedabad (2585 ha), Kutch (1128 ha), Mehsana (1010 ha), Surendranagar (769 ha) are the major dill growing districts of Gujarat. The productivity of these districts ranged from 500 kg/ha (Ahmedabad) to 1413 kg/ha (Kutch). Dill in Ahmedabad district is totally grown under conserve moisture conditions. Whereas, in other districts it is grown under both irrigated as well as rainfed conditions. Though the productivity of dill has increased from 813 to 850 kg/ha (2001-02 to 2004-05) after release of GD-1 and GD-2 (2001), the rate of increase is very low, because of lack of variety suitable for both rainfed as well as irrigated conditions and high susceptibility to powdery mildew. Dill seeds are used as condiments in soups, salads, processed meats, sauces and pickling. Both seed and oil are used in formulation of various ayurvedic medicines. The essential oil is used as flavour and in soap industry and provides relief in flatulence, abdominal and colic pain. The gripe water is prepared from its seeds which are used to improve digestion and control of vomiting in infants and children. Seeds are also very useful for women and milking animals in delivery for expulsion of placenta and promotion of milk.

Materials and Methods

Since last few years, there is drastically decrease in area and production in India. There are some causes for declining the crop status. Obviously, having shattering habit of the grains, non-synchronous maturity of crop, as delaying in harvesting of early umbels automatically and shattered on plot, next year the seed grows as weeds deteriorates the remaining crop. So farmers keep away from this serious problem. Moreover, lack of variety suitable for under irrigated as well as unirrigated conditions hence, there is urgent need to enhance the area as well as production and improve the quality of the produce to make a dill crop more profitable.

Keeping these in view, the breeding programme was concentrated on developing varieties with high yield potential and more adaptability, disease resistance and better grain quality. Selection plays a predominant role in crop improvement and most of the high yielding varieties are made out of the selection of germplasm. Therefore, 77 numbers of germplasms collected and evaluated at Jagudan. Guj.Dill-3 (JD-95-161) was selected from the germplasm collected from Manud village of Patan district and maintained as JD-95-161 at Jagudan centre.The proposed variety GD-3 (JD-95-161) alongwith other germplasm entries was evaluated for grain yield and other attributes at Jagudan with Mehsana local and Jamnagar selection as there was not any released variety in Gujarat up to 2000. Considering its yield, plant type, maturity, it was first entered in PET during 2001-02 at Vallabhipur for testing under rainfed conditions. It was included in co-ordinated trial for multi location testing and tested at Vallabhipur (3 years), Arnej (2 years) as well as Dhandhuka (one year) under conserved moisture and Jagudan for three years (2003-04 to 2005-06) under irrigated conditions.

The trial was conducted in Randomised block Design with detail treatment keeping eight rows of 4 meter length with row to row spacing of 45 cm and plant to plant spacing was 30 cm. All the recommended crop protection measures were followed timely to control pests and diseases for successful raising the crop. A uniform fertilizer dose of 20 kg N and 40 kg P_2O_5 ha^{-1} was applied and normal agronomic practices were followed. Observations were taken on five randomly selected plants from each of the replications for quantitative and qualitative characters are as per minimal descriptor developed by Malhotra and Vashishtha (2006). Mean data from each of the replications were used to analyzed, statistics.

Results and Discussion

High Yield Potential

In PET during 2001-02 at Vallabhipur for testing under rainfed conditions. It was tested at Vallabhipur (3 years), Arnej (2 years) as well as Dhandhuka (one year) and out yielded GD-1 and GD-2 by producing 23.04 and 15.04 per cent higher yield (Tables 27.1–27.3). It was also tested at Jagudan for three years (2003-04 to 2005-06) under irrigated conditions. It has produced on an average 1273 kg/ha and significantly out yielded the check varieties GD-1 (1128 kg/ha) and GD-2 (1080 kg/ha) by producing 12.85 and 17.87 per cent higher yield. On an average of 9 trials conducted, the proposed variety GD-3 produced 1062 kg/ha, which was 18.79 and 16.19 per cent higher than the check variety GD-1 (Irrigated) and GD-2 (Rainfed), respectively, clearly indicating its high yield potential under both the varying production conditions (Tables 27.1–27.5). The yield potential of GD-3 under irrigated and rainfed conditions is 1659 and 1296 kg/ha, respectively.

Better Grain Quality

The grains are uniform in size, attractive, plumpy having dark brown colour and non splitting habit. The essential oil (2.7 per cent) is higher than both the checks. The stalk of the fruit (seed) easily detached during threshing, whereas, it remains attached in GD-1 and GD-2 and deteriorate the quality of produced (Table 27.6).

Table 27.1: Yield Performance of GD-3 (JD-95-161) Compared to Checks and Other Varieties Under Rainfed Conditions at Vallabhipur

Sl.No.	Treatment	Yield (kg/ha)				Per cent IOC	
		2001–02 (PET)	2002–03 (PET)	2004–05 (LST)	Average	GD–1	GD–2
1.	JD–95–45	1108@	932	988	1009	16.92	3.49
2.	JD–95–72	753	633	1123	836	–	–
3.	JD–95–124	1111@	985	1148	1081*	25.26	10.87
4.	JD–95–161	1296*@	917	1204	1139*	31.98	16.82
5.	GD–1 (Ch)	827	824	938	863	–	–
6.	GD–2 (Ch)	1028	867	1030	975	–	–
	S.Em±	93	57	122	66	–	–
	C.D. at 5 per cent	267	169	NS	186	–	–
	C.V. per cent	20.90	21.91	20.32	19.98	–	–

Note: The LST during 2003–04 was vitiated because very poor germination due to heavy moisture stress.

Table 27.2: Yield Performance of GD-3 (JD-95-161) Compared to Checks and Other Varieties Under Rainfed Conditions at Arnej

Sl.No.	Treatment	Yield (kg/ha)			Per cent IOC	
		2004–05 (LST)	2005–06 (LST)	Average	GD–1	GD–2
1.	JD–95–45	648	822	735	12.38	4.40
2.	JD–95–72	843	743	793	21.25	12.64
3.	JD–95–124	596	779	688	5.20	–
4.	JD–95–161	690	841	736	12.54	4.54
5.	GD–1 (Ch)	515	793	654	–	–
6.	GD–2 (Ch)	654	754	704	–	–
	S.Em.±	76	58	50	–	–
	C.D. at 5 per cent	NS	NS	NS	–	–
	C.V. per cent	20.29	12.67	–	–	–

Synchronous Plant Type and Uniform Maturity

The proposed variety matures in 142 days which is 14 and 6 days earlier than GD-1 and GD-2, respectively. The plant type is medium dwarf having synchronous branching and uniform maturity resulting into less shattering of grains. The average per cent of shattering in GD-3 recorded only 6.9 per cent compared to 18.5 and 15.4 per cent of GD-1 and GD-2, respectively (Table 27.6).

Table 27.3: Yield Performance of GD-3 (JD–95–161) Compared to Checks and Other Varieties Under Rainfed Conditions at Dhandhuka

Sl.No.	*Treatment*	*Yield (kg/ha)*	*Per cent IOC*	
		2005–06 (LST)	*GD–1*	*GD–2*
1.	JD–95–45	848 @	10.99	30.7
2.	JD–95–72	667	–	2.77
3.	JD–95–124	698	–	7.55
4.	JD–95–161	848 @	10.99	30.7
5.	GD–1 (Ch)	764	–	17.72
6.	GD–2 (Ch)	649	–	–
	S.Em.±	42		
	C.D. at 5 per cent	122		
	C.V. per cent	11.10		

Table 27.4: Yield Performance of GD-3 (JD-95-161) Compared to Checks and Other Varieties Under Irrigated Conditions at Jagudan

Sl.No.	*Treatment*	*Yield (kg/ha)*				*Per cent IOC*	
		2003–04 (LST)	*2004–05 (LST)*	*2005–06 (LST)*	*Average*	*GD–1*	*GD–2*
1.	JD–95–45	774@	1570	1081	1142	1.24	5.74
2.	JD–95–72	968*@	1646	1161	1258*@	11.52	16.48
3.	JD–95–124	869 @	1600	1169	1213 @	7.53	12.31
4.	JD–95–161	887*@	1659	1274*@	1273*@	12.85	17.87
5.	GD–1 (Ch)	772	1543	1070	1128	–	–
6.	GD–2 (Ch)	694	1451	1096	1080	–	–
	S.Em.±	34	78	46	33	–	–
	C.D. at 5 per cent	98	NS	134	93	–	–
	C.V. per cent	8.70	10.03	8.22	9.70	–	–

Moderately Resistant to Powdery Mildew and Less Prone to *Ramularia* Blight

Powdery mildew caused by *Erisyphe polygoni* is the most destructive disease causing heavy loss of yield and deterioration of grain quality and decrease in oil content. The proposed variety is moderately resistant to this disease. The per cent disease intensity (PDI) recorded at Jagudan, Arnej and Dhandhuka ranged from 19 to 25 per cent (average 20.6) in GD-3 as compared to 62 to 80 per cent (average 67.4) and 80 to 90 per cent (average 85.8) of GD-1 and GD-2, respectively. Whereas the PDI (35.5) for *Ramularia* blight recorded at Jagudan indicated that proposed variety is less prone to this disease. The average PDI recorded for GD-3, GD-1 and GD-2 was 35.5, 54.3 and 68.5, respectively (Table-7). The stem is more solid than GD-1 and GD-2. Dwarf plant type and solid stem make the variety less prone to lodging.

Table 27.5: Overall Performance

Location	No. of Trials	Average Yield (kg/ha)						Pooled C.D.
		JD-95-45	JD-95-72	JD-95-124	GD-3 (JD–95–161)	GD–1 (Ch)	GD–2	
(I) Under Irrigated Condition								
Jagudan	3	1142	1258*@	1213*	1273*@	1128	1080	93
(II) Under Rainfed Condition								
Vallabhipur	3	1009	836	1081*	1139*	863	975	186
Dhandhuka	1	848@	667	698	848@	764	649	122
Arnej	2	735	793	688	736	654	704	NS
Mean	6	891	794	886	956	777	831	
Overall Mean	9	975	948	995	1062	894	914	
Per cent IOC (Irri.) (3)	GD–1	1.24	11.52	7.83	12.85			
	GD–2	5.74	16.48	12.31	17.87			
Per cent IOC (Rainfed) (6)	GD–1	14.67	2.19	14.02	23.04			
	GD–2	7.22	–	6.62	15.04			
Per cent IOC Overall (9)	GD–1	9.06	6.04	11.30	18.79			
	GD–2	6.67	3.72	8.86	16.19			
Frequency of out yielding	GD–1	9/9	5/9	8/9	9/9			
	GD–2	5/9	6/9	9/9	9/9			

Yield data significantly superior than GD–1 (*) and GD–2 (@).

Table 27.6: Yield and Quality Attributes of GD-3 in Comparison to Checks

Sl.No.	Character	Location/Mean	GD-3 I and R	GD-1(Ch) Irrigated	GD-2(Ch) Rainfed
1.	Days to 50 per cent flowering	*Jagudan (3)	72	85	77
		**Vallabhipur (3)	70	81	75
		Mean	71	83	76
2.	Days to maturity	Jagudan (3)	145	165	154
		Vallabhipur (3)	139	146	141
		Mean	142	156	148
3.	Plant height (cm)	Jagudan (3)	113	136	131
		Vallabhipur (3)	89	115	96
		Mean	101	126	114

Contd...

Table 27.6–Contd...

Sl.No.	*Character*	*Location/Mean*	*GD-3 I and R*	*GD-1(Ch) Irrigated*	*GD-2(Ch) Rainfed*
4.	No. of Branches/plant	Jagudan (3)	8.0	6.4	5.0
		Vallabhipur (3)	5.9	5.3	5.5
		Mean	7.0	5.7	5.3
5.	No. of umbels/plant	Jagudan (3)	25.0	21.2	21.0
		Vallabhipur (3)	20.5	18.5	20.5
		Mean	22.8	19.9	20.8
6.	No. of Umbellates/umbel	Jagudan (3)	26.0	20.4	20.2
		Vallabhipur (3)	20.0	19.0	20.0
		Mean	23.0	19.7	20.1
7.	No. of seeds/umbellate	Jagudan (3)	29.5	28.3	23.6
		Vallabhipur (3)	–	–	–
		Mean	29.5	28.3	23.6
8.	1000 grains weight (gms)	Jagudan (3)	4.30	3.60	4.1
		Vallabhipur (3)	3.08	3.15	3.5
		Mean	3.69	3.40	3.8
9.	Volatile oil (per cent)	Jagudan (3)	2.7	2.5	2.2
		Vallabhipur (3)	–	–	–
		Mean	2.7	2.5	2.2
10.	Grain Shattering (per cent)	Range (Jagudan)	5.5-9.0	14.5-21.0	13.3-19.6
		Mean (Jagudan)	6.9	18.5	15.4

*: Mean of 2003-04, 2004-05 and 2005-06; **: Mean of 2001-02, 2002-03 and 2004-05.

Table 27.7: Reaction to Diseases and Pests

Location	*Year*	*Per cent Disease Intensity (PDI)*				
		JD-95-72	*JD-95-124*	*JD-95-161*	*GD-1 (Ch)*	*GD-2 (Ch)*
A. Powdery *Mildew* (*Erisyphe polygoni*)						
Jagudan	2003-04	90.0	89.0	25.0	60.0	80.0
	2004-05	88.0	87.5	20.0	65.0	86.0
	2005-06	92.5	91.5	19.0	70.0	88.0
Arnej	2005-06	87.0	89.0	21.0	62.0	85.0
Dhandhuka	2005-06	85.0	88.0	18.0	80.0	90.0
	Mean	88.4	89.0	20.6	67.4	85.38

Contd...

Table 27.7–Contd...

Location	Year	Per cent Disease Intensity (PDI)				
		JD-95-72	JD-95-124	JD-95-161	GD-1 (Ch)	GD-2 (Ch)
B. *Ramularia* blight (*Ramularia* sp.)						
Jagudan	2003-04	75.00	68.75	37.75	56.25	68.75
	2004-05	56.25	62.50	37.50	50.00	68.25
	2005-06	75.00	68.75	31.25	56.25	68.25
	Mean	68.8	66.80	35.50	54.30	68.50

C. Aphid (*Myzus persicae*): Not observed in any of varieties during 2003-04 to 2005-06.

Table 27.8: Market Price and Economics of GD-3 Compared to Checks

Sl.No.	Trader's name	APMC Markets	Market Price Rs./20 kg		
			GD-3	GD-1 (Ch)	GD-2 (Ch)
1.	M/s D. A. Patel	Unjha	625	600	600
2.	M/s Suresh Trading Company	Unjha	620	590	595
3.	M/s R. C. Enterprise	Unjha	630	611	621
4.	M/s Rasikbhai A. Patel	Unjha	614	592	595
5.	B. Somabhai Patel and Co.	Mehsana	630	600	610
6.	M/s.Patel Babubhai Joitaram and Co.	Mehsana	621	595	600
7.	Secretary, APMC	Harij	585	565	530
	Total...		4325	4153	4151
		Average	618	593	593

Table 27.9a: Profitability of GD-3 Over Checks

Sl.No.	Particulars	Under Irrigated Condition		Under Rainfed Condition		Over all		
		GD-3	GD-1	GD-3	GD-2	GD-3	GD-1	GD-2
1.	Production (kg/ha)	1273	1128	956	831	1062	894	914
2.	Price (Rs. per 20 kg)	618	593	618	593	618	593	593
3.	Total Income (Rs./ha)	39336	33445	29540	24639	32816	26507	27100
4.	Net profit over checks (Rs./ha)	–	5891	–	4901	–	6309	5716
5.	Profit over checks (per cent)	–	17.61	–	19.89	–	23.80	21.09

More Profitability

The variety GD-3 fetches premium price (Rs. 25/20 kg more) in market because of its attractive and uniform grains having non splitting habit and detached stalk of grains. By growing GD-3, farmers can earn on an average Rs.39336 per ha under irrigated conditions which is Rs.5891/ha more than

GD-1. Similarly, in rainfed conditions earning and net profit over GD-2 would be Rs.29540 and Rs.4901 per ha, respectively (Table 27.9a). If GD-3 adopted at rate of 25 per cent by farmers in the Dill growing area in the state they may be benefited by 1 crore Rupees more (Table 27.9b).

Table 29.9b: Projection of Productivity, Production and Income Increase by Adoption of New Dill Variety Guj. Dill-3

District	*Area (ha)*	*Productivity (kg/ha)*	*Production (Mt)*	*Income Rs/ha*		*Total Rs (crore)*	
		*A**	*B***	*A*	*B @*	*A*	*B @*
Patan	4499	901	1054	4055	4227	12.02	12.68
Kutch	1128	1913	1653	1594	1662	4.72	4.98
Mehsana	1010	1000	1170	1010	1053	2.99	3.16
Ahmedabad	25850	500	600	1293	1358	3.83	4.07
Surendranagar	769	724	847	557	581	1.50	1.63
State	10079	850	995	8567	8932	25.40	26.30

Base year: 2005-06 A*= Present B*= After adoption of GD-3.

@ Calculated on the basis of 25 per cent adoption rate.

Table 27.10: Descriptor of Dill Variety GD-3

Sl.No.	*Characters*	*Descriptor*
1.	Early plant vigour	Good
2.	Plant Growth habit	Dwarf
3.	No. of branches/plant	7.0
4.	Leaf Size	Small
5.	Foliage Colour	Lush green
6.	Days to 50 per cent flowering	71
7.	Inflorescence Colour	Yellow
8.	Umbel diameter (cm)	16.1
9.	No. of umbels/plant	22.8
10.	No. of umbellates/Umbel	23.0
11.	No. of seeds/umbellate	29.5
12.	Fruit shape	Ellipsoidal
13.	Seed Size	Medium
14.	Seed type	Compact and non-splitting
15.	Seed Colour	Dark brown
16.	Plant height (cm)	101 cm
17.	Days to maturity	142
18.	1000 grains weight (gms)	3.7
19.	Volatile oil content (per cent)	2.7

Conclusion

Variety GD-3 with high stable yield under irrigated and rainfed conditions, better grain quality moderately resistant to powdery mildew will provide very good alternative to both GD-1 and GD-2. Cultivation of GD-3 will not only enhance the productivity but also improve profitability of dill under varying environmental conditions.

References

DOA, 2009. *District-Wise Area and Production of Seed Spices in Gujarat*. Directorate of Agriculture, Govt. of Gujarat, Gandhinagar.

Malhotra, S.K. and Vashishta, B.B., 2006. *A Manual on Minimal Descriptors of Seed Spices*. National Research Centre on Seed Spices, Ajmer, Rajasthan.

Patel, I.D. and Patel, M.S., 2001. *Status of Seed Spices in Gujarat: Seed Spices Production, Quality and Export*. Pointer Publishers, Jaipur, Rajasthan, India, pp. 41–50.

Recent Advances in Seed Spices (2011) *Pages* **273–279**
Editors: **Y. Ravindrababu, R.K. Jaiman, K.D. Patel, N.R. Patel and S.B.S. Tikka**
Published by: **DAYA PUBLISHING HOUSE, NEW DELHI**

28

Gujarat Methi-2: A Newly Bold Seeded Fenugreek Variety having High Yield Potential and Disease Resistance

D.B. Prajapati, A.V. Agalodiya, G.M. Patel, K.D. Patel and N.R. Patel
Centre for Research on Seed Spices, S.D.A.U, Jagudan, Dist. Mehsana, Gujarat

ABSTRACT

Fenugreek (*Trigonella foenum graecum* L.) is one of the important condiments largely cultivated in North India during rabi season. It is grown for both grain as well as vegetable purpose. During 2004-05, India produced 62800 MT from an area of 50400 hectares with productivity of 1245 kg/ha. Gujarat grows fenugreek in 4900 hectares and produces 6280 MT with an average productivity of 1144 kg/ha (2004-05). The contribution of North Gujarat and Kutch in area and production is 47.8 and 48.8 pre cent, respectively. Though the productivity of fenugreek has increased from 1100 kg (2000-01) to 1178 (2003-04) after release of GM-1 (2000) the rate of increase is very low. The reasons are high susceptibility of GM-1 to powdery mildew disease slow rate of adoption of variety and lack of better alternative to GM-1. Therefore, looking to the importance of fenugreek, there is urgent need to enhance productivity as well as production to make the crop more profitable. The investigation was carried out with twelve genotypes obtained local check variety of respective co-coordinating centres *viz.* Jobner (Rajasthan), Guntur (A.P.), Kumarganj (U.P.) and Coimbatore (Tamil Nadu) of AICRP. On an average GM-2 yielded 1920 kg/ha and showed its superiority by producing 10.03 and 20.32 per cent higher than GM-1 and Hissar Sonali respectively. Results indicating its high stability under varying range of environmental conditions. The yield potential of this variety is 2079 kg/ha. The grains are bold, uniform in size, attractive and have more test weight. Therefore, it fetches Rs.200/qtl more than GM-1. It possesses high number of pods and more pod length as well as grains per pod than GM-1 and Hissar Sonali. Less prone to powdery mildew and other diseases causing heavy loss of yield and deterioration of grain quality.

Introduction

Fenugreek (*Trigonella foenum graecum* L.) is the third largest seed spice in India (after coriander and cumin).This is one of the oldest cultivated plant. It was a part of Indian diet even 3000 years ago. In India this crop is grown on an area of about 32600 hectares with production of about 35700 tonnes. Rajasthan is considered as 'fenugreek bowl' of the country and contribute about 80 per cent in the total production (Tiwari and Agrawal, 2004).

Table 28.1: Area, Production and Productivity of Seed Spices in India (2005–06)

Crop	Area '000 ha		Production '000 MT		Productivity kg/ha	
	India	Gujarat	India	Gujarat	India	Gujarat
Cumin	403.0	267.9 (66.5)	199.9	147.6 (73.9)	496	551
Coriander	356.2	8.67 (2.4)	254.9	8.42 (3.3)	716	971
Fennel	45.0	36.3 (80.6)	61.3	57.23 (93.4)	1499	1577
Fenugreek	32.6	4.9 (15.0)	35.7	5.9 (16.6)	1095	1204
Ajwain	14.5	2.3 (15.6)	4.4	1.3 (29.2)	303	565
Dill	09.4	4.8 (51.7)	5.7	3.8 (67.5)	606	792
Total	860.7	324.87	561.9	224.25		

(...) Share of Gujarat

Materials and Methods

Low yield potential of present day varieties, susceptibility to disease powdery mildew caused by the fungus *Erysiphe poligoni, Oidium* spp. and downy mildew caused by the fungus *Pernospora trigonellae* Gaum (Tiwari and Agrawal, 2004) and poor quality of the grain are the major constraints in fenugreek. Keeping these in view, the breeding programme was concentrated on developing varieties with high yield potential and adaptability, disease resistance and better grain quality. Selection plays a predominant role in crop improvement and most of the high yielding varieties are made out of the selection of germplasm. Most of the selections are made on the basis number of branches, number of pods per plant, pod length, grains per pod, and test weight have high genotypic coefficient of variation, heredity estimates and genetic advance. Above characters also exhibits maximum correlation with grain yield, hence they serve as useful selections (Mehta *et al.*, 1992, Hariharan and Vijaykumar, 1996, Kailash Chandra *et al.*, 2000, Tiwari and Agrawal, 2004, Prajapati, D.B., 2004). Therefore, large number of germplasm collected and evaluated at Jagudan. GM-2 was selected from the germplasm collected from Vadgam village of Banaskantha district and maintained as AC-279 at Jagudan centre. It was evaluated with other accessions in PRT and PYT and found promising for yield and other attributes. During 2001-02, it was included in co-ordinated trial for multi location testing.

Results and Discussion

High Stable Yield

The data presented in Table 28.2 revealed that at Jagudan (Gujarat) variety GM-2 has significantly out yielded GM-1 and Hissar Sonali by producing 10.03 and 20.32 per cent higher yield, respectively. This variety was also tested at Jobner (Rajasthan), Guntur(A.P), Kumarganj (U.P), Dholi(Bihar), Coimbatore (Tamil Nadu) and Hissar (Haryana). The proposed variety GM-2 significantly out yielded the state released/local check varieties in all the years of testing at Jobner, Guntur and Hissar and proved its superiority by producing on an average 48.29, 36.91 and 25.05 per cent higher yield, respectively. At Coimbatore, it yielded (5.49 q/ha) 16.31 per cent higher than local check. Further, GM-2 produced 59.65 and 58.96 per cent higher than National check Hissar Sonali at Guntur and Coimbatore, respectively. This clearly indicates high adaptability of GM-2 under varying environmental conditions. (Table 28.3). This variety possesses more number of branches, pods per plant, pod length and high test weight which are important yield contributing traits (Table 28.4).

Table 28.2: Yield Performance of Fenugreek Variety J.Fenu-244 (GM-2) Compared to Check Varieties at Jagudan (2001–02 to 2005–06)

Sl.No.	Year	Trial	Yield kg/ha						C.D.	C.V. Per cent	Trial Mean Kg/ha	Per cent IOC	
			J.Fenu–244 (GM–2)		GM–1		H.Sonali (N.C.)					GM–1	Hissar Sonali
			Y	R	Y	R	Y	R					
1	2001–02	CVT	1557	2	1532	3	–		182	8.48	1448	4.21	–
2	2002–03	CVT	2049*	1	1838	4	–		205	8.08	1753	11.48	–
3	2003–04	CVT	1906	3	1760	5	–		231	8.98	1783	8.30	–
4	2004–05	CVT	2079⊗	1	1846	4	1748	9	295	11.22	1820	12.62	18.94
5	2005–06	CVT	1971*⊗	1	1748	5	1617	10	174	7.10	1704	12.76	21.89
Overall Mean (5 trials)			1920	1	1745	5	1683		–	–	–	10.03	20.32
Superiority over check			5/5	–	–	–			–	–	–	–	–
Frequency of first rank			3/5	–	0/5	–			–	–	–	–	–
Pooled over 4 years (2002–03 to 2005–06)			2001*	1	1798	5			116	9.06	–	11.29	–

* ⊗:The proposed variety is significantly superior than checks GM–1 and Hissar Sonali, respectively.

Table 28.3: Yield Performance of GM-2 in Different States

Centre	Year	Yield (Qtl./ha)		Per cent IOC	Yield (Qtl./ha)		Per cent IOC	C.D. at 5 per cent
		GM–2	Local Check		GM–2	Hissar Sonali		
Kumarganj (UP)	2003–04	16.90	17.00		–	–		2.2
	2004–05	17.60	16.60		17.60	18.17		1.8
	Mean	17.25	16.80	2.68	17.60	18.17	(–) 3.24	–
Guntur (A.P.)	2002–03	5.68 *	4.79		–	–		0.73
	2003–04	11.02 *	9.53		–	–		1.42
	2004–05	13.57 *	7.78		13.57 *	8.50		2.62
	Mean	10.09	7.37	36.91	13.57	8.50	59.65	

Contd...

Table 28.3–Contd...

Centre	Year	Yield (Qtl./ha) GM–2	Yield (Qtl./ha) Local Check	Per cent IOC	Yield (Qtl./ha) GM–2	Yield (Qtl./ha) Hissar Sonali	Per cent IOC	C.D. at 5 per cent
Coimbatore	2003–04	5.48	4.53		–	–		–
(T.Nadu)	2004–05	5.50	4.90		5.50	3.46		
	Mean	5.49	4.72	16.31	5.50	3.46	58.96	–
Jobner (Raj.)	2001–02	16.00 *	10.00		16.00	16.67@		1.37
	2002–03	17.36 *	12.76		17.36	16.50@		2.24
	2003–04	14.84 *	9.72		14.84	13.82@		1.53
	Mean	16.06	10.83	48.29	16.06	15.66@	2.55	–
Hissar (Haryna)	2002–03	16.46 *	14.58		16.46	18.08		1.21
	2003–04	19.60 *	15.50		19.60	19.15		2.19
	2004–05	24.37 *	18.25		24.37	25.31		3.14
	Mean	20.14	16.11	25.02	20.14	20.85	(–) 3.52	–
Dholi (Bihar)	2002–03	11.30	13.40		–	–	–	–
	2003–04	13.40	14.00		–	–	–	–
	2004–05	12.10	14.00		–	–	–	–
	Mean	12.27	13.80	(–) 12.47	–	–	–	–

* Significant at 5 per cent level @ RMT–1 was used as National check.

Table 28.4: Duration and Other Characteristic Features of GM-2 in Comparison to Check

Sl.No.	Character	Mean Values (02–03 to05–06) GM–2	Mean Values (02–03 to05–06) GM–1(ch)	Mean Values (04–05 to 05–06) GM–2	Mean Values (04–05 to 05–06) Hissar Sonali (N.Ch.)
1.	Days to 50 per cent flowering	52.5	51.7	53.5	56.5
2.	Days to maturity	116	113.5	116	114.5
3.	Plant height (cm)	65.9	65.1	60.0	69.6
4.	No. of Branches/plant	5.5	5.3	5.7	4.9
5.	No. of pods/plant	62.2	51.9	39.0	33.9
6.	Length of pod (cm)	11.3	10.5	12.8	11.7
7.	No. of grains/pod	17.5	16.8	17.9	16.3
8.	1000 grain weight (gms)	16.49	14.32	15.35	11.60

Disease Resistance

The proposed variety GM-2 along with other promising entries and check varieties was screened in natural condition for powdery mildew disease in the field during 2003-04 to 2005-06. The per cent disease intensity (PDI) ranged from 35.0 to 40.0 per cent (Average 37.5 per cent) in GM-2 as compared to 49.15 to 81.25 per cent (Average 68.47 per cent) of GM-1 and 78.25 and 75.00 (average 76.62) of Hissar Sonali. Further the average (2001-02 and 2002-03) intensity of root rot (per cent) and downy mildew (0-9 grade) disease recorded at Jobner (Rajasthan) was 3.0 per cent and 3 respectively, indicating high degree of resistance of GM-2 against both the diseases (Table 28.5).

Table 28.5: Reaction to Diseases

Disease	*Year*	*Per cent Disease Intensity (PDI per cent)*		
		GM–2	*GM–1 (C)*	*H.Sonali (N.C.)*
Powdery Mildew	2003–04	40.00	49.15	–
	2004–05	35.00	81.25	78.25
	2005–06	37.50	75.00	75.00
	Mean	37.50	68.47	76.62
Root rot (per cent)*	Mean (01–02 and 02–03)	3.00	10 **	–
Downy Mildew* (0–9)	Mean (01–02 and 02–03)	3	6 **	–

* Root rot and downy mildew recorded at Jobner. ** Local Check variety.

Better Grain Quality and Medicinal Properties

The grains are bold, uniform in size, attractive and have more test weight. Therefore, it fetches Rs.200/qtl more than GM-1. It possesses high number of pods and more pod length as well as grains per pod than GM-1 and Hissar Sonali.

Table 28.6: Quality (Medicinal) Parameters of GM-2 and GM-1

Sl.No.	*Parameters*	*GM–2*	*GM–1*
1.	Bitterness Value (Unit)	1155	1040
2.	Swelling Index	11	11
3.	Foaming Index	100	<100
4.	Extractable Oil residue (per cent w/w)	5.3	5.1
	Fixed Oil Content (per cent v/w)	6.00	5.55
5.	Oleoresin Content (per cent w/w) (Alcohol Soluble Extractive)	12	12

According to pharmacopoeia minimum requirement of swelling index for fenugreek is 6.

Foaming Index indicates saponin content, higher the index more the saponin content, while bitterness indicates alkaloids (*Trigonelline*) content.

The fenugreek has high medicinal value. Medicinal "papyri" from ancient Egyptian tombs reveal that it was used both to reduce fever and also as a food. The seeds are used in colic flatulence, dysentery, diarrhoea and diabetes. Although the use of most spices in medicine has been declined substantially in recent years. Fenugreek is an important exception to the rule. Recent study in England indicates that Fenugreek seeds contain substantially the steroidal substance diosgenin which is used as a starting material in the synthesis of sex hormones and oral contraceptive. This diosgenin is derived from seponin. The chemical analysis carried out at Department of Pharmocognosy and Phytochemestry, S.K.Patel College of Pharmaceutical Education and Research, Ganpat Vidhyanagar

(Dist.Mehsana) as per pharmacopoeial methods and WHO guidelines reveal that, GM-2 possesses high bitterness value which is responsible for alkaloids (*Trigonelline*) content. The foaming index that indicates saponin contents in fenugreek is more in GM-2 than GM-1. The fixed oil content (V/W per cent) is 5.3 and 6.0 in GM-2 compared to 5.1 and 5.55 of GM-1, respectively. It clearly indicates that GM-2 has better medicinal properties than GM-1 (Table 28.6).

Profitability

The variety GM-2 fetches premium price in market because of bold grains attractive and uniform in size. Farmer can earn 43488 Rs/ha by growing GM-2 variety which is 7454 Rs/ha (20.69 per cent) more than GM-1 variety (Tables 28.7a and b)

Table 28.7a: Market Price and Economics of GM–2 Compared to Check

Sl.No.	Trader's name	APMC Markets	Market Price Rs./20 kg	
			GM–2	GM–1 (c)
1.	B. Somabhai Patel and Co.	Mehsana	425	390
2.	Patel Lalajibhai Jayantilal and Co.	Mehsana	425	400
3.	Patel Mahendrabhai Joitaram and Co.	Unjha	460	415
4.	Patel Bhogilal Ishwarlal	Unjha	460	415
5.	K. Mafatlal and Co.	Unjha	480	450
6.	Pravinbhai Amrutlal and Co.	Unjha	470	410
		Average	453	413

Table 28.7b: Profitability of GM–2 Over Check

Sl.No.	Particulars	GM–2	GM–1 (ch)
1.	Production (kg/ha)	1920	1745
2.	Price (Rs. per 20 kg)	453	413
3.	Total Income (Rs./ha)	43488	36034
4.	Net profit over GM–1 (Rs./ha)	7454	–
5.	Profit over GM–1 (per cent)	20.69	–

Table 28.8: Descriptors of Fenugreek Variety GM–2

Sl.No.	Characters	
1.	Plant Growth habit	Medium
2.	Leaf Margin	Serrated
3.	Leaf Size	Medium
4.	Leaf Tip	Rounded
5.	Category of Plant	Seed type
6.	Foliage Colour	Green

Contd...

Table 28.8–Contd...

Sl.No.	Characters	
7.	Stem pubescence	Absent
8.	Stem Shape	Cylindrical
9.	Inflorescence Colour	White
10.	No. of branches/plant	5.5
11.	Days to 50 per cent flowering	52.5
12.	Branching Habit	Divergent
13.	Pod colour	Light Green
14.	Pod shape	Curved
15.	Pod Length (cm)	11.3
16.	No. of pods/plant	62.2
17.	No. of seeds/pod	17.5
18.	Seed Colour	Light yellow
19.	Seed Size	Bold
20.	Plant height	65.9
21.	Days to maturity	116
22.	Seed yield/plant (gms)	17.95
23.	1000 seed weight	16.49

Conclusion

Variety GM-2 with high stable yield, better grain quality as well as medicinal properties and tolerance against diseases will provide very good alternative to GM-1 and Hissar Sonali. Cultivation of GM-2 will not only enhance the productivity but also improve profitability of fenugreek under varying environmental conditions.

References

Chandra, Kailash, Sastry, E.V.D. and Singh, D., 2000. Genetic variation and character association of grain yield and its component characters in fenugreek. *Agric. Sci. Digest.*, 20: 93–95.

Hariharan, K. and Vijaykumar, M., 1996. Studies on genetic variability in fenugreek. *South Indian Horti.*, 45(3 and 4): 143–147.

Mehta, K.G., Patel, R.H. and Kachhadia, B.T., 1992. Genetic variability and path analysis in fenugreek. *Indian Cocoa, Areca nut and Spices Journal*, 15(4): 114–116.

Prajapati, D.B., 2004. Genetic variability, character association and path analysis in fenygreek (*Trigonella foenum graecum* L.). *M.Sc.(Agri.) Thesis*, Sardarkrushinagar Dantiwada Agricultural University, Sardarkrushinagar, Gujarat (Unpublished).

Tiwari, R.S. and Agrawal, Ankur, 2004. *Production Technology of Spices*. International Book Distributing Co. Pub., Lucknow, pp. 250, 296.

Recent Advances in Seed Spices (2011) *Pages* **280–284**
Editors: **Y. Ravindrababu, R.K. Jaiman, K.D. Patel, N.R. Patel and S.B.S. Tikka**
Published by: **DAYA PUBLISHING HOUSE, NEW DELHI**

29

Studies on Character Association in Fenugreek (*Trigonella foenum-graecum* L.)

D.B. Prajapati[1], S.N. Badaya[2] and B.H. Prajapati[2]
[1]Centre for Research on Seed Spices, S.D.A.U., Jagudan, Dist.Mehsana, Gujarat
[2]Head of Department Botany (Ex), C.P. College of Agriculture, S.K.Nagar, Gujarat

ABSTRACT

Sixty four genotypes of fenugreek (*Trigonella foenum-graecum* L.) from geographic and genetic origin, among these 10 genotypes were included from the All India Co-ordinated Research Project Centre for Research on Seed Spices, S.D.A.U, Jagudan and 54 genotypes from different states of India were studied during *rabi* 2002-2003 adopting simple lattice design with 2 replications. Association of grain yield with its component characters was studied. Direct and indirect effects of the component characters on grain yield were estimated. Grain yield per plant had positive and significant correlations with number of pods per plant, number of seeds per pod and 1000-seed weight.

Path coefficient analysis revealed that characters *viz.*, number of pods per plant, days to 50 per cent flowering and test weight had highest positive direct effects on grain yield, while, days to maturity and number of branches per plant had negative and direct effects on grain yield.

Based upon the present investigations it is suggested that in future breeding programmes major emphasis should be given for number of pods per plant, 1000-seed weight and number of seeds per pod, as these traits had positive correlation with grain yield per plant with high direct effects. The genotypes, UM-323, J Fenu-244 and UM-18 recorded higher grain yield and possessing all the desirable morphological traits. These genotypes may be utilized as parents in hybridization programme.

Keywords: *Fenugreek, Correlation, Path coefficient, Trigonella foenum-graecum.*

Introduction

Low productivity of fenugreek (*Trigonella foenum graecum* L.) in India is due to non-availability of suitable high yielding varieties for various agro-climatic regions in the country and poor crop husbandry. Studies on variability and character associations for different agronomic characters in fenugreek have been made by Singh and Raghuvanshi (1984), Sharma *et al.* (1990), Dash and Kole (2000) and Saha and Kole (2001). Such studies are prerequisites for genetic improvement of any crop. The present study was undertaken to assess variability and interrelations among different quantitative characters in fenugreek in Gujarat.

Materials and Methods

The experimental material consisted of 64 diverse genotypes from geographic and genetic origin, among these 10 genotypes were included from spices research station, Gujarat Agricultural University, Jagudan and 54 different genotypes from Jobner (Rajasthan), Hissar (Haryana) and Lam (Andhra Pradesh). The genotypes were grown in a simple lattice design with two replications at Main Spices Research Station, Gujarat Agricultural University, (23°52' N, 72°43' E, 70 m MSL), (Gujarat) Jagudan during rabi 2002-03. Each plot consisted of two rows of 4 m length with inter and intra-row spacing of 30 cm and 10 cm, respectively. A uniform fertilizer dose of 20 kg N, 40 kg P_2O_5 and 10 kg S ha^{-1} was applied and normal agronomic practices were followed. Observations were taken on five randomly selected plants from each of the replications for quantitative characters. Mean data from each of the replications were used to estimate phenotypic and genotypic coefficients of variation (Burton 1952), heritability and genetic advance (Johnson *et al.*, 1955), correlation (Al. Jibouri *et al.* 1958), and path coefficient (Dewey and Lu 1959) following standard methods.

Results and Discussion

The analysis of variance revealed highly significant differences among the genotype for all the 9 characters studied. The estimates of phenotypic and genotypic coefficients of variation (PCV and GCV) were high for grain yield per plant followed by number of pods per plant and number of primary branches per plant and plant height. This is in agreement with earlier reports for number of primary branches per plant by Pant *et al.* (1984) and Singh (2000). Similar results have been reported by Hariharan and Vijaykumar (1996). Whereas, moderate GCV and PCV was recorded for 1000 seed weight (Table 29.1). The PCV in general, were higher than the GCV for all the characters which indicated considerable effect of environment on the character expression, although the differences between PCV and GCV being low. Hence, the effect of environment for expression of these characters will be less. High estimates of GCV were recorded for grain yield, number of pods per plant and number of primary branches per plant, which indicated the presence of high amount of genetic variability for these characters. These results are in agreement with the earlier findings for number of pods per plant reported by Shukla and Sharma (1978), Pant *et al.* (1984), Kailash Chandra *et al.* (2000).

High heritability estimates coupled with higher GCV and high to moderate expected genetic advance were observed for grain yield per plant. Similar results were recorded for grain yield per plant by Sharma *et al.* (1990), Meena (1994) and Singh (2000) and for seed yield per plant high heritability with high genetic advance were recorded by Pant *et al.* (1984). Moderate heritability estimates with the highest genetic advance was observed for number of pods per plant and reported by Pant *et al.* (1984). The results of phenotypic and genotypic coefficient of variability, heritability and genetic advance revealed that selection of quantitative characters like number of pods per plant, 1000 seed weight and number of grains per pod would be effective for improvement of grain yield in this population.

Table 29.1: Mean, Range and Genetic Parameters for Different Characters in Fenugreek

Sl.No.	Characters	Mean	Range		Coefficient of Variation (Per cent)		Broad Sense Heritability (Per cent)	Genetic Advance	Genetic Advance as Percentage of Mean
			Min	Max	GCV	PCV			
1.	Days to 50 per cent flowering	50.70	43.50	62.00	5.118	5.637	82.40	4.853	9.574
2.	Days to maturity	111.18	96.00	123.00	2.867	3.115	84.70	6.043	5.435
3.	Plant height (cm)	76.69	29.00	95.80	11.212	12.324	82.80	16.114	21.012
4.	No. of primary branches/plant	5.69	3.90	8.00	13.644	17.670	59.60	1.235	21.702
5.	No. of pods/plant	84.26	29.30	110.00	18.159	21.728	69.80	26.342	31.263
6.	Length of pod (cm)	8.88	7.40	11.50	7.405	8.901	69.20	1.127	12.690
7.	No. of grains/pod	16.24	11.90	22.00	9.359	11.075	71.40	2.645	16.293
8.	1000-seed weight (g)	13.96	8.620	17.125	10.778	11.251	91.80	2.970	21.268
9.	Grain yield/plant (g)	13.79	5.00	21.80	25.788	29.409	76.90	6.424	46.582

Positive and highly significant correlation of grain yield were found for number of pods per plant, 1000 seed weight and number of seeds per pod and thus these characters are considered as main components for grain yield (Table 29.2). The results are in agreement with Shukla and Sharma (1978), Sade *et al.* (1996), Kailash Chandra *et al.* (2000) and Dash and Kole (2000), with number of seeds per pod reported by Meena (1994) and test weight reported by Kohli *et al.* (1988).

Table 29.2: Phenotypic (P) and Genotypic (G) Correlation Among Nine Characters of Fenugreek

Characters	Days to 50 per cent Flower	Days to Maturity	Plant Height (cm)	No. of Primary Branches/ Plant	No. of Pods/ Plant	Length of Pod (cm)	No. of Grains/ Pod	1000-Seed Weight (gm)	Grain Yield/ Plant (gm)
Days to 50 per cent flower	1.000	0.8541**	0.1463	–0.0081	0.0445	0.0267	–0.0130	–0.0659	0.0117
Days to maturity	0.8436**	1.000	0.4258**	–0.1578	0.1730	0.0896	.0.0719	–0.0617	–0.0075
Plant height (cm)	0.1431	0.3885**	1.000	–0.1622	0.1423	0.0848	0.1512	–0.1024	–0.0113
No. of primary branches/plant	–0.1111	–0.0521	–0.2524**	1.000	0.2726**	0.0113	–0.0734	0.0274	0.0056
No. of pods/plant	0.0423	0.1408	0.1272	0.1763*	1.000	0.1219	0.2494**	–0.0030	0.5286**
Length of pod (cm)	0.0079	0.0952	0.1475	–0.2646**	–0.0177	1.000	0.4130**	–0.0549	0.1479
No. of grains/pod	–0.0384	0.0725	0.2059**	–0.3597**	0.1391	0.3235**	1.000	0.0838	0.3087**
1000–seed weight (gm)	–0.0760	–0.0617	–0.1190	–0.0672	–0.0675	–0.1330	0.0372	1.000	0.3966**
Grain yield/plant	–0.0316	–0.0411	–0.0319	–0.1353	0.5467**	0.0498	0.2548**	0.4326**	1.000

*, ** significant at 5 per cent and 1 per cent level respectively.

Upper diagonal shows phenotypic (P) and lower diagonal shows genotypic (G) correlation.

Among other inter character correlations, length of pod showed positive and non significant association with grain yield per plant at both the levels. While, days to flowering, plant height, number of branches per plant and days to maturity had negative and non significant association with grain yield per plant. Similar results were reported by Singh and Raghuvanshi (1984), Shukla and Sharma (1978), Berwal *et al.* (1996).

Therefore, the genetic correlation points out relatively strong associations of grain yield per plant with number of pods per plant, 1000 seed weight and number of grains per pod. Hence, these traits can be used as selection parameters.

Path analysis at genotypic level was carried out taking eight characters as independent variables and grain yield as dependent variables (Table 29.3). Srivastava and Sharma (1976) suggested that only direct component should be used for path analysis. The residual effect at genotypic level indicated that the eight characters included in this study explained the high percentage of variation in grain yield in this population. Days to maturity had negative direct effect on grain yield as been earlier reported by Singh and Raghuvanshi (1984). Numbers of pods per plant, 1000 seed weight and number of grains per pod have highest positive direct effects on grain yield. Highest positive effects on grain yield have been earlier reported by Meena (1994) and Sade *et al.* (1996), for number of pods per plant Berwal *et al.* (1996), for number of seeds per pod Singh (2000) and for test weight Kailash Chandra *et al.* (2000).

Table 29.3: Estimate of Path Coefficient Showing Direct and Indirect Effects on Seed Yield of Fenugreek at Genotypic Level

Characters	*Days to 50 per cent Flower*	*Days to Maturity*	*Plant Height (cm)*	*No. of Primary Branches/ Plant*	*No. of Pods/ Plant*	*Length of Pod (cm)*	*No. of Grains/ Pod*	*1000-Seed Weight (gm)*	*Correlation with Grain Yield*
Days to 50 per cent flower	*0.3260***	0.2784**	0.0477	–0.0026	0.0145	0.0087	–0.0042	–0.0215	0.0117
Days to maturity	–0.3141**	*–0.3678***	–0.1429	0.0192	–0.0518	–0.0350	–0.0267	0.0099	–0.0075
Plant height (cm)	–0.0015	–0.0039	*–0.0100*	0.0016	–0.0014	–0.0009	–0.0015	0.0190	–0.0113
No. of primary branches/plant	0.0014	0.0090	0.0280	*–0.1728**	–0.0471	–0.0019	0.0127	0.0010	0.0056
No. of pods/plant	0.0257	0.0812	0.0821	0.1572	*0.5764***	0.0702	0.1437	–0.0047	0.5286**
Length of pod (cm)	0.0021	0.0076	0.0068	0.0009	0.0097	*0.0798*	0.0329	–0.0017	0.1479
No. of grains/pod	–0.0015	0.0086	0.0179	–0.087	0.0295	0.0489	*0.1184*	–0.0044	0.3087**
1000–seed weight (gm)	–0.0263	–0.0206	–0.0408	0.0109	–0.0012	–0.0219	0.0334	*0.3989***	0.3966**

Residual effect: 0.6950.

Italic figures shows direct effect.

*, ** significant at 5 per cent and 1 per cent level respectively.

Conclusion

Results reveal that selection through the character viz; Numbers of pods per plant, 1000 seed weight and number of grains per pod rewarding to lay more emphasis for improvement of grain yield in fenugreek. Among these traits number of pods per plant had highest positive significant correlation on grain yield, hence; it should be given more concentration during selection for obtaining high yielding lines.

References

Al-Jibouri, H.A., Miller, P.A. and Robinson, H.F., 1958. Genotypic and environmental variance and co-variance in an upland cotton cross of interspecific origin. *Agron. J.*, 50: 3633–3637.

Berwal, K.K., Singh, J.V., Jhorar, B.S., Lodhi, G.P. and Kishor, C., 1996. Character association studies in fenugreek (*Trigonella foenum-graecum* L.). *Ann. Agric. Bio. Res.*, 1: 93–99.

Burton, G.W., 1952. Quantitative inheritance in grasses. *Proc. Sixth Intern. Grassland Cong.*, 1: 227–283.

Dash, S.R. and Kole, P.C., 2000. Association analysis of grain yield and its components in fenugreek (*Trigonella foenum-graecum* L.). *Crop Research (Hissar)*, 20: 449–452.

Dewey, D.R. and Lu, K.H., 1959. A correlation and path coefficient analysis of components of crested wheat grass seed production. *Agron. J.*, 51: 515–518.

Hariharan, K. and Vijaykumar, M., 1996. Studies on genetic variability in fenugreek. *South Indian Horti.*, 45(3 and 4): 143–147.

Johnson, H.W., Robinson, H.F. and Comstock, R.E., 1955. Estimates of genetic and environmental variability in soybean. *Agron. J.*, 47: 314–318.

Kailash Chandra, Sastry, E.V.D. and Singh, D., 2000. Genetic variation and character association of grain yield and its component characters in fenugreek. *Agric. Sci. Digest*, 20: 93–95.

Kohli, U.K., Sharma, O.P. and Singh, J., 1988. Genetic variability, correlation and path analysis in fenugreek. *Indian J. Hort.*, 45: 119–125.

Meena, S.S., 1994. Genetic divergence in germplasm of fenugreek. *M.Sc. (Agri.) Thesis*, Rajasthan Agricultural University, Bikaner (Unpublished).

Pant, K.C., Chandel, K.P.S. and Pant, D.C., 1984. Variability and path coefficient in fenugreek. *Indian J. Agric. Sci.*, 54: 655–658.

Sade, B., Akinerdem, F., Tomkoc, A., Topal, A., Aéar, R. and Soylu, S., 1996. Correlation and path analysis of yield and components in fenugreek (*Trigonella foenum-graecum* L.) lines. *Turkish J. Agric. and Fores.*, 20: 153–156.

Saha, A. and Kole, P.C., 2001. Genetic variability in fenugreek grown in sub-humid lateritic belt of West Bengal. *Madras Agric. J.*, 88: 345–348.

Sharma, K.C., Sharma, N.M. and Sharma, R.K., 1990. Nature of variability and association in fenugreek. *Indian J. Genet.*, 50: 260–262.

Shrivastava, M.N. and Sharma, K.K., 1876. Analysis of path coefficients in rice. *Z. Pflanzenzuchtung*, 77: 174–177.

Shukla, G.P. and Sharma, R.K., 1978. Genetic variability, correlation and path analysis in fenugreek. *Indian J. Agric. Sci.*, 48: 518–521.

Singh, A., 2000. Estimation of variability for grain yield and its component characters in fenugreek (*Trigonella foenum-graecum* L.). *M.Sc. (Ag.). Thesis*, Rajasthan Agricultural University, Bikaner (Unpublished).

Singh, R.R. and Raghuvanshi, S.S., 1984. Correlation and path coefficient analysis in fenugreek. *Indian J. Hort.*, 41: 294–298.

Recent Advances in Seed Spices (2011) *Pages* **285–290**
Editors: **Y. Ravindrababu, R.K. Jaiman, K.D. Patel, N.R. Patel and S.B.S. Tikka**
Published by: **DAYA PUBLISHING HOUSE, NEW DELHI**

30

Evaluation of Half Sib Progenies in Population of Fennel (*Foeniculum vulgare* Mill.)

Dhirendra Singh*, Brijendra Pareek, E.V. Divakara Sastry and K.C. Sharma
Department of Plant Breeding and Genetics, S.K.N. College of Agriculture, Rajasthan Agriculture University, Jobner, Bikaner, Rajasthan

ABSTRACT

The present investigation was aimed to evaluate half sib progenies in population of RF-101 of fennel for yield attributes. Half sib progenies were developed by open pollination. One hundred twenty five half sib progenies generated using the population RF-101 by open pollination were evaluated in Randomized Block Design with three replications at S.K.N. College of Agriculture, Jobner (Rajasthan). The variance was partitioned into between families and within families. The mean squares due to between families were significant over within families for all the characters indicating the inherent differences between the families for most of the characters. The between families components was again partitioned into mean squares due to replications, families and error component. The mean squares due to replication indicated non-significant differences for 50 per cent flowering. Plant height, seed yield per plant, umbels per plant and 1000 seed weight in half sib progenies, showing that all the three replication had uniformity for these characters. The differences between estimates of σ^2_A and σ^2_P was more in all the characters except branches per plant, umbellets per umbel and seeds per umbel in half sibs progenies indicated the presence of dominance variance in the inheritance of these characters.

* Corresponding Author: E-mail: dhirendrasinghjobner@gmail.com.

Introduction

The success of any breeding programme depends on the availability of genetic variability. Therefore, a method, which exposes more of the variability and gives greater genetic advance, would obviously be more effective in providing material for selection of good genotypes. In cross pollinated crops, selection schemes using S_1 progenies, open pollinated (half sibs) and full sibs have been very effective in exposing hidden variability and to improve base populations, so that selection can be made effective (Hallaur and Mirinda, 1988). Indian fennel (*Foeniculum vulgare* Mill) is an annual, stout and aromatic herb of 100–180 cm height having slender, branched smooth stem which become hollow at maturity with distinct veins. Fennel is a cross pollinated crop with high heterozygosity prevailing in the natural population (Ramanujam *et al.*, 1964). Reports regarding population improvement and comparative study of half sibs progenies is lacking in fennel, therefore the present investigation was carried out to estimate the extent of variation in a set of half-sibs families generated in fennel variety RF-101.

Materials and Methods

The experimental material in the present study was derived from the population of fennel *viz.*, RF-101, developed from a local collection from Sohela (Tonk). The plants of this population are erect and medium tall with long attractive seeds and belong to medium maturing group.

Half sib progenies were developed by open pollination. One hundred twenty five half sib progenies generated using the population RF-101 by open pollination were evaluated in Randomized Block Design with three replications at S.K.N. College of Agriculture, Jobner (Rajasthan). Data on plant height, branches per plant, umbels per plant, umbellets per umbel, seeds per umbel seed yield per plant and 1000 seed weight was obtained on a random sample of ten plants in each progeny/replication while, for 50 per cent flowering the data was recorded on whole plot basis. Ten single plants were randomly tagged before initiation of flowering so as to reduce the biasness in the plant selection in each progeny/replication for recording the above observations. Data obtained on various characters were subjected to appropriate statistical analyses.

Results and Discussion

Analysis of variance of plant height, branches per plant, umbels per plant, umbellets per umbel, seeds per umbel seed yield per plant and 1000 seed weight was done on the random sample of ten plants in each replication while, for the character 50 per cent flowering the analysis was done on plot means (Table 30.1). The mean squares due to between families were found significant over within families for all the characters indicating that the between families variance was higher than the within families variance. The between families mean squares were partitioned into MSS due to replications, families and error components. Results presented in Table 30.1 indicated that the mean squares due to replications were non-significant for days to 50 per cent flowering, plant height, umbels per plant, seed yield and 1000 seed weight and found significant for branches per plant, umbellets per umbel and seeds per umbel indicating the effect of environment on these characters. Perusal of Table 30.1 showed that the mean squares due to families were significant for all the traits indicating existence of significant difference among families for the characters. The error mean squares were found to be significant over within families mean squares for all the traits, the error mean squares was used to test the significance of mean squares due to families of all traits. The mean squares due to families was found significant for all the traits, this indicates that the variation among the families is significant. Such types of results have been obtained by Sastry and Yadav (1999) in taramira and Singh *et al.* (2003) in fennel.

Table 30.1: Mean Sum of Squares of Different Characters of Half Sib Progenies

Source	*d.f.*	*Days to Flowering*	*Plant Height (cm)*	*Branches per Plant*	*Umbels per Plant*	*Umbellets per Umbel*	*Seeds per Umbel*	*Seed Yield (gm)*	*1000-Seed Weight (gm)*
Between families	374	–	54103.13**	593.15**	5013.60**	2623.67**	949955.70**	883.89**	1332.46**
Replications	2	93.39	5172.80	2860.70**	447.97	1196.70**	870890.28**	156.46	203.86
Families	124	41.99*	12547.13**	95.22**	1205.30**	672.55**	230308.27**	157.61**	259.54**
Error	248	31.23	1843.82	18.77	149.82	49.74	21082.01	53.23	69.53
Within Families	3375	–	134.61	1.90	12.07	9.69	2887.72	11.03	6.72

*: Significant at $P = 0.05$ and **: Significant at $P=0.01$.

The Estimates of Mean, Range and Phenotypic Coefficient of Variation

The general mean and range for each character estimated from the family means over replications is presented in the Table 30.2. The mean number of days taken by the half sibs for 50 per cent flowering was 122.29 days and range for this character was 117.33 (HS 40 and HS 41) to 129.00 (HS 32) days. The general mean height of open pollinated plants was 100.16 cm and plant height ranges from 84.83 cm (HS 79) to 119.33 cm (HS 82). The branches per plant ranged from 4.23 (HS 31) to 6.93 (HS 68) whereas, the general mean was 5.60. The general mean of the umbels per plant was 10.67 and the lower limit of range was 7.17 (HS 31) whereas, the upper limit was 16.80 (HS 111). The general mean of half sib progenies for umbellets per umbel was 15.13 and the range was 12.20 (HS 9) to 21.77 (HS 22). The general mean for seeds per umbel was 177.21. The range for seeds per umbel was 111.40 (HS 9) to 280.43 (HS 97). The observed general mean of half sib progenies for seed yield was 6.67g and the minimum seed yield per plant was 4.01 g (HS 43) while maximum seed yield per plant was 10.35 g (HS 108). The general mean for seed weight was 3.89 g per 1000 seed. The minimum weight for 1000 seeds was 2.81 g (HS 29) and the maximum was 6.59 g (HS 91) for 1000 seeds. In half sib progenies the variability was higher for 1000 seed weight, umbels per plant, seeds per umbel and seed yield per plant on the basis of phenotypic coefficient of variation. The PCV values indicated that the variation among the progenies was higher for most of the traits. The magnitude being highest in 1000 seed weight, seed yield, umbels per plant and seeds per umbel.

Table 30.2: General Mean, Range and Estimates of Additive (σ^2_A), Phenotypic Variance (σ^2_P) in Half Sib Progenies

Characters	*General Mean*	*Range*	σ^2_F	σ^2_{HS}	σ^2_A	σ^2_P	*PCV (Per cent)*
Days to flowering*	122.29	117.33–129.00	–	–	3.58 (σ^2_g)	34.82	4.83
1.55 (GCV, per cent)							
Plant height (cm)	100.16	84.83–119.33	356.78	931.00	1427.11	3270.93	57.10
Branches per plant	5.6	4.23–6.93	2.55	8.23	10.19	28.96	96.10
Umbels per plant	10.67	7.17–16.80	35.18	81.50	140.73	290.55	159.75
Umbellets per umbel	15.13	12.20–21.77	20.76	34.43	83.04	132.78	76.16
Seeds per umbel	177.21	111.40–280.43	6974.21	13135.23	27896.83	48978.85	124.89
Seed yield (gm)	6.67	4.01–10.35	3.48	17.91	13.92	67.85	122.85
1000 seed weight (gm)	3.89	2.81–6.59	6.33	29.48	25.33	94.87	263.95

*: The estimates presented are based on plot means.

Estimates of Components of Variance, Heritability and Genetic Advance

Table 30.3 indicated that heritability in broad sense for days to 50 per cent flowering and genetic advance were low (10.30 per cent and 1.03 per cent, respectively). The heritability in narrow sense was found to be high in umbellets per umbel (62.54 per cent), seeds per umbel (56.96 per cent), umbels per plant (48.44 per cent), plant height (43.63 per cent) and branches per plant (35.20 per cent), while moderate in 1000 seed weight (26.71 per cent) and seed yield per plant (20.73 per cent). The expected genetic advance due to selection within families was higher in comparison to expected genetic advance due to selection among families. This indicates that selection within families is more effective than selection among families. The expected genetic advance due to among families was highest in 1000 seed weight (6.43 per cent) and lowest in the seeds per umbel (0.31 per cent). The estimate of genetic advance due to selection within families was highest in umbels per plant (84.42 per cent) and was lowest in seeds per umbel (4.21 per cent). The genetic advance due to among families + genetic advance due to within families was highest in umbels per plant and lowest in seeds per umbel. Plant height and seeds per umbel in half sib progenies showed high heritability accompanied with low genetic advance that indicated preponderance of non-additive gene action. The high heritability is being exhibited due to favourable influence of environment rather than genotype and for such traits may not be rewarding. Such results also reported by Arha *et al.* (1990) and Debnath and Azad (1993) in maize. Days to 50 per cent and seed yield in half sib progenies showed low heritability with low genetic advance that indicated that the character was highly influenced by environmental effects and selection should be ineffective. Low heritability was reported by Reis *et al.* (1982) and Shahi *et al.* (1986) and low genetic advance was reported by Choudhary and Singh (1983) in maize.

Table 30.3: Heritability (h^2) and Genetic Advance (GA) in Half Sib Progenies of Fennel

Characters	h^2 as Per cent	GA Among Families	GA Among Families as Per cent Mean	GA within Families	GA within Families as Per cent Mean	GA Among + GA within	GA Among + GA within as Per cent Mean
Days to flowering*	10.30(BS)	1.25 (GA)	1.03 (GA)				
Plant height (cm)	43.63	0.39	0.39	8.19	8.18	8.58	8.57
Branches per plant	35.20	0.32	5.69	4.15	74.09	4.47	79.78
Umbels per plant	48.44	0.44	4.17	9.01	84.42	9.45	88.59
Umbellets per umbel	62.54	0.62	4.10	6.62	43.75	7.24	47.85
Seeds per umbel	56.96	0.55	0.31	7.46	4.21	8.01	4.52
Seed yield (gm)	20.73	0.20	3.00	0.98	14.62	1.18	17.62
1000 seed weight (gm)	26.71	0.24	6.43	2.91	78.72	3.15	85.35

*: The estimates presented are based on plot means.

References

Arha, M.D., Sarda, R.P. and Agrawal, K.N., 1990. Studies on maize gene pools: Heritability and expected genetic advance. *Acta Agronomica Hungarica*, 39(1–2): 115–120.

Chaudhary, S.K. and Singh, J., 1983. Genetic architecture of an early maturing composite following five cycles of full sib selection. In: *Abstract of Contributed Papers of 15th Int. Congress of Genetics*, (Ed.) M.S. Swaminathan, December, New Delhi, 2(936): 528.

Debnath, S.C., 1987. Genetic variability in maize (*Zea mays* L.). *Bangladesh Journal of Agriculture*, 12(4): 217–221.

Hallauer, A.R. and Miranda, J.B., 1988. *Quantitative Genetics in Maize Breeding*, 2nd Edn. Iowa State University Press, Ames, I.A., pp. 75–82.

Ramanujam, S., Joshi, B.S. and Saxena, M.B.L., 1964. Extent and randomness of cross pollination in some umbelliferous spices of India. *Indian Journal of Genetics and Plant Breeding*, 24: 62–67.

Reis, F.P., Silva, J.C., Regazzi, A.J. and Oliveira, L.M., 1982. Heritabilities, correlations and selection indices in maize (*Zea mays* L.). *Revista Ceres*, 29: 268–283 (*Plant Breeding Abstracts* 53: 6240).

Sastry, E.V.D. and Yadav, B.S., 1999. Evaluation of half sib progenies developed from RTM-314 in *Eruca sativa* for seed yield and its components. *Annals of Arid Zone*, 38(1): 49–53.

Shahi, J.P. and Singh, I.S., 1988. Genetic variability in maize composite. *Crop Improvement*, 15(1): 101–104.

Singh, V.V., Singhania, D.L. and Sastry, E.V.D., 2003. Evaluation of half-sib progenies in fennel (*Foeniculum vulgare* Mill.). *Indian Journal of Genetics and Plant Breeding*, 63(3): 277–278.

Recent Advances in Seed Spices (2011) *Pages* **291–294**
Editors: **Y. Ravindrababu, R.K. Jaiman, K.D. Patel, N.R. Patel and S.B.S. Tikka**
Published by: **DAYA PUBLISHING HOUSE, NEW DELHI**

31

Organogenesis in Callus Cultures of Cumin (*Cuminum cyminum* L.)

M.L. Jakhar
Department of Plant Breeding and Genetics, SKN College of Agriculture, Rajasthan Agricultural University, Jobner – 303 329, Rajasthan

ABSTRACT

Shoot apex explants of cumin genotypes were cultured on various concentrations of cytokinin and auxin used singly or in combination for induction of callus in cumin. Profuse callus induction was observed in shoot apex explant at 0.3 mg/l BAP and this callus remained potent for longer time. This callus was sub-cultured in varying concentrations of cytokinins and auxins either alone or in combination for regeneration. However, it did not regenerated in complete plant even after various manipulations in all genotypes except RZ-19 where shoot organogenesis was observed when MS medium supplemented with BAP(0.1mg/l) + NAA (0.1mg/l). Regenerated shoots were separated and sub-cultured on plain MS medium induced roots at the base and complete plant were obtained. This investigation suggests need of genotypic specific protocol with interacting phytohormone and physical factor combination for improvement of cumin through non-conventional breeding methods.

Introduction

Cumin (*Cuminum cyminum* L.) an annual herb, is one of the most extensively used seed spice crop. The average productivity of cumin is low. Rendering between 350-500 kg/ha depending upon season particularly on disease and insect pest problems. The crop has limited variability for most of the traits which limits the scope for crop improvement (Peter *et al.,* 1994). Thus to increase the production and productivity of the crop research has to be concentrated on development of high yielding varieties. Development of high yielding varieties in cumin is possible only by widening genetic base. Thus induction of genetic variability is very essential in cumin. *In vitro* organogenesis of many plant species has been the result of growth medium and environmental manipulation (Fosket, 1979). In addition to manipulating the media and culture condition which are essential to help express the ability of right genotypes, the improvement for plantlet regeneration may also be made by selecting and recombining the suitable genotypes (Butenko, 1985). Therefore, the variable organogenesis response of different genotypes cultivated *in vitro* must be work out to develop perfect culture system for individual crop before they can be any value in routine breeding programmes related to crop improvement.

Materials and Methods

Seeds of cumin (var. RZ-19) were surface sterilized with 0.1 per cent mercuric chloride for 3 minutes then washed thoroughly by sterilized distilled water for 5-6 times. The seeds were then transferred aseptically to sterilized culture tubes containing paper bridge partly dipped in ½ Murashige and Skoog [6] medium for germination. Shoot apex explants were excised from 15 days old seedlings. These explants were inoculated in 100 ml wide neck Erlenmeyer conical flask dispensed with 40 ml of MS medium supplemented with various levels of cytokinins (BAP/Kn-0.1-2.0 mg/) and auxins (IAA/NAA-0.1-2.0 mg/l) added singly or in combination for induction of callus and were exposed to 14:10 photoperiod (3000 lux) at 25 ± 2° C temperature. Each treatment comprised a minimum of ten cultures and was repeated on particular growth regulator level twice and observed regularly. Observations were taken at seven day interval. At the end of each experiment (35 days after incubation) observations were taken for callus initiation and callus was visually assessed on a scale of slight, medium and profuse callus.

Stock callus cultures developed on MS medium supplemented with 0.3 mg/l BAP were assessed for their organogenetic capacity on various plant growth regulators added singly or in combination in medium. The levels of various growth regulators and their combination for the regeneration study were same as those use for callus induction. Mainly the genotype RZ-19 was used for various studies and rest were used at responsive levels only.

Results and Discussion

Induction of Callus

The shoot apex explants responded differently to different concentrations of cytokinins and auxins used either singly or in combination for induction of callus. After 15-18 days of incubation on medium supplemented with cytokinins predominantly multiple shoots induced from shoot apex explants at lower levels (0.3–0.05mg/l). High levels inhibited multiple shoots induction completely. After induction and growth of multiple shoots medium to profuse callus appeared at the base of shoots on BAP supplemented medium, whereas very little callus was noticed from shoots developed under the influence of Kn. Various levels of auxins did not evoke a significantly different response on various concentration of IAA/NAA in shoot apex explants. IAA supplemented medium induced only roots in shoot apex explants, whereas little callus induction was observed from the explants incubated on NAA.This observation in closed agreement of Bhojwani and Raidan (1996) and Narayanaswamy (1994) observation suggesting differential response of plant growth regulators with kind of explants.

Greenish, Semi-compact callus at the base of shoot apex explant was initiated when medium was supplemented with 0.3 mg/l BAP with 100 per cent frequency. Both lower and higher levels of BAP gave lower response. This observation is in agreement to Skoog and Miller (1957) who stated that response of *in vitro* cultures depends upon exogenous ratio of auxins/cytokinins. In combination treatments (0.3 mg/l BAP + 0.1 mg/l IAA and 0.3 mg/l BAP + 0.1 mg/l NAA) profuse callus proliferation was also observed. Callus induced from shoot apex explant at 0.3 mg/l BAP remained fresh for longer time and retained potency through subcultures. Similar results were reported by Jakhar *et al.* (2003). All genotypes induced profuse callus at 0.3 mg/l BAP except local genotype. The frequency of callus induction was 100 per cent in all genotypes except RZ-19, RZ-209and UC-223 (Table 31.1). This also indicates strong effect of genotypes on callus induction and freqency as observed by Dietert *et al.* (1982).

Table 31.1: Callus Induction in Different Genotypes of Cumin

Genotype	*Callus Induction in MS Medium Supplemented with 0.3 mg/l BAP*
RZ-19	+++, (80)
RZ-209	+++, (80)
EC-279082	+++, (100)
UC-220	+++, (100)
UC-198	+++, (100)
JC-147	+++, (100)
UC-217	+++, (100)
UC-218	+++, (100)
UC-223	+++, (100)
Local	++, (80)

+++: Profuse callus; ++: Medium callus.

( ): Values in parenthesis represents frequency of callus induction.

Shoot Morphogenesis in Callus Cultures

Semi-compact callus proliferating from the base of shoot apex explants on medium supplemented with BAP (0.3 mg/l) was also assessed for its organogenetic capacity. Upon sub-culturing on cytokinins (BAP/Kn 0.1-2.0 mg/l) added singly in medium and in combination with auxins (IAA/NAA 0.1-2.0 mg/l). However, it did not regenerated in complete plant even after various manipulations in all genotypes only callus proliferation was observed. The Shoot morphogenesis (40 per cent) was observed in shoot apex derived callus cultures (0.3 mg/l BAP) of genotype RZ-19 upon sub-cultures on MS medium supplemented with BAP (0.1mg/l) + NAA (0.1mg/l). Similar results were not obtained in other genotypes at same level of plant growth regulators only callus proliferation was observed. In current investigation, regeneration from callus cultures could not be established truly. There could be many reasons for this. Since 10 genotypes, including one exotic and one land race, were tested for their morphogenetic potential, it appears that *Cumium cyminum* may be inherently less regenerative or recalcitrant. Thus a close molecular scrutiny is needed for such contrasting system for a predictable outcome in future efforts on *in vitro* regeneration in this crop.

Induction of Roots in *in vitro* Regenerated Shoots

Shoots regenerated from callus in cumin variety RZ-19 on the medium supplemented with BAP (0.1mg/l) + NAA (0.1mg/l) were separated and sub-cultured on plain MS medium induced roots at the base and complete plant was obtained. These plants grew well under *in vitro* conditions and flowered profusely but seed setting was not observed. When these seedlings were cultured on 1 mg/l NAA, eighty per cent plantlets produced roots within three weeks. Forty-sixty per cent of rooted shoots produced flowers after 4 more weeks on this medium. Seed formation was also recorded after 65-70 days of inoculation.

References

Bhojwani, S.S. and Rajdan, M.K., 1996. Cellular totipotency. In: *Plant Tissue Culture: Theory and Practice.* Elsevier, Netherlands, pp. 95–123.

Butenko, R.B., 1985. Some features of cultured plant cells. In: *Plant Cell Culture*, (Ed.) R.G. Butenko. Mir Publishers, Moscow, pp. 1–11.

Dieterat, M.F., Barron, S.A. and Yoder, O.C., 1982. Effect of genotypes on *in vitro* cultures in the geneus Brassica. *Plant Science Letters*, 26: 233–240.

Fosket, D.E., 1979. Hormonal control of morphogenesis in cultured tissues. In: *Plant Growth Substances*, (Ed.) F. Skoog. Springer-Verlag, Heidelberg, pp. 362–369.

Jakhar, M.L., Sastry, E.V.D., Dhayal, L.S. and Rathore, V.S., 2003. Callusing response of hypocotyl, shoot apex and leaf explant of cumin (*Cuminum cyminum* L.) to plant growth regulator treatments. *Indian J. Plant Physiol. (Special Issue)*, pp. 685–688

Murashige, T. and Skoog, F., 1962.A revised medium for rapid growth and bioassay with tobacco tissue cultures. *Physiol. Plant*, 15: 473–497.

Narayanaswamy, S., 1994. Tissue (callus) cultures. In: *Plant Cell and Tissue Culture*. Madras Science Foundation, Madras, pp. 51–93.

Peter, G.B., Vedamuthan, M.D., Ablut Khader and Solarajan, F., 1994. Improvement of seed species. In: *Advances in Horticulture, Vol. 9: Plantation and Spice Crops Plant*, (Eds.) Chadha and P. Rethinum. Malhotra Publishing House, New Delhi, pp. 345–374.

Skoog, F. and Miller, C.O., 1957. Chemical regulation of growth and organ formation in plant tissue cultured *in vitro*. *Symposium of Society of Experimental Biologists*, 11: 118– 131.

Recent Advances in Seed Spices (2011) *Pages* **295–298**
Editors: **Y. Ravindrababu, R.K. Jaiman, K.D. Patel, N.R. Patel and S.B.S. Tikka**
Published by: **DAYA PUBLISHING HOUSE, NEW DELHI**

32

Effect of Temperature and Photoperiod on Organogenesis of Cumin (*Cuminum cyminum* L.)

M.L. Jakhar
Department of Plant Breeding and Genetics, SKN College of Agriculture, Rajasthan Agricultural University, Jobner – 303 329, Rajasthan

ABSTRACT

Cumin (*Cuminum cyminum* L.) is an important spice crop belonging to family apiaceae. The productivity of the crop is very low due to narrow genetic variability thus, enhancement of genetic variability is essential for crop improvement, which may be enchanced through *in vitro* techniques. In organogenesis cells are redifferentiated from dedifferentiated state leading to somaclonal variation. *In vitro* organogenesis of many plant species is influenced by the role of phytohormones, light and temperature of the medium. Thus the present investigation was carried out to find out the role of photoperiod and temperature is organogenesis of cumin. The profuse callus induced at 0.3 mg/l from shoot apex explants of cumin was subjected in different regimes of temperature (15±2, 20±2, 25±2 and 30±2) and photoperiod (16:8, 14:10, 12:12 and 8:16) with regenerating medium (M.S. medium supplemented with 0.1 mg/l BAP + 0.1 mg/l NAA). The maximum organogenesis were observed at 25±2 ° C, the number of shoots decreased with decreasing temperature and totally inhibited at higher temperature indicating effects of temperature in exploitation of morphogenesis. The significant differences were also observed at different photoperiod regimes. Maximum shoot organogenesis was observed at 14:10 photoperiod regime and organogenesis was completely inhibited at 8:16 photoperiod regimes. Thus behavior of organogenesis of cells and tissues is largely dependent on light and temperature even on responsive level of plant growth regulator and medium. Thus the optimum organogenesis in cumin can be achieved only when cultures are subjected with proper temperature (25±2) and photoperiod (14:10) regimes along with responding levels of plant growth regulators.

Introduction

Cumin (*Cuminum cyminum* L., 2n = 14), a seed-spice crop species of Apiaceae family, is an important foreign exchange earner. Gujarat and Rajasthan are the important cumin growing states of India. High yield, increase in volatile oil content, large seed size and attractive dark brown colour of grains are the principal objectives of cumin improvement programme. The germplasm evaluation reports have indicated existence of low genetic variation for seed yield, seed yield components, volatile oil content and resistance to diseases (Peter *et al.*, 1994). Thus, induction of genetic variability is very important in cumin.

Morphogenesis *in vitro* culture is dependent on large number of factors such as inorganic ions, amino acids, nucleic acid base adenine, plant growth regulators like cytokinins, auxins and gibberellic acid as well as physical factors like culture conditions *viz.*, light, temperature, physiological state of donor plant, age of explant and genotypes (Fosket, 1979). Besides, very little is known about endogenous hormone levels of cultured tissues or how these levels change in response to culture conditions. The response of explants to initiate callus varies with the type and concentration of plant growth regulators (Fosket, 1979). However, no such systematic studies have been conducted in cumin to initiate callus from various explants and to induce *in vitro* regeneration. Also, there are no reports on induction of *in vitro* organogenesis under the influence temperature and photoperiod regimes. Thus the present investigation was carried out to find out the role of photoperiod and temperature on organogenesis of cumin.

Materials and Methods

Seeds of cumin (var. RZ-19) were surface sterilized with 0.1 per cent mercuric chloride for 3 minutes then washed thoroughly by sterilized distilled water for 5-6 times. The seeds were then transferred aseptically to sterilized culture tubes containing paper bridge partly dipped in ½ Murashige and Skoog (1962) medium for germination. Shoot apex explants were excised from 15 days old seedlings. These explants were inoculated in 100 ml wide neck Erlenmeyer conical flask dispensed with 40 ml of MS medium supplemented with BAP (0.3 mg/l) were exposed to 14:10 photoperiod (3000 lux) at 25 ± 2° C temperature for callus induction. This potent callus sub-cultured on regeneration medium (BAP- 0.1 mg/l + NAA- 0.1 mg/l) to see the effect of different photoperiod and temperature regimes on in vitro cultures, specially in relation to shoot organogenesis under different temperature(15±2, 20±2, 25±2 and 30±2.) and photoperiod (16:8, 14:10, 12;12 and 16:8) regimes. Each treatment comprised a minimum of ten cultures and was repeated on particular growth regulator level twice at each temperature and photoperiod regimes to assess their effects on organogenesis. Observations were taken at seven day interval. At the end of each experiment (35 days after incubation) observations were taken for shoot morphogenesis and days to shoot induction. The data were analyzed using CRD design.

Results and Discussion

The profuse semi compact callus induced on the base of shoot apex explants inoculated on MS medium supplemented with 0.3 mg/l BAP. This potent callus sub- cultured for morphogenetic response on different regimes of temperature and photoperiod in regeneration medium (Ms medium containing 0.1 mg/l BAP+0.1 mg/l NAA).

Effect of Temperature

When callus sub-cultured on the regeneration medium in different temperature regimes at 14:10 photoperiod. The maximum number of shoots (10.2) induced when culture incubated at 25±2° C. The number of shoots drastically reduced with decrease in temperature regimes and completely inhibited at higher temperature regimes (30±2° C) (Table 32.1). Similar observation was also observed by Bhargava (1986) in regeneration from leaf explants of *vigna aconitifolia* where, maximum regeneration was observed at 32° C. Such temperature requirement matched to field temperature requirement of normal crop growth. This statement is also true in case of cumin which requires low temperature.

Table 32.1: Effect of Temperature Regimes on Organogenesis of Callus Cultures of Cumin

Temperature Regime		*Number of Shoot Regenerated*	*Days Taken for Shoot in Duration*
Days	*Night*		
15±2	10±2	4.6	29.20
20±2	15±2	6.2	25.00
25±2	20±2	10.2	21.60
30±2	25±2	–	–

Effect of Photoperiod

When callus sub-cultured on regeneration medium in different photoperiod regimes at 25±2°C temperature. Shoot organogenesis response was best when the dark period was shortest and the reverse when the dark period was longest. Regeneration was not observed in cultures incubated at 8:16 hr photoperiod. The days taken for the shoot induction were non-significant at responsive levels of photoperiod (Table 32.2). Maximum shoot organogenesis (8.80) was observed when cultures were incubated at 14:10 hr photoperiod. The complete inhibition of shoot organogenesis at 8:16 hr. photoperiod due to the role of phytochrome in shoot induction Bagga *et al.* (1985). This for enhance the organogenesis in cumin physical factor like temperature and photoperiod should be kept at 25±2°C and 14:10 hr.

Table 32.2: Effect of Photoperiod Regimes on Organogenesis of Callus Cultures of Cumin

Photoperiod	*Number of Shot Regenerated*	*Days Taken for Shoot Induction*
16: 8	6.80	16.40
12: 12	6.20	17.40
8: 16	-	-
14: 10	8.80	16.00

References

Bagga, S., Rajasekhar, V. K., Guha-Mukherjee, S. and Sopory, S.K., 1985. Enhancement in the formation of shoot initials by phytochrome in stem callus cultures of *Brassica oleracea* var. *botrytis*. *Plant Science*, 38: 61–64.

Bhargava, S.C., 1986. Studies on growth and morphogenesis in cell and tissue cultures of *Vigna aconitifolia* (Jacq.) Marechal. *Ph.D. Thesis*, University of Rajasthan, India.

Butenko, R.B., 1985. Some features of cultured plant cells. In: *Plant Cell Culture*, (Ed.) R.G. Butenko, Mir Publishers, Moscow, pp. 1–11.

Fosket, D.E., 1979. Hormonal control of morphogenesis in cultured tissues. In: *Plant Growth Substances*, (Ed.) F. Skoog. Springer-Verlag, Heidelberg, pp. 362–369.

Murashige, T. and Skoog, F., 1962. A revised medium for rapid growth and bioassay with tobacco tissue cultures. *Physiol. Plant*, 15: 473–497.

Peter, G.B., Vedamuthan, M.D., Ablut Khader and Solarajan, F., 1994. Improvement of seed species. In: *Advances in Horticulture, Vol. 9: Plantation and Spice Crops Plant*, (Eds.) Chadha and P. Rethinum. Malhotra Publishing House, New Delhi, pp. 345–374.

Recent Advances in Seed Spices (2011) *Pages* **299–310**
Editors: **Y. Ravindrababu, R.K. Jaiman, K.D. Patel, N.R. Patel and S.B.S. Tikka**
Published by: **DAYA PUBLISHING HOUSE, NEW DELHI**

33

Effect of Levels and Sources of Sulphur on Yield and Chemical Composition of Cumin (*Cuminum cyminum* L.) in Loamy Sand

P.K. Patel, B.B. Patel, N.J. Jadav, R.P. Pavaya, M.V. Patel, V.R. Patel and C.H. Raval
Department of Agricultural Chemistry and Soil Science, C.P. College of Agriculture, S.D. Agricultural University, Sardarkrushinagar – 385 506, Gujarat

ABSTRACT

A field experiment entitled "Effect of levels and sources of sulphur on yield and chemical composition of cumin (*Cuminum cyminum* L.) in loamy sand" was conducted at Agronomy Instructional Farm, C. P. College of Agriculture, S. D. Agricultural University, Sardarkrushinagar (Gujarat) in randomized block design (factorial) with three replications during the *Rabi* season of 2002-03. The treatments consist of three levels (10, 20 and 30 kg ha^{-1}) and three sources of sulphur (gypsum, elemental sulphur and ammonium sulphate). The results indicated that irrespective to sources and levels of sulphur, an application of sulphur significantly increased seed and straw yield, volatile oil and protein content of cumin seed and uptake of N, P, K, S, Mg, Fe, Mn, Zn and Zu by cumin crop over the corresponding values with no sulphur. The mean effect of sulphur significantly increased the amount of available S in the soil after harvest of crop. The different sources of sulphur were found significant in influencing the volatile oil content of seed, S and Fe uptake by seed. The maximum S and Fe uptake by seed were recorded with gypsum, where as maximum S content of soil was found due to the application of elemental sulphur, while maximum volatile oil content was recorded with ammonium sulphate application. The levels of sulphur produced significant effect on seed and straw yield, volatile oil and protein content of seed, N, K, S, Fe and Cu uptake by the crop as well as available status of phosphorus and sulphur in soil after harvest. The maximum seed yield, protein content and available S in the soil after harvest were recorded with 30 kg S ha^{-1}, whereas straw yield, volatile oil content of seed and Fe uptake by the crop were recoded maximum under 20 kg S ha^{-1}. The significant interaction effect of sources and levels of sulphur were observed in case of volatile oil content of seed. Maximum volatile oil content was observed with ammonium sulphate @ 20 kg ha^{-1}. Application of gypsum was found to be the most economical sources and recorded the highest net return of Rs. 29054 ha^{-1}. Level 30 kg S ha^{-1} gave the highest net return of Rs. 29346 ha^{-1}. As far as combined effect is concerned, combination of gypsum @ 30 kg S ha^{-1} (Gy x S_{30}) reported the highest net return.

Keywords: *Cumin, Sulphur, Gypsum, Ammonium sulphate, Elemental sulphur.*

Introduction

India is the largest producer (25 lakh tones), consumer (22.8 lakh tonnes) and exporter (2.10 lakh tonnes) of spices in the world. According to Bureau of Indian standard, 63 spices are grown in the world. India leads with 60 per cent of world production of spices and 40 per cent world exports. Spices exports earning is worth about Rs. 2000 crores a year. India's share in world market is about 35 per cent. Among all the major spices grown, "cumin" (*Cuminum cyminum* L.) locally known as "*Jeeru*" is one of the important spice crop. It is an annual herbaceous plant belonging to family umbelifereae of dicotyledon group. In India, it is widely grown in Rajasthan, Gujarat, Uttar Pradesh and Punjab with total area of about 2.8 lakh hectares and 91,700 M.T. production. In Gujarat, area under cumin crop is around 1.087 lakh hectares and production is about 35,400 M.T. with average productivity 326 kg ha^{-1} in the districts *viz.*, Ahmedabed, Banaskantha, Mehsana, Sabarkantha, Kheda, Kachchh and Saurashtra region (Directorate of Agriculture, 2001).

Sulphur is an essential element for all plants and it is indispensable for the synthesis of certain amino acids, proteins, oils, chlorophyll as well as for various metabolic and enzymatic processes of the plants. Sulphur is involved in the activation of enzymes, in the formation of glucosides or glucosinolates and the SH-sulphydryl linkage which provide the source of pungency in oils etc. It is a part of co-enzyme pyrophosphate, vitamins such as biotin and thiamine. Considering the requirement of sulphur, it can be rightly called as the fourth major element in plant nutrition, after nitrogen, phosphorus and potassium (Tandon and Messick, 2002). Reports on the incidence of sulphur deficiency are appearing mainly due to intensification of agriculture, use of high analysis fertilizers and increasing use of high yielding crop varieties in coarse textured soil and soils low in organic matter. Analysis of 90 samples from dryland fields of North Gujarat showed that more than 90 per cent of the samples were deficient in heat soluble sulphur (Patel and Kalyanasundaram, 1988). According to Tandon (1989), the soils of Banaskantha and Sabarkantha districts are deficient in sulphur. Cumin is extensively cultivated in North Gujarat, which is considered as a more potential area for cumin cultivation. As it contains 17.7 per cent protein and 23.8 per cent fat, it may respond to sulphur fertilization to improve seed quality and seed yield. However, very little scientific information is available on sulphur nutrition in cumin crop, particularly under North Gujarat conditions. Keeping in view of above facts, the present investigation was carried out to study the effects of levels and sources of sulphur like gypsum, elemental sulphur and ammonium sulphate in loamy sand soil of North Gujarat.

Materials and Methods

A field experiment was conducted at Agronomy Instructional farm, C. P. College of Agriculture, Sardarkrushinagar Dantiwada Agricultural University, Sardarkrushinagar during the *rabi* 2002-03. The experiment was laid out in randomized block design comprising of three levels (10, 20 and 30 kg ha^{-1}) and three sources of sulphur (gypsum, elemental sulphur and ammonium sulphate) replicating thrice and was conducted in Typic Ustipsament with loamy sand texture having pH 7.6 and Ec 0.12 dsm^{-1}. The soil was low in available nitrogen (167 kg ha^{-1}), phosphorus (P_2O_5 kg ha^{-1}) and sulphur (8.2 mg S kg^{-1}) but medium in available potassium (180 kg K_2O ha^{-1}) contents. The DTPA extractable Fe, Mn, Zn and Cu content of soil were 6.8, 10.9, 0.86 and 0.57 mg kg^{-1}, respectively. Recommended dose of 30 kg N (Through DAP and Urea) + 15 kg P_2O_5 (through DAP) ha^{-1} was applied as common dose. Phosphorus was applied as basal, the nitrogen was applied in two splits, half as basal and remaining half was applied at 30 days after sowing as top dressing. Among the sources of sulphur, gypsum and elemental sulphur were applied 25 days before sowing of crop and the experimental plot was irrigated to maintain the moisture for oxidation of sulphur, while ammonium sulphate was applied at the time of sowing. The processed plant and soil samples were analysed for chemical properties by employing standard procedures. The economics of the treatments were worked out on the basis of purchase price of different chemical fertilizers and other costs involved, while the income was calculated on the basis of prevailing selling price of the cumin seeds.

Results and Discussion

Yield and Quality of Cumin

Effect of Sulphur

The data given in Table 33.1 showed that seed and straw yield, 1000 seed weight as well as volatile oil and protein content were recorded significantly higher with sulphur fertilization. The sulphur application registered significantly higher seed (5.18 q ha^{-1}) and straw (6.50 q ha^{-1}) yield than control. The increased in yield was 16.9 per cent higher over control. Significantly higher test weight (3.82 g) was recorded with sulphur application than control, which was 5.5 per cent over the control. Sulphur application significantly recorded higher volatile oil (3.59 per cent) and protein (16.61 per cent) content over control. The application of sulphur significantly increased seed yield, straw yield, 1000 seed weight, volatile oil and protein content by 16.9, 16.7, 5.5, 12.2 and 13.1 per cent, respectively over control. These increases in yield and quality of cumin may be attributed to the balanced nutrition to crop with the application of sulphur in sulphur deficient soils. Increase in yield attributes may also because of improvement in availability of sulphur and favourable nutritional environment in *rhizosphere* and plant system which resulted in vigorous growth and branching, thereby increasing the sink size in terms of flowering and fruit setting.

Effect of Sources of Sulphur

Different sources of sulphur did not show significant variation in cumin seed yield, straw yield, 1000 seed weight and protein content (Table 33.1). Similar findings were observed by Patel *et al.* (1994) in castor and Gupta *et al.* (1998) in soybean. However, the effect of different sources was found significant on volatile oil content of cumin seed. The maximum volatile oil content (3.70 per cent) in cumin seed was recorded with ammonium sulphate. The minimum volatile oil content (3.54 per cent) by cumin seed was noted under gypsum and elemental sulphur. This might be due to higher solubility of ammonium sulphate and presence of sulphur as $SO_4^{=}$. Venkatesh *et al.* (2002) observed effectiveness of sources in the order of gypsum > SSP > elemental sulphur with groundnut crop in respect of oil content.

Table 33.1: Effect of Source and Levels of Sulphur on Yield, Volatile Oil and Protein Content of Cumin

Treatment	*(Yield q ha^{-1})*		*1000 Seed Weight (g)*	*Volatile Oil Content (per cent)*	*Protein Contend (per cent)*
	Seed	*Straw*			
No sulphur Vs Sulphur					
S_0: No sulphur	4.43	5.57	3.62	3.20	14.68
S_1: Sulphur	5.18	6.50	3.82	3.59	16.61
S.Em.±	0.15	0.21	0.04	0.06	0.061
C. D. at 5 per cent	0.45	0.60	0.11	0.17	1.76
CV per cent	6.04	6.43	1.99	3.33	7.39
Sources of sulphur					
Gypsum	5.27	6.66	3.84	3.54	16.18
Elemental sulphur	5.13	6.46	380	3.54	16.89
Ammonium sulphate	5.13	6.37	3.81	3.70	16.76
S.Em.±	0.094	0.125	0.022	0.033	0.365
C. D. at 5 per cent	NS	NS	NS	0.097	NS
Levels of Sulphur					
S_{10}: 10 kg S ha^{-1}	4.85	6.13	3.75	3.48	15.71
S_{20}: 20 kg S ha^{-1}	5.33	6.75	3.84	3.67	16.91
S_{30}: 30 kg S ha^{-1}	5.35	6.63	3.86	3.62	17.20
S.Em.±	0.094	0.125	0.022	0.033	0.365
C. D. at 5 per cent	0.27	0.365	0.064	0.097	1.067
Interaction					
S.Em.±	0.162	0.217	0.038	0.058	0.633
C. D. at 5 per cent	NS	NS	NS	0.169	NS
CV per cent	+6.26	6.67	2.00	3.22	7.62

Sources x Levels Mean table (Volatile oil)

Sources of Sulphur	*Levels of Sulphur*		
	S_{10}	*S_{20}*	*S_{30}*
Gypsum	3.68	3.55	3.40
Elemental sulphur	3.48	3.53	3.63
Ammonium sulphate	3.30	3.95	3.85

Effect of Levels of Sulphur

The data in Table 33.1 indicated that the sulphur levels significantly influenced yield, test weight and quality of cumin. The maximum seed yield (5.35 q ha^{-1}) of cumin was obtained with 30 kg S ha^{-1} level and minimum (4.85 q ha^{-1}) with 10 kg S ha^{-1} level. From the statistical point of view, application of 20 and 30 kg S ha^{-1} gave significantly higher seed yield than that of 10 kg S ha^{-1} but, they were statistically at par with each other. Application of sulphur @ 30 and 20 kg ha^{-1} increased the seed yield to the tune of 10.3 and 9.9 per cent, respectively over the level of 10 kg S ha^{-1}. The highest straw yield (6.75 q ha^{-1}) was recorded with 20 kg S ha^{-1} and the lowest (6.13 q ha^{-1}) with 10 kg S ha^{-1} level. Application of sulphur @ 20 kg ha^{-1} increased the straw yield to 6.75 q ha^{-1} compared to 6.12 q ha^{-1} at 10 kg S ha^{-1}. The application of 20 kg S ha^{-1} was statistically superior to 10 kg S ha^{-1} but it was at par with 30 kg S ha^{-1} level with respect to straw yield. The highest test weight (3.86 g) was observed with 30 kg S ha^{-1} level whereas, minimum test weight (3.75 g) was recorded with 10 kg S ha^{-1}. Application of 20 and 30 kg S ha^{-1} were at par with each other with respect to test weight of cumin but significantly superior to 10 kg level of S ha^{-1}. Sulphur application @ 20 kg ha^{-1} noticed the highest (3.67 per cent) volatile oil content, which was statistically at par with that obtained under 30 kg S ha^{-1} (3.62 per cent). Significantly the lowest (3.48 per cent) volatile oil content was observed under 10 kg S ha^{-1} level. The level, 30 kg S ha^{-1} showed higher protein content (17.20 per cent) as compared to 20 and 10 kg S ha^{-1} levels. From the statistical point of view, level of 30 kg S ha^{-1} was proved significantly superior to 10 kg S ha^{-1}, but was at par with 20 kg S ha^{-1}.

Marked effect on increasing the seed yield was noted due to sulphur application @ 20 kg ha^{-1}. This might be due to increased 1000 seed weight, which is important yield attributes having significant correlation. Moreover, improved nutritional environment in root zone and plant system might have favorably influenced the energy transformation, activation of enzymes and chlorophyll synthesis as well as increased carbohydrate metabolism, ultimately increased seed yield. These results are in agreement with the findings of Bhati *et al.* (1989) and Sivkumaran *et al.* (1996) in coriander, Singh and Singh (1990) in linseed, Shekhavat (1996) in cluster bean and Patel *et al.* (2004) in lucerne. Just like seed yield, straw yield was also increased remarkably up to 20 kg S ha^{-1}. Increased content of sulphur in plants helped in better development and thickening of xylem, collenchyma tissues, such as favorable effects might have resulted in stronger stem and increasing photosynthetic and meristematic activities which might have promoted vegetative growth consequently the more straw yield. Sulphur response to different crop has been studied very widely by Singh and Singh (1990) in linseed, Singh and Kumar (1996) in lentil and Mishra *et al.* (2003) in mustard. The increase in oil content might be attributed to the increased availability of nutrients owing to the favorable environment created by sulphur and also it plays a significant role in overall biosynthetic processes. Thus, sulphur supply seems to be involved in an increased conversion of primary fatty acids metabolites to end product of fatty acids. Similar beneficial effect of sulphur on oil content was also observed by Jaggi (1994) in raya, Ali *et al.* (1996) in rape seed, Ganeshmurthy (1996) in soybean and Singh *et al.* (2002) in mustard. The beneficial effect of sulphur fertilization may be because of the fact that sulphur is a constituent of amino acid *i.e.* biotin, cystine and methionine which present in protein and further, it also participates in several biochemical reactions in eventually resulting in increased protein content. The results are in accordance with the findings of Tripathi (1992) in sorghum, Dubey *et al.* (1993) in mustard, Ali *et al.* (1996) in rapeseed and Patel *et al.* (2004) in lucerne crop.

Interaction

The interactive effect of sources and levels of sulphur on volatile oil content and oil yield are presented in Table 33.1 showed that the combination of ammonium sulphate with 20 kg S ha^{-1} level recorded maximum volatile oil content (3.95 per cent) and oil yield (21.06 kg ha^{-1}) closely followed by ammonium sulphate and 30 kg S ha^{-1} treatment combination, whereas volatile oil content and oil yield were minimum with 10 kg S ha^{-1} through ammonium sulphate.

Nutrients Uptake by Cumin

Effect of Sulphur

The data presented in Table 33.2 revealed that the N uptake by cumin was significantly affected with sulphur application. The maximum uptake of nitrogen (20.51 kg ha^{-1}) was registered with sulphur fertilization, which was significantly higher than that of control. Sulphur fertilization significantly affected P uptake in cumin. The maximum uptake (3.05 kg ha^{-1}) was recorded with sulphur application as against the minimum uptake (2.22 kg ha^{-1}) under control. A significant higher K uptake (12.68 kg ha^{-1}) was observed with S fertilization than that of control (10.06 kg ha^{-1}). The K uptake due to S application was 26.0 per cent higher over control. The S uptake by cumin was affected significantly due to sulphur fertilization. The highest value of S uptake of 15.04 kg ha^{-1} was observed with sulphur application as against the minimum of 8.91 kg ha^{-1} S uptake registered under control which was 68.9 per cent higher over control. The results revealed that there were not significant differences in the Ca uptake by cumin due to sulphur application, whereas Mg uptake was altered up to the level of significance due to sulphur application. The maximum uptake (12.58 kg ha^{-1}) of Mg was observed with sulphur application, which was 23.9 per cent higher over control. It is apparent from Table 33.2 that the maximum Fe uptake (245.25 g ha^{-1}) by cumin was observed with sulphur application as against the minimum uptake under control. The data on the mean effect of sulphur revealed that the application of sulphur significantly increased Mn uptake (316.16 g ha^{-1}) by the crop, which was 33.5 per cent higher over control. Sulphur application significantly increased the uptake of Zn in cumin. The highest Zn uptake (44.15 g ha^{-1}) was registered with sulphur fertilization which was 29.4 per cent higher over control. The Cu uptake was significantly affected due to sulphur fertilization. The maximum Cu uptake (3.41 g ha^{-1}) was recorded with sulphur application its uptake (2.55 g ha^{-1}) under control. On percent basis, application of sulphur increased the Cu uptake by 33.7 per cent over control.

Table 33.2: Effect of Source and Levels of Sulphur on Nutrient Uptake by Cumin Crop

Treatment	*Nutrient Uptake (kg ha^{-1})*						*Nutrient Uptake (g ha^{-1})*			
	N	*P*	*K*	*S*	*Ca*	*Mg*	*Fe*	*Mn*	*Zn*	*Cu*
No Sulphur Vs Sulphur										
S_0: No sulphur	15.82	2.22	10.06	8.91	11.52	10.15	149	237	34.13	2.55
S_1: Sulphur	20.51	3.05	12.68	15.04	13.61	12.58	245	316	44.15	3.41
S.Em.±	0.88	0.24	0.60	0.71	0.88	0.66	20.2	17.1	2.18	0.16
C. D. at 5 per cent	2.55	0.72	1.90	2.27	NS	2.10	58.6	49.5	6.32	0.45
CV per cent	8.75	15.9	9.61	9.89	13.2	10.7	17.1	11.1	10.1	9.41
Sources of Sulphur										
Gypsum	20.49	3.08	12.67	16.21	14.24	12.95	270	320	44.52	3.55
Elemental sulphur	20.77	2.96	12.72	14.96	13.38	12.66	238	314	45.90	3.29
Ammonium sulphate	20.26	3.11	12.64	14.06	13.21	12.14	228	315	42.02	3.38
S.Em.±	0.53	0.14	0.36	0.43	0.53	0.40	11.8	10.4	1.33	0.09
C. D. at 5 per cent	NS	NS	NS	1.25	NS	NS	34.4	NS	NS	NS

Contd...

Table 33.2–Contd...

Treatment	Nutrient Uptake (kg ha^{-1})						Nutrient Uptake (g ha^{-1})			
	N	P	K	S	Ca	Mg	Fe	Mn	Zn	Cu
Levels of Sulphur										
S_{10}: 10 kg S ha^{-1}	18.66	2.83	11.90	12.95	12.71	11.80	208	324	43.22	3.12
S_{20}: 20 kg S ha^{-1}	21.28	3.12	13.04	15.95	13.77	13.16	265	317	44.64	3.44
S_{30}: 30 kg S ha^{-1}	21.58	3.20	13.09	16.23	14.36	12.79	262	308	44.58	3.65
S.Em.±	0.526	0.14	0.36	0.43	0.53	0.40	11.8	10.4	1.33	0.09
C. D. at 5 per cent	1.536	NS	1.06	1.25	NS	NS	34.4	NS	NS	0.27
Interaction										
S.Em.±	0.91	0.24	0.63	0.74	0.92	0.68	20.4	18.1	2.30	0.16
C. D. at 5 per cent	**NS**	**NS**	**NS**	**NS**	**NS**	**NS**	**NS**	**NS**	**NS**	**NS**
CV per cent	8.89	15.6	9.88	9.85	13.5	10.9	16.6	11.4	10.4	9.47

Effect of Sources of Sulphur

From the data given in Table 33.2, it was observed that the effect of different sources of sulphur on N, P, K, Ca, Mg uptake were found non- significant. However, sulphur uptake by cumin was influenced significantly by different sulphur sources, wherein, gypsum was significantly superior over remaining two sources of sulphur. Among sources of sulphur tried, significant difference was observed with respect to Fe uptake by cumin. The highest Fe uptake (270.08 g ha^{-1}) was registered with gypsum, whereas the minimum uptake of Fe (228.16 g ha^{-1}) by cumin was recorded under ammonium sulphate. Gypsum was significantly superior to ammonium sulphate but was at par with elemental sulphur with respect to Fe uptake by cumin. The effect of different sources were found non-significant in case of Mn, Zn and Cu uptake by cumin (Table 33.2). Application of sulphur sources not only act as a source of sulphur but also influence the physical, chemical and biological properties of soil resulting in changes like drop in soil reaction, release of nutrient in available forms and improvement of physical properties might have influenced the availability of other nutrients leading to higher absorption. Jaggi (1994) reported that sulphur uptake was significantly higher with gypsum than elemental sulphur. Similar results for lucerne crop were also observed by Patel *et al.* (2004).

Effect of Levels of Sulphur

A perusal of the data (Table 33.2) indicated that the N uptake by cumin crop was significantly affected due to different applied levels of sulphur, wherein the N uptake values under levels of 10, 20 and 30 kg S ha^{-1} were 18.66, 21.28 and 21.58 kg ha^{-1}, respectively. The P uptake by cumin increased with increasing levels of sulphur application however, it did not reach to the level of significance. The effect of different levels of sulphur on K uptake was found significant, wherein the highest uptake of K (13.09 kg ha^{-1}) in cumin was registered under 30 kg S ha^{-1} level as against the minimum of 11.90 kg ha^{-1} under 10 kg S ha^{-1}. The K uptake by cumin was increased by 10.0 and 9.6 per cent in cumin due to 30 and 20 kg S ha^{-1} level, respectively over 10 kg S ha^{-1} level. From the statistical point of view, level 30 kg S ha^{-1} was significantly superior to 10 kg S ha^{-1} level with respect to K uptake by cumin but it was at par with 20 Kg S ha^{-1} level. The data in Table 33.2 showed that the application of sulphur progressively enhanced S uptake by cumin. The maximum value of S uptake (16.23 kg ha^{-1}) was noted under 30 kg

S ha^{-1}, whereas the lower values of S uptake (12.95 kg ha^{-1}) was observed at 10 kg S ha^{-1}. The levels of 20 and 30 kg S ha^{-1} were at par with each other and significantly superior to 10 kg S ha^{-1} with respect to S uptake by cumin. The results further showed that S uptake was increased by 23.2 and 25.3 per cent due to 20 and 30 kg S ha^{-1}, respectively over 10 kg S ha^{-1} level. The levels of sulphur were not effective to alter the values of Ca and Mg uptake by cumin up to significant level. It is evident from the data presented in Table 33.2 that the effect of sulphur levels on Fe uptake by cumin was found significant. The maximum Fe uptake (265.25 g ha^{-1}) was observed with 20 kg S ha^{-1} which was significantly higher than that (208.41 g ha^{-1}) recorded with 10 kg S ha^{-1} but it was at par with that (262.08 g ha^{-1}) observed under 30 kg S ha^{-1} (Table 33.2). Varying levels of sulphur application failed to produce the significant effect on Mn and Zn uptake by cumin crop. Different levels of sulphur showed their significant effect on Cu uptake. The maximum (3.65 g ha^{-1}) and minimum (3.12 g ha^{-1}) Cu uptake was observed with levels of 30 and 10 kg S ha^{-1}, respectively. The increase uptake of the most of plant nutrients could be attributed to the enhanced vigor of the crop growth due to application of sulphur in sulphur deficient soil. The higher uptake of N by cumin due to S application may be explained on the basis of synergistic effect between two elements. Singh and Singh (1990) reported that S uptake was significantly increased up to 40 kg S ha^{-1} in linseed crop. Similar findings were also observed by Patel *et al.* (1994) in castor and Patel and Patel (1994) in lucerne crops. Singh *et al.* (2004) noted that an application of 40 kg S ha^{-1} resulted in better S uptake than no and 20 kg S ha^{-1} in chickpea crop.

Interaction

The combined effect of sources and levels of sulphur on major, secondary and micro nutrients uptake by cumin was found non-significant.

Nutrient Status After Harvest of Crop

Soil fertility status after the harvest of crop depends on nutrient applied, its removal by the crop and the extent of losses during the crop growth. Results of the present investigation have been shown that the application of sulphur significantly increased the status of available sulphur in soil at harvest than that of no application of sulphur (Table 33.3). Among different levels 30 kg S ha^{-1} gave maximum available sulphur and 20 kg S ha^{-1} application gave maximum available phosphorus status in soil after harvest of crop. A significant increase in available sulphur due to sulphur application has been reported by Singh and Singh (1975), Vakharia (1979), Vakharia and Dangarwala (1983). Raikhy *et al.* (1985) and Randhava and Arora (1997) reported that the available phosphorus status increased due to application of sulphur. Application of sulphur did not show significant effect on organic carbon content and availability of N, P and K in soil after harvest.

Among different sources of sulphur, elemental sulphur significantly increased the available sulphur status of soil after harvest that might be due to slow release of S from elemental sulphur than other two sources. The results were in accordance with findings of Dhilon and Dev (1976), Singh and Kumar (1982).

Yield Economics

The data pertaining to economics of applied sulphur sources and levels to cumin are presented in Table 33.4. It is evident from the data that different sources, gypsum, elemental sulphur and ammonium sulphate recorded additional net return of Rs. 6039, 4818 and 4699 ha^{-1} over control, respectively, which gave additional net return of Rs. 1221 and Rs. 1341 over elemental sulphur and ammonium sulphate, respectively. Looking to the economics of sulphur @ 10, 20 and 30 kg ha^{-1} gave additional net return of Rs. 2938, 6287 and 6331 ha^{-1} over control, respectively. Application of S @ 20 kg ha^{-1}

ranks first (1: 24.80) in ICBR over rest of the levels. Higher economic return with 30 kg S ha^{-1} compared to 20 kg and 10 kg S ha^{-1} could be justified on the basis of higher seed yield.

Table 33.3: Effect of Source and Levels of Sulphur on Nutrient Status in Soil

Treatment	*Available Nutrients (kg ha^{-1})*			*Available S (mg kg^{-1})*	*Organic Carbon (Per cent)*
	N	*P*	*K*		
No Sulphur Vs Sulphur					
S_0: No sulphur	163.70	19.39	173.50	5.48	0.20
S_t: Sulphur	177.72	20.86	174.97	8.55	0.22
S.Em.±	10.74	0.97	10.56	0.27	0.03
C. D. at 5 per cent	NS	NS	NS	0.78	NS
CV per cent	12.19	9.33	12.08	6.59	23.71
Sources of Sulphur					
Gypsum	176.41	19.92	173.33	8.10	0.21
Elemental sulphur	179.41	21.53	175.00	8.94	0.22
Ammonium sulphate	177.33	21.14	176.58	8.61	0.22
S.Em.±	6.03	0.58	5.33	0.16	0.02
C. D. at 5 per cent	NS	NS	NS	0.47	NS
Levels of Sulphur					
S_{10}: 10 kg S ha^{-1}	178.00	19.96	172.75	7.15	0.22
S_{20}: 20 kg S ha^{-1}	179.58	22.05	168.83	8.54	0.21
S_{30}: 30 kg S ha^{-1}	175.58	20.58	183.30	9.95	0.21
S.Em.±	6.03	0.58	5.33	0.16	0.02
C. D. at 5 per cent	NS	1.70	NS	0.47	NS
Interaction					
S.Em.±	10.44	1.01	9.23	0.28	0.03
C. D. at 5 per cent	NS	NS	NS	NS	NS
CV per cent	11.76	9.67	10.55	6.46	24.62

Table 33.4: Effect of Source and Levels of Sulphur on Economics of Cumin

Treatment	*Seed Yield (q ha^{-1})*	*Gross Return (Rs. ha^{-1})*	*Additional Cost (Rs. ha^{-1})*	*Total Cost of Production (Rs. ha^{-1})*	*Net Return (Rs. ha^{-1})*
Control	4.43	32353	–	9339	23014
Sources of Sulphur					
Gypsum	5.27	38493	100	9439	29054
Elemental sulphur	5.13	37464	292	9631	27833
Ammonium sulphate	5.13	37420	367	9704	27713

Contd...

Table 33.4–Contd..

Treatment	Seed Yield ($q\ ha^{-1}$)	Gross Return ($Rs.\ ha^{-1}$)	Additional Cost ($Rs.\ ha^{-1}$)	Total Cost of Production ($Rs.\ ha^{-1}$)	Net Return ($Rs.\ ha^{-1}$)
Levels of Sulphur					
S_{10}: 10 kg S ha^{-1}	4.85	35434	143	9482	25952
S_{20}: 20 kg S ha^{-1}	5.33	38894	253	9592	29302
S_{30}: 30 kg S ha^{-1}	5.35	39048	363	9702	29346

	Additional Net Return Over Control ($Rs.\ ha^{-1}$)	Incremental Cost Benefit Ration (ICBR)
Sources of Sulphur		
Gypsum	6039	1: 60.39
Elemental sulphur	4818	1: 16.50
Ammonium sulphate	4699	1: 12.79
Levels of Sulphur		
S_{10}: 10 kg S ha^{-1}	2938	1: 20.56
S_{20}: 20 kg S ha^{-1}	6287	1: 24.80
S_{30}: 30 kg S ha^{-1}	6331	1: 17.44

Cost of seed Rs. 7300 q^{-1}.

References

Ali, M.H., Zaman, S.M.H. and Hossain, S.M.A., 1996. Variation in yield, oil and protein content of rapeseed (*Brassica campestris* L.) in relation to levels of nitrogen, sulphur and plant density. *Indian Journal of Agronomy*, 41(2): 290–295.

Bhati, D.S., Rathore, P.S. and Mali, A.L., 1989. Response of coriander to soil and foliar applied sulphur. *Journal of Farming System*, (1–2): 49–51.

Dhillon, N.S. and Dev, G., 1976. Studies on extractability of sulphur from soil applied through its various compounds. *Indian Journal of Agricultural Research*, 10(4): 229–232.

Director of Agriculture, 2001. *District-Wise Area, Production and Yield per Hectare of Important Food and Non-Food Crops in Gujarat State*, p. 43.

Dubey, O.P., Sahu, T.R., Garg, D.C. and Khan, R.A., 1993. Response of mustard to sulphur and nitrogen under irrigated Vertisol conditions. *Journal of Oilseeds Research*, 10(1): 11–15.

Ganeshmurthy, A.N., 1996. Critical plant sulphur content and effect of S application on grain and oil yield of rainfed soybean in Vertic Ustochrepts. *Journal of the Indian Society of Soil Science*, 44(2): 290–294.

Gupta, R.K. and Dubey, S.K., 1998. Sulphur management in rainfed cropping system in Madhya Pradesh. *Fertilizer News*, 43(7): 57–60.

Jaggi, R.C., 1994. Response of raya to sulphur through different sources in a acid Alfisol. *Journal of the Indian Society of Soil Science*, 42(2): 287.

Misra, S.K., 2003. Effect of sulphur and potassium on yield, nutrient uptake and quality characteristics of mustard (*Brassica juncea* L.) in Udic Haplustepts of Kanpur. *Journal of the Indian Society of Soil Science*, 51(4): 544–548.

Patel, A.T. and Kalyansundram, N.K., 1988. *Research Highlights AICRP for Dryland Agriculture*, Dantiwada Centre, Gujarat Agricultural University.

Patel, M.K., Patel, J.J., Patel, P.G., Fatteh, U.G. and Kalyansundram, N.K., 1994. Effect of sources and rate of sulphur fertilization on castor. *GAU Research Journal*, 20(1): 42–45.

Patel, P.C. and Patel, J.R., 1994. Effect of phosphorus and sulphur on forage yield and nutrient uptake by lucerne. *Journal of the Indian Society of Soil Science*, 42(1): 154–156.

Patel, P.C., Yadavendra, J.P. and Kotech, A.V., 2004. Effect of sources and level of sulphur on seed yield and nitrogen and sulphur uptake by lucerne (*Medicago sativa*). *Indian Journal of Agronomy*, 49(2): 128–130.

Raikhy, N.P. and Agrawal, R.K., 1985. Performance of pyrites and gypsum as sulphur source in an arid sandy soil. *Journal of the Indian Society of Soil Science*, 33: 447–449.

Randhawa, P.S. and Arora, C.C., 1997. Effect of phosphorus and sulphur on their availability in soils. *Journal of the Indian Society of Soil Science*, 45(2): 306–310.

Shekhavat, P.S., Rathore, A.S. and Singh, M., 1996. Effect of source and level of sulphur on yield attributes and seed yield of clusterbean (*Cyamopsis tetragonoloba*). *Indian Journal of Agronomy*, 41(3): 424–426.

Shivkumaram, S., Gururaj Hansigi, Basavaraju, H.K. and Sridhara, S., 1996. Study on nutrient uptake, seed and oil yield of coriander as influenced by nitrogen, phosphorus and sulphur. *Indian Agriculture*, 40(2): 89–92.

Singh, B. and Kumar, V., 1996. Effect of phosphorus and sulphur on lentil under rainfed condition. *Indian Journal of Agronomy*, 41(3): 420–423.

Singh, B. and Singh, M., 1975. Sulphur status of Hissar soils and availability of sulphur and phosphorus affected by fertilizer application. *Haryana Agriculture University Journal of Research*, 5(3): 222–230.

Singh, B.P., Om Prakash, Singh, B. and Singh, S.K., 2002. Comparative performance of Indian mustard (*Brassica juncea*) genotypes in relation to sulphur fertilization. *Indian Journal of Agronomy*, 47(4): 531–536.

Singh, M. and Kumar, V., 1982. Oxidation of elemental S in soil and effect of organic moisture and calcium carbonate on the mineralization of the native sulphur. *Haryana Agricultural University Journal of Research*, 12(3): 405–412.

Singh, R.P. and Singh, V., 1990. Effect of S, Mg and K on yield and uptake by linseed. *Journal of the Indian Society of Soil Science*, 38: 169–170.

Tandon, H.L.S., 1989. *A Guidebook on Sulphur Fertilizers for Indian Agriculture*. FDCO, New Delhi, pp. 1–22.

Tandon, H.L.S. and Messick, D.L., 2002. *Practical Sulphur Guide*. The Sulphur Institute, Washington, D.C., pp. 1–20.

Tripathi, S.B., Singh, R.S. and Tripathi, R.K., 1992. Effect of S fertilization on quality constituents of fodder sorghum cv M. P. Chari. *Forage Research*, 18(1): 9–14.

Vakharia, D.N., 1979. Effect of S sources on its availability as well as on yield and chemical composition of onion (*Allium cepa* L.). *M.Sc. (Agri.) Thesis*, Gujarat Agricultural University, Anand Campus, Anand.

Vakharia, D.N. and Dangarwala, R.J., 1983. Effect of sulphur sources on pH, EC and sodium chloride extractable sulphur content of *goradu* soil. *Gujarat Agricultural University Research Journal*, 8(2): 79–83.

Venkatesh, M.S., Majumdar, B., Lal, B. and Kumar, K., 2002. Relative performance of sulphur sources of sulphur nutrition of groundnut (*Arcachis hypogea*) in acid Alfisols of Meghalaya. *Indian Journal of Agricultural Sciences*, 72(4): 216–219.

Recent Advances in Seed Spices (2011) *Pages* **311–318**
Editors: **Y. Ravindrababu, R.K. Jaiman, K.D. Patel, N.R. Patel and S.B.S. Tikka**
Published by: **DAYA PUBLISHING HOUSE, NEW DELHI**

34

Response of Isabgol (*Plantago oveta* Forsk.) to Nitrogen and Iron Application in Loamy Sand

B.B. Patel, T.V. Gangwal, N.J. Jadav, R.P. Pavaya, M.V. Patel and V.R. Patel
Department of Agricultural Chemistry and Soil Science, C.P. College of Agriculture, S.D. Agricultural University, Sardarkrushinagar – 385 506, Gujarat

ABSTRACT

Field experiment entitled "Response of Isabgol (*Plantago oveta* Forsk.) to Nitrogen and Iron Application in Loamy Sand" was conducted at Agronomy Instructional Farm, C. P. College of Agriculture, S. D. Agricultural University, Sardarkrushinagar (Gujarat) in randomized block design (factorial) with three replications during the *Rabi* season of 2001-02. The experiment comprised three levels each of nitrogen (20, 40 and 60 kg N ha^{-1}) and iron (0, 12, 24 kg $FeSO_4$ ha^{-1}). The result revealed that application of 40 kg N ha^{-1} recorded the maximum plant height, tillers $plant^{-1}$, spike length, grain (1.36 Mg ha^{-1}) and straw (3.30 Mg ha^{-1}) yields. In case of Fe application, the maximum yield of grain (1.35 Mg ha^{-1}) and straw (3.37 Mg ha^{-1}) were observed with the application of 12 kg $FeSO_4$ ha^{-1}. However, the combined application of 40 kg N and 12 kg $FeSO_4$ ha^{-1} recorded the maximum grain (1.48 Mg ha^{-1}) and straw (3.84 Mg ha^{-1}) yield of isabgol. The highest uptake of N, P, K, Fe, Mn, Zn and Cu by seeds of isabgol at harvest in grain were observed under 40 kg N ha^{-1}. The maximum Fe uptake of all the nutrients by seed of isabgol were noticed under combined application of 40 kg N and 12 kg $FeSO_4$ ha^{-1}. It could be concluded that for getting higher yield of isabgol in light textured iron deficient soil of North Gujarat, the crop should be fertilized with 40 kg N along with 12 kg Fe ha^{-1}.

Keywords: *Isabgol, Nitrogen, Iron.*

Introduction

Bland psyllium or isabgol is an important medicinal and traditional cash crop of North Gujarat. India is the sole export to the world markets. Being economically important cash crop, it is essential to increase the productivity. With increasing trend of intensive cultivation, high yielding varieties, scare use of organic manures and heavy usage of high analysis fertilizers have resulted in micronutrient deficiency in many areas. In Fe and Zn deficient light textured soils of North Gujarat, respective micronutrient application responded well. As there is a paucity of information on the response of micronutrient on isabgol, present investigation was undertaken to find out the optimum production of isabgol.

Materials and Methods

A field experiment was conducted at Agronomy Instructional farm, C. P. College of Agriculture, Sardarkrushinagar Dantiwada Agricultural University, Sardarkrushinagar during the *rabi* 2001-02 The experiment was laid out in randomized block design (factorial) comprising of three levels each of nitrogen (20, 40 and 60 kg N ha^{-1}) and iron (0, 12 and 24 kg $FeSO_4$$^{-1}$) replicating four times and was conducted in Typic Ustipsament with loamy sand texture having pH 7.6 and Ec 0.09 dsm^{-1}. The soil was low in available nitrogen (221 kg ha^{-1}) medium in available phosphorus (36.4 kg P_2O_5 ha^{-1}) and rich in available potassium (274 kg K_2O ha^{-1}) contents. The DTPA extractable Fe, Mn, Zn and Cu content of soil were 4.63, 7.13, 0.73 and 0.92 mg kg^{-1}, respectively. Half of N in the form of urea and full dose of Fe in form of ferrous sulphate were given as basal and half of N was top dressed after one month of sowing. The crop was sown with a seed rate of 5 kg ha^{-1} using variety "Gujarat Isabgol-2". The crop was fertilized with 24 kg P_2O_5 ha^{-1} as a basal dose in the form of single super phosphate and six irrigations were applied. The processed plant and soil samples were analysed for chemical properties by employing standard procedures. The economics of the treatments were worked out on the basis of purchase price of different chemical fertilizers and other costs involved, while the income was calculated on the basis of prevailing selling price of the cumin seeds.

Results and Discussion

Growth, Yield and Yield Attributes

Effect of Nitrogen

The results revealed that nitrogen application affected the entire yield attributing characteristics, grain and straw yield significantly (Tables 34.1 and 34.2). Application of 40 kg N ha^{-1} gave significantly higher values of plant height, No. of tillers per plant, No. of spikes per plant and spike length. The maximum grain (1355 kg ha^{-1}) and straw yield (3298 kg ha^{-1}) were noted with the application of 40 kg N ha^{-1}. Further, an application of nitrogen @ 60 kg N ha^{-1} showed decline in yield and yield attributing characters as compared to 40 kg N ha^{-1}. The beneficial effect of nitrogen on growth and yield attributing characters might have contributed towards higher grain and straw yields. Such beneficial effects of nitrogen on plant growth was also reported by Singh and Chouhan (1994), Parihar and Singh (1995) and Sharma *et al.* (1999).

Effect of Iran

A significant response to Fe application was obtained with respect to grain and straw yield of isabgol. Yield obtained under 12 kg $FeSO_4$ ha^{-1} was higher, which evidently has resulted from plant and spike length. Favourable response of isabgol to foliar application of 0.5 per cent $FeSO_4$ was observed by Solanki *et al.* (1997).

Table 34.1: Effect of N and Fe Levels on Growth and Yield Attributes of Isabgol

Nitrogen (kg ha^{-1})	*$FeSO_4$ (kg ha^{-1})*			*Mean*
	00	*12*	*24*	
		Plant height (cm)		
20	22.05	28.93	27.15	26.04
40	27.00	31.07	28.20	28.76
60	25.72	30.07	26.04	27.28
Mean	24.92	30.03	27.13	
	N	*Fe*	*N x Fe*	
S. Em.±	0.40	0.40	0.70	
C. D. at 5 per cent	1.17	1.77	2.04	
CV per cent		5.10		
		Number of tillers per plant		
20	3.43	3,62	3.55	3.53
40	4.27	4.70	4.58	4.52
60	3.87	4.05	3.87	3.93
Mean	3.86	4.13	4.00	
	N	*Fe*	*N x Fe*	
S. Em.±	0.03	0.04	0.04	
C. D. at 5 per cent	0.07	0.07	0.13	
CV per cent		2.16		
		Number of spikes per plant		
20	19.70	24.63	21.88	22.07
40	25.35	29.00	27.00	21.12
60	23.10	28.05	26.24	25.79
Mean	22.72	27.23	25.04	
	N	*Fe*	*N x Fe*	
S. Em.±	0.19	0.19	0.33	
C. D. at 5 per cent	0.55	0.55	NS	
CV per cent		2.63		
		Spike length (cm)		
20	3.18	3.61	3.56	3.45
40	3.92	4.26	3.98	4.05
60	3.73	4.06	3.24	3.68
Mean	3.61	3.98	3.59	
	N	*Fe*	*N x Fe*	
S. Em.±	0.07	0.07	0.13	
C. D. at 5 per cent	0.22	0.22	0.37	
CV per cent		6.85		

Interaction Effect

Interaction of nitrogen and iron in respect to growth and yield attributing characters were found significant except number of spikes per plant and straw yield. All the growth and yield attributing characters were higher with the treatment combination 40 kg N ha^{-1} + 12 kg $FeSO_4$ ha^{-1}. The grain yield of 1476 kg ha^{-1} and straw yield of 3841 kg ha^{-1} recorded under the treatment combination 40 kg N ha^{-1} + 12 kg $FeSO_4$ ha^{-1} were the highest over all other treatment combinations.

Table 34.2: Effect of N and Fe Levels on Yield and Major Nutrients Uptake by Isabgol

Nitrogen (kg ha^{-1})	*$FeSO_4$ (kg ha^{-1})*			*Mean*
	00	*12*	*24*	
		Grain yield (kg ha^{-1})		
20	891	1221	1189	1100
40	1283	1476	1306	1355
60	1193	1364	1035	1195
Mean	1222	1354	1177	
	N	*Fe*	*N x Fe*	
S. Em.±	39.1	39.1	67.7	
C. D. at 5 per cent	114.1	114.1	198	
CV per cent		11.12		
		Straw yield (kg ha^{-1})		
20	2471	3057	2679	2736
40	2852	3841	3201	3298
60	2732	3216	2563	2857
Mean	2685	3371	2814	
	N	*Fe*	*N x Fe*	
S. Em.±	91.5	91.5	158.5	
C. D. at 5 per cent	264.1	264.1	NS	
CV per cent		10.72		
		N uptake by seed (kg ha^{-1})		
20	16.50	20.18	25.90	23.86
40	25.74	39.10	35.20	33.35
60	27.68	33.44	22.70	27.94
Mean	23.30	33.90	27.93	
	N	*Fe*	*N x Fe*	
S. Em.±	1.02	1.02	1.77	
C. D. at 5 per cent	2.98	2.98	5.17	
CV per cent		12.47		

Contd...

Table 34.2–Contd...

Nitrogen (kg ha^{-1})	*$FeSO_4$ (kg ha^{-1})*			*Mean*
	00	*12*	*24*	
		P uptake by Seed (kg ha^{-1})		
20	7.45	11.18	10.61	9.74
40	11.72	15.09	12.73	13.18
60	11.35	13.41	8.85	11.20
Mean	10.18	13.22	10.73	
	N	*Fe*		*N x Fe*
S. Em.±	0.47	0.47		0.81
C. D. at 5 per cent	1.36	1.36		2.36
CV per cent		14.18		
		K uptake by seed (kg ha^{-1})		
20	8.27	13.02	11.91	11.06
40	13.82	16.90	14.33	15.01
60	11.58	15.30	9.90	12.26
Mean	11.22	15.07	12.05	
	N	*Fe*		*N x Fe*
S. Em.±	0.60	0.60		1.04
C. D. at 5 per cent	1.75	1.75		NS
CV per cent		16.22		

Nutrients Uptake by Isabgol Seeds

Effect of Nitrogen

The results clearly indicated that nitrogen application significantly increased the uptake of major (N, P, K) and micronutrients (Fe, Mn, Zn and Cu) by isabgol seed (Tables 34.2 and 34.3). Application of 40 kg N ha^{-1} registered significantly higher values of N (33.35 kg ha^{-1}), P (13.18 kg ha^{-1}), K (15.01 kg ha^{-1}), Fe (161.00 g ha^{-1}), Mn (24.98 g ha^{-1}), Zn (35.21 g ha^{-1}) and Cu (15.98 g ha^{-1}) uptake by seeds of isabgol on account of higher production of grain. The nutrients uptake registered under 40 kg N ha^{-1} was statistically at par with that of 60 kg N ha^{-1} but significantly higher as compared to control. The results of present study are in agreement with those reported by Shakhela (1980) and Patel (1982).

Effect of Iran

A significant response to Fe application was obtained with respect to nutrients uptake by seed of isabgol. The uptake of nutrients *viz.*, N (33.9 kg ha^{-1}), P (13.22 kg ha^{-1}), K (15.07 kg ha^{-1}), Fe (155.98 g ha^{-1}), Mn (23.73 g ha^{-1}), Zn (35.63 g ha^{-1}) and Cu (16.18 g ha^{-1}) were recorded higher under 12 kg $FeSO_4$ ha^{-1}. The nutrients uptake noticed from under 12 kg $FeSO_4$ ha^{-1} was statistically at par with that of 24 kg $FeSo_4$ ha^{-1} but significantly higher as compared to control. It may be due to the higher production of grain of isabgol under 12 kg $FeSO_4$ ha^{-1}. Such type of effect of Fe application on nutrient uptake was also reported by Singh and Yadav (1980) and Singh *et al.* (1995).

Table 34.3: Effect of N and Fe Levels on Micronutrients Uptake by Isabgol

Nitrogen (kg ha⁻¹)	*FeSO₄ (kg ha⁻¹)*			*Mean*
	00	*12*	*24*	
		Fe uptake by seed (g ha⁻¹)		
20	92.25	132.46	138.87	121.19
40	141.69	180.81	160.51	161.00
60	127.09	154.66	112.33	131.36
Mean	120.34	155.98		
	N	*Fe*	*N x Fe*	
S.Em. ±	5.21	5.21	9.03	
C.D. at 5 per cent	15.21	15.21	26.35	
CV per cent		13.10		
		Mn uptake by seed (g ha⁻¹)		
20	16.01	21.04	20.41	19.15
40	23.99	27.34	22.12	24.98
60	21.57	22.83	15.38	19.92
Mean	20.52	23.73	19.30	
	N	*Fe*	*N x Fe*	
S.Em. ±	0.93	0.93	1.61	
C.D. at 5 per cent	2.71	2.71	4.70	
CV per cent		15.20		
		Zn uptake by seed (g ha⁻¹)		
20	18.22	29.58	28.24	25.34
40	31.35	41.29	33.01	35.21
60	28.40	36.03	23.84	29.42
Mean	25.99	35.63	23.36	
	N	*Fe*	*N x Fe*	
S.Em. ±	1.08	1.08	1.88	
C.D. at 5 per cent	3.17	3.17	5.48	
CV per cent		12.52		
		Cu uptake by Seed (g ha⁻¹)		
20	10.81	14.81	14.50	13.37
40	15.86	18.43	13.65	15.98
60	14.15	15.31	10.47	13.31
Mean	13.61	16.18	12.87	
	N	*Fe*	*N x Fe*	
S.Em. ±	0.51	0.51	0.88	
C.D. at 5 per cent	1.48	1.48	2.57	
CV per cent		14.62		

Interaction Effect

Interaction of N and Fe in respect to uptake of nutrients by isabgol seeds were found significant except K uptake. In general, combined application of 40 kg N ha^{-1} along with 12 kg $FeSO_4$ ha^{-1} registered the maximum nutrients uptake. The results obtained under the present investigation are in conformity with those observed by Patel (1982), Dahiya and Bhatia (1982), Pavaya (1983) Singh *et al.* (1995).

Nutrient Available in Soil After Harvest of Crop

Effect of Nitrogen

The results clearly indicated that nitrogen application significantly increased the available nitrogen at the higher level *i.e.* 60 kg N ha^{-1}. Whereas, DTPA extractable Fe was increased with the application of 40 kg N ha^{-1} (Table 34.4). It might be due to fact that application of nitrogen decrease soil pH and increase iron availability in soil. Srivastava and Verma (1985) observed that the application of nitrogen to field pea increased residual soil fertility status of soil.

Table 34.4: Effect of N and Fe Levels on Available Nitrogen and Iron in Soil

Nitrogen (kg ha^{-1})	*$FeSO_4$ (kg ha^{-1})*			*Mean*
	00	*12*	*24*	
	Available N in soil after harvest of crop (kg ha^{-1})			
20	234	240	237	237
40	242	254	249	248
60	251	260	253	255
Mean	242	251	246	
	N	*Fe*	*N x Fe*	
S. Em.±	2.26	2,26	3.93	
C. D. at 5 per cent	6.62	6.62	NS	
CV per cent		3.18		
	DTPA Extractable Fe in soil after harvest of crop (mg kg^{-1})			
20	4.71	7.02	9.21	6.98
40	5.89	9.17	10.18	8.41
60	5.02	8.20	9.19	7.47
Mean	5.21	8.13	9.53	
	N	*Fe*	*N x Fe*	
S. Em.±	0.05	0.05	0.09	
C. D. at 5 per cent	0.14	0.14	0.25	
CV per cent		2.24		

Effect of Iran

Iron application increased the available nitrogen of soil after harvest was higher in the plot receiving 12 kg $FeSO_4$ which, was at par with that of 24 kg $FeSO_4$. The results may be due to increase in nitrate reductase activity by the iron application. Application of 24 kg $FeSO_4$ ha^{-1} recorded higher DTPA extractable Fe in soil. Similar result was also reported by Francis *et al.* (1979).

Interaction Effect

The combined application of 60 kg N ha^{-1} and 12 kg $FeSO_4$ ha^{-1} registered higher available nitrogen. While higher value of DTPA extractable was registered with combined application of 40 kg N ha^{-1} and 12 kg $FeSO_4$ ha^{-1}. This might be due to N and Fe have synergistic effect on each other. Devasenapathy and Subbarayalu (1986) indicated that application of recommended doses of N, P, K increased the availability of N and Fe in soil

References

Dahiya, S.S. and Bhatia, B.K., 1982. Effect of iron and nitrogen levels on dry matter yield and nutrient uptake by maize. *Haryana Agri. Univ. J. Res.*, 12 (2): 254–258.

Devasenapathy, P. and Subbarayalu, M., 1986. Effect of nitrogen, FYM and iron on uptake of nutrients by sorghum. *Madras Agric. J.*, 73(1): 53–56.

Francis, H.J., Rajagopal, C.K. and Krishnamoorthy, K.K., 1979. Effect of organically complexed iron on the available iron content in soil and uptake by sorghum CSH-5 in two different soils at successive growth stages. *Mysore J. Agric. Sci.*, 13: 131–137.

Parihar, G.N. and Singh, R., 1995. Response of Psyllium (*Plantago ovata*) to nitrogen and phosphorus fertilization. *Indian J. Agron.*, 40(3): 529–531.

Patel, M.C., 1982. Influence of zinc, boron and iron application on yield and chemical composition of isabgol. *M.Sc. (Agri.) Thesis*, GAU, Anand.

Pavaya, R.P., 1983. Influence of nitrogen-phosphorus with iron sources and levels on yield and uptake of nutrients by wheat. *M.Sc. (Agri.) Thesis*, GAU, Anand.

Shakhela, R.R., 1980. Yield and chemical composition of isabgol as influenced by nitrogen levels and methods of application. *M.Sc. (Agri.) Thesis*, GAU, Sardarkrushinagar.

Sharma, P.K., Yadav, G.L. and Kumar, S., 1999. Effect of row spacing and nitrogen levels on the yield of isabgol varieties. *Haryana J. Agron.*, 15(4): 210–211.

Singh, I. and Chouhan, G.S., 1994. Effect of mixtalol and different combination of nitrogen and phosphorus fertilizer on growth and yield of blond psyllium (*Plantago ovata*). *Indian J. Agron.*, 39(4): 711–712.

Singh, J.P., Dahiya, D.J. and Kumar, V., 1995. Effect of nitrogen and iron supply on the growth and nutrient uptake by wheat on sandy soil. *Crop Res.*, 10(3): 271–276.

Singh, M. and Yadav, D.S., 1980. Effect of copper, iron and liming on the growth, concentration and uptake of Cu, Fe, Mn and Zn in sorghum. *J. Indian Soc. Soil Sci.*, 28(1): 113–118.

Solanki, N.S., Sahu, M.P. and Shaktawat, R.P.S., 1997. Effect of sources of phosphorus, sulphur and foliar spray of zinc and iron on growth and yield of isabgol. *Haryana J. Agron.*, 13(2): 97–99.

Srivastava, S.N.L. and Verma, S.C., 1985. Effect of nitrogen, phosphorus and molybdenum fertilization on growth, nodulation and residual fertility in field pea. *Indian J. Agric. Res.*, 19(3): 131–137.

Recent Advances in Seed Spices (2011) *Pages* **319–325**
Editors: **Y. Ravindrababu, R.K. Jaiman, K.D. Patel, N.R. Patel and S.B.S. Tikka**
Published by: **DAYA PUBLISHING HOUSE, NEW DELHI**

35

Cyclic Mode of Sewage and Well Water Irrigation with Different Levels of Nitrogen in Dill Seed (*Anethum graveolens*)

M.M. Patel, P.T. Patel, M.M. Patel and S.M. Patel
Department of Agronomy, C.P. College of Agriculture, S.D. Agricultural University, Sardarkrushinagar – 385 506, Gujarat

ABSTRACT

With a view to find out economic feasibility of sewage water for irrigation purpose, a field experiment was taken up on dill seed (*Anethum graveolens* L.) during *rabi* season of 2004-05 and 2005-06 involving tube well and sewage water in a cyclic mode of four irrigations. Eighteen treatment combinations comprising of six irrigation schedules from both the sources *i.e.* sewage (S) and tube well water (T) *i.e.* I_1: TTTT; I_2:SSSS; I_3: TSTS; I_4: TTSS; I_5: TSSS and I_6: STTT as main plots and three levels of nitrogen *i.e.* 0, 20, and 40 kg N/ha as sub plots were tested in split plot design with four replications. All four irrigations with sewage water produced significantly higher seed yield (1.30 t/ha) than rest of the treatments except a treatment involving one tube well + three of sewage water. It accounted 20 per cent higher seed yield over local practice *i.e.* four irrigations of tube well water. Higher yield recorded with sewage water application is attributed to increase in number of branches, umbels and seeds/umbellate. Inclusion of two irrigations of sewage water in a cycle also helped to enhance seed yield than only sole schedule of tube well water.

Nitrogen application brought linear increase in seed yield. Nitrogen 40 kg/ha realized 8.26 and 22.43 per cent higher seed yield than 20 kg N/ha and control. Significant improvement in number of branches, umbellates, seeds per umbellate and higher test weight due to application of 40 kg N/ha contributed to enhance seed yield.

Maximum net returns (Rs. 11,818/ha) and B:C ratio (2.31) were secured with application of four irrigations with sewage water. Similarly 40 kg N/ha showed its superiority accruing the net return of Rs. 11,902/ha and B:C ratio of 2.32.

In water scare area, inclusion of two or more irrigations of sewage water in a schedule along with 40 kg N/ha, brought significant improvement in yield of dill seed.

Keywords: *Dill seed, Irrigation, Nitrogen, Seed yield, Sewage water, Nutrients uptake and status.*

Introduction

Dill seed (*Anethum graveolens* L.) is one of the under utilized seed spices of India. The area under dill seed crop is increasing in North Gujarat due to its increasing demand in national and international markets and higher profitability with minimum risk. It has also low monetary input requirement particularly irrigation and fertilizers, and less pest and disease incidence. The cost of nitrogenous fertilizers is increasing; hence, it becomes imperative to substitute nitrogen by some other cheaper sources like sewage water, sludge etc. Such cheaper sources increase the yield and economize the cost of fertilizers to some extent (Tiwari *et al.*, 1996). The efficient use of sewage water reduces the requirement of chemical fertilizers and enriches soil nitrogen (Malarvzhi and Rajamannar 2001). Thus, the present investigation was undertaken to find out the efficient combination of irrigation schedule of sewage and well water along with nitrogen levels on growth and yield of dill seed under North Gujarat condition.

Materials and Methods

The field experiment was conducted during *rabi* season of 2004-05 and 2005-06 at Sardarkrushinagar Dantiwada Agricultural University, Sardarkrushinagar. Sardarkrushinagar is situated at 24°-19' North latitude, 72°-19' East longitude and has an elevation of 154.52 m above the mean sea level. The soil was loamy sand in texture, low in organic C (0.26 and 0.30 per cent) and available nitrogen (148 and 159 kg/ha), medium in available P (19 and 22 kg/ha), K (124 and 142 kg/ha) and Zn (0.64 ppm) and deficient in Fe (4.04 ppm) with 7.6 and 7.7 pH in 2004-05 and 2005-06, respectively. It is characterized by semi arid climate with extreme cold winter, hot and dry windy summer. Eighteen treatment combinations comprising of six irrigation schedules from both the sources *i.e.* sewage (S) and tube well water (T) *i.e.* I_1-T T T T; I_2-S S S S;, I_3-T S T S; I_4-T T S S; I_5-T S S S and I_6-S T T T as main plots and three levels of nitrogen *i.e.* N_0-0 kg/ha, N_1 20 kg/ha and N_2 40 kg/ha as sub plots were tested in split plot design with three replications. Dill seed GD-1 was sown on 9th November, 2004 and 30th October, 2005 using seed rate of 6 kg/ha at 45 cm spacing. Half of the total nitrogen and full dose of phosphorus (20 kg/ha) for each treatment was applied in previously opened furrows just before sowing as basal dose and remaining half nitrogen at 45 DAS. Dill seed crop was harvested on 7th and 1st April, respectively during 2005 and 2006. Based on the analysis, average composition of sewage and tube well water is given in Table 35.1.

Results and Discussion

Growth and Yield

Over the seasons, irrigation schedules and nitrogen levels had significant influence on growth, yield attributes and yield (Table 35.2). Dill seed irrigated with sewage water produced significantly taller plants with more number of branches, umbels/plant and seeds per umbellate over tube well water irrigated crop. However it remained on par with the treatment involving one tube well + three sewage water irrigations. The improvement in yield attributes under this treatment led to increase 20 per cent seed yield. Higher N and P uptake by seeds and stover as well as buildup of available N and P status after harvest of crop under sewage water application (Table 35.3) are the principal reasons for better performance of dill seed under this treatment. Moreover increase in Fe and Zn status (Table 35.3) had helped to meet micronutrient requirement of the crop. Similar effects of sewage water irrigation have also been found for rice, wheat, spinach, cauliflower and sorghum as reported at CSSRI, Karnal (2001-02).

Table 35.1

Characteristics	*Sewage Water*	*Well Water*	*Method Employed*
pH	7.38	7.65	Potentiometric method
Ec (dSm^{-1})	0.92	0.68	Schofield method
Carbonate (meL^{-1})	0.89	0.67	Volumetric titration
Bicarbonate (meL^{-1})	7.27	4.55	Volumetric titration
Chloride (meL^{-1})	4.88	3.50	Volumetric titration
Sulphate (meL^{-1})	0.98	0.81	Turbidometry
Ca (meL^{-1})	2.90	1.75	Varsenate titration
Mg (meL^{-1})	4.15	4.70	Varsenate titration
Na (meL^{-1})	4.05	1.50	Flame photometry
K (meL^{-1})	0.31	0.11	Flame photometry
Fe (mgL^{-1})	3.25	2.08	Atomic absorption Spectophotometry
Mn (mgL^{-1})	0.03	0.01	Atomic absorption Spectophotometry
Zn (mgL^{-1})	0.25	0.11	Atomic absorption Spectophotometry
Cu (mgL^{-1})	0.09	0.04	Atomic absorption Spectophotometry
Lead (mgL^{-1})	0.09	0.08	Atomic absorption Spectophotometry
Cd (mgL^{-1})	0.02	0.02	Atomic absorption Spectophotometry
Ni (mgL^{-1})	0.06	0.06	Atomic absorption Spectophotometry
CO (mgL^{-1})	0.04	0.03	Atomic absorption Spectophotometry
NO_3-N(ppm)	85	20	Phenol Di-Sulphonic acid Method

Inclusion of two irrigations of sewage water in the schedule also improved seed yield to the tune of 11.24 and 9.48 per cent over tube well water irrigation (Table 35.2). Higher yield with these treatments was owing to higher value of yield attributes *viz.* branches and umbles per plant, umbellates per umbel (Table 35.2) as well as nutrients removal (N and P) by the seeds (Table 35.3).

A linear increase in seed and stover yield was observed with increase in N levels. Higher level of N (40 kg/ha) showed its superiority recording 8.02 and 22.17 per cent higher yield over N_1 and N_0, respectively. Rise in yield with 40 kg N/ha is due to improvement in yield attributes (Table 35.2) and higher removal of nitrogen and phosphorus by seeds and stover (Table 35:3). Kewalanand *et al.* (2001) and Patel (2002) also reported response to higher dose of N by dill seed.

Table 35.2: Growth, Yield Attributes and Yield of Dill Seed as Influenced by Different Irrigation Schedules and Nitrogen Levels (mean of 2 years)

Treatment	*Plant Height (cm)*	*Number of Primary Branches*	*Number of Umbels per Plant*	*Number of Umbellates per Umbel*	*Number of Seeds per Umbellate*	*Test Weight (g)*	*Seed Yield (t/ha)*	*Stover Yield (t/ha)*
				Irrigation schedules				
I_1 TTTT	111.29	4.35	10.52	13.13	11.07	4.00	1.08	2.45
I_2 SSSS	126.99	5.21	11.92	13.77	12.61	4.31	1.30	2.94
I_3 TSTS	118.53	4.77	11.24	13.45	11.57	4.10	1.20	2.75
I_4 TTSS	116.22	4.74	11.15	13.50	11.52	4.10	1.18	2.74
I_5 TSSS	123.32	4.96	11.66	13.65	12.51	4.20	1.27	2.88
I_6 STTT	115.02	4.63	10.75	13.31	11.32	4.07	1.15	2.61
SEm±	3.33	0.11	0.21	0.28	0.23	0.08	0.03	0.07
C.D. (P=0.05)	9.83	0.31	0.61	NS	0.68	NS	0.09	0.02
				Nitrogen levels (Kg/ha)				
N_0 0	111.01	4.38	10.69	13.11	11.36	4.03	1.07	2.49
N_1 20	116.73	4.74	11.06	13.26	11.64	4.12	1.21	2.74
N_2 40	127.94	5.21	11.87	14.03	12.30	4.24	1.31	2.30
SEm±	2.33	0.07	0.14	0.17	0.14	0.04	0.02	0.05
C.D. (P=0.05)	6.64	0.19	0.40	0.49	0.41	0.10	0.06	0.13
Interaction	NS	NS	NS	NS	NS	NS	NS	NS

T: Tubewell water; S: Sewage water.

Table 35.3: Nutrients Uptake and Postharvest Balance in Soil as Influenced by Different Treatments (Mean of 2 years)

Treatment	*Nutrients Yptake, kg/ha*				*Postharvest Nutrients Status of Soil*			
	N		*P*		*Available Nutrients, kg/ha*		*DTPA Extractable Content, mg/kg*	
	Seeds	*Stover*	*Seeds*	*Stover*	*N*	*P*	*Fe*	*Zn*
				Irrigation schedules				
I_1 TTTT	14.53	22.83	3.31	7.40	141.4	18.90	3.61	0.66
I_2 SSSS	16.83	27.14	4.09	9.00	170.7	23.25	5.95	0.90
I_3 TSTS	15.61	24.73	3.49	8.43	154.6	21.10	4.78	0.78
I_4 TTSS	15.41	24.65	3.51	8.53	154.4	20.88	4.78	0.77
I_5 TSSS	16.36	26.16	3.77	8.76	165.9	22.36	5.29	0.83
I_6 STTT	14.89	23.93	3.36	7.90	149.6	20.26	4.39	0.71
SEm±	0.29	0.55	0.09	0.23	3.9	0.46	0.13	0.03
C.D. (P=0.05)	0.85	1.61	0.27	0.67	11.4	1.35	0.40	0.08

Contd...

Table 35.3–Contd...

Treatment	*Nutrients Yptake, kg/ha*				*Postharvest Nutrients Status of Soil*			
	N		*P*		*Available Nutrients, kg/ha*		*DTPA Extractable Content, mg/kg*	
	Seeds	*Stover*	*Seeds*	*Stover*	*N*	*P*	*Fe*	*Zn*
				Nitrogen levels (Kg/ha)				
N_0 0	13.92	21.74	2.88	6.67	148.5	20.69	4.60	0.75
N_1 20	15.37	24.81	3.63	8.58	157.0	21.16	4.81	0.78
N_2 40	17.53	28.18	4.26	9.76	162.8	21.53	4.99	0.80
SEm±	0.15	0.34	0.06	0.15	2.4	0.25	0.05	0.01
C.D. (P=0.05)	0.43	0.97	0.17	0.42	6.9	NS	0.14	NS
Interaction	NS	NS	NS	NS	NS	NS	NS	NS

T: Tubewell water; S: Sewage water.

Nutrients Uptake

Nitrogen and phosphorus uptake by seeds and stover were significantly influenced by irrigation schedules. Maximum removal of N and P by both the components was observed with sewage water application followed by a combination of one tube well + three sewage water irrigations. Higher removal of the nutrients has reflected in improvement of yield. Tiwari *et al.* (1996) have also reported such higher removal of N and P with sewage water in different crops.

Removal of nitrogen by seeds due to application of 40 kg N/ha was increased to the extent of 25.93 per cent over control. Likewise, phosphorus uptake by seeds was enhanced to the tune of 47.9 per cent. Similar trend was also followed for N and P uptake by stover. These results are in accordance with the findings of Patel (2002).

Soil Fertility Improvement

Significant buildup in soil status of available N and P was noted with sewage water application followed by one tube well and three sewage water irrigations. The improvement in soil N and P was also observed with supplementation of two irrigation with sewage water in the schedule of total four irrigations. Similar trend was also observed for micronutrients *viz.* iron and zinc (Table 35.3). Higher content of NO_3-N as well as Fe and Zn in sewage water had contributed to buildup soil fertility. These findings are in agreement with the Abdou and Nehhah (1980), Jeyabaskaran and Ramulu (1996), Datta *et al.* (2000) and Saraswat *et al.* (2005).

Remarkable rise in available N and P status of soils was observed with application of 40 kg N/ha. Similarly, iron and zinc content of soil also got improved with increasing level of nitrogen.

Economics

All four irrigations with sewage water irrigation schedule registered the highest net return of Rs. 11,818 per hectare with 2.31 B:C ratio. Among N levels, maximum net realization of Rs. 11,902 per hectare with 2.32 B:C ratio was obtained with 40 kg N/ha.

Table 35.3: Economics of Dill Seed as Influenced by Different Irrigation Schedules and Nitrogen Levels (Mean of 2 years)

Treatment	*Cost of Production (Rs ha^{-1})*	*Gross Realization (Rs ha^{-1})*	*Net Realization (Rs ha^{-1})*	*B:C Ratio*
		Irrigation schedules		
I_1 TTTT	8723	17221	8499	1.97
I_2 SSSS	8998	20816	11818	2.31
I_3 TSTS	8860	19163	10302	2.16
I_4 TTSS	8860	18853	9993	2.13
I_5 TSSS	8929	20299	11370	2.27
I_6 STTT	8792	18336	9544	2.08
		Nitrogen levels (kg/ha)		
N_0 0	8708	17107	8398	1.96
N_1 20	8873	19336	10463	2.18
N_2 40	9000	20901	11902	2.32

T: Tubewell water; S: Sewage water.

Cost: Sewage water Rs. 300/irrigation, Urea Rs. 5.02/kg, Seed Rs. 50/kg.

Tube well water Rs. 240/irrigation, SSP Rs. 2.8/kg.

Selling price: Seed Rs. 40/kg.

From the present investigation, it can be concluded that dill seed crop raised with four irrigations using sewage water or one tube well + three sewage water irrigations fertilizing with 40 kg N/ha helped to increase seed yield and net return over sole schedule of tube well water irrigations.

References

Abdou, F.M. and El-Nehhah, M., 1980. Effects of irrigating loamy soil by liquid sewage sludge on its content of some micronutrients. *Plant Soil*, 56: 53–57.

CSSRI, 2001. *Annual Progress Report NATP-MM Project on Use of Urban and Industrial Effluent in Agriculture, 2001-02*, Central Soil Salinity Research Institute, Karnal.

Datta, S.P., Biswas, D.R., Saharan, N., Ghosh, S.K. and Rattan, R.K., 2000. Effect of long term application of sewage effluents on organic carbon, bio-available phosphorous, potassium and heavy metal status of soils and contents of heavy metals in crops grown thereon. *J. Indian Soc. Soil. Sci.*, 48(4): 836–839.

Jeyabaskaran, K.J. and Ramulu, U.S., 1996. Distribution of heavy metals in soils of various sewage farms in Tamil Nadu. *J. Indian Soc. Soil Sci.*, 44(3): 401–404.

Kewalanand, Pandey, C.S., Bisht, L.D. and Singh, S., 2001. Influence of umbel harvesting and nitrogen levels on European dill (*Anethum graveolens* L.). *Journal of Medicinal Aromatic Plant Science*, 23(3): 361–364.

Malarvzhi, P. and Rajamannar, A., 2001. Soil fertility management in fodder cultivated area through sewage water irrigation. *Madras Agricultural Journal*, 88(7–9): 472–477.

Patel, V.S., 2002. Response of dill seed to different sowing dates under varying levels of nitrogen and phosphorus. *M.Sc. (Agri.) Thesis*, Gujarat Agricultural University, Sardarkrushinagar.

Saraswat, P.K., Tiwari, R.C., Agrawal, H.P. and Kumar, S., 2005. Micronutrient status of soils and vegetable crops irrigated with treated sewage water. *J. Indian Soc. Soil Sci.*, 53(1): 111–115.

Tiwari, R.C., Kumar, A. and Misra, A.K., 1996. Influence of treated sewage and tube well water irrigation with different fertilizer levels on rice and soil properties. *Journal of Indian Society of Soil Science*, 44: 547–549.

Recent Advances in Seed Spices (2011) *Pages* **326–328**
Editors: **Y. Ravindrababu, R.K. Jaiman, K.D. Patel, N.R. Patel and S.B.S. Tikka**
Published by: **DAYA PUBLISHING HOUSE, NEW DELHI**

36

Effect of Weedicides on Cumin Crop and its Residual Effect on Greengram

S.M. Patel, J.C. Patel, M.M. Patel and H.D. Patel
Department of Agronomy, C.P. College of Agriculture, S.D. Agricultural University, Sardarkrushinagar – 385 506, Gujarat

Cumin is an irrigated *rabi* crop which remains in the field about four months. It can be grown on well drained sandy loam soils. Cold and dry weather is favorable for crop. Looking to the potential of cumin in global market, farmers are more inclined to grow this crop in the North Gujarat. As being broadcast sowing of this crop, it can not be intercultured. The control of weeds by manually is always troublesome. Unavailability of labours and high rate of wages during critical crop weed competition period are major constraints for the farmers. Under such circumstances only alternative left to farmers is to use chemicals for control of weeds. To find out suitable herbicide for the control of weeds in cumin crop and the residual effects of herbicides on succeeding green gram crop, an experiment was planned to study the "Effect of weedicides on cumin crop and its residual effect on green gram".

Objectives

1. To find out the efficacy of weedicide in cumin crop.
2. To compare the economics of different weedicides.
3. To know the residual effect of different weedicides on germination of green gram.

Materials and Methods

A field experiment on cumin crop was conducted at Agronomy Instructional Farm, C.P. College of Agriculture, S.D.A.U., Sardarkrushinagar during 2003-04 and 2004-05 in randomized block design with four replications. Total twelve treatments *viz.*, T_1: Pendimethalin @ 1000 g/ha as pre-emergence,

T_2: Trifluralin @ 500 g/ha as PPI, T_3: Trifluralin @ 1000 g/ha as pre-emergence, T_4: Oxadiazyl @ 75 g/ha as pre-emergence, T_5: Oxadiazyl @ 100 g/ha as pre-emergence, T_6: Oxadiazyl @ 125 g/ha as pre-emergence, T_7: Oxadiazyl @ 75 g/ha as post-emergence(25 DAS), T_8: Oxadiazyl @ 100 g/ha as post-emergence(25 DAS), T_9: Oxadiazyl @ 125 g/ha post-emergence(25 DAS) T_9: Hand weeding at 25 and 35 DAS, T_{11}: Weed free and T_{12}: Weedy check were tested.

The soil was sandy loam having pH -7.6, EC – 0.07 mmhoscm^{-1},O.C.-0.26 per cent and available P and K were 49.00 and 294.00 kg ha $^{-1}$, respectively. Cumin *Cv* Guj.Cumin-3 was grown during two years. The crop was sown at a distance of 22.5 cm apart. Recommended agronomical practices were adopted for raising crop. The crop was failed in the year 2005-06 due to aberrant weather condition.

Results

Weed Flora

Monocot weeds *viz., Cynodon dactylon* L., *Cyprus rotundus* L., *Eragrostis doageoides* Beauv., *Cenchrus biflorus* L. and Dicot weeds *viz., Chenopodium album* L. *Chenopodium murale* L., *Phyllanthus niruri* L, *Tribulus terrestris* L., *Portulaca oleracea* L., *Lanuca nudicaulis* L., *Asphodelus tenufolius* Cavan, *Melilotus alba* Des, *Acanthosperum hispidum* De., *Agremone mexicana* L. were observed in experimental site. Among all the weeds C*henopodium album* L. was dominant weed during *rabi* season.

Weed Counts and Dry Weight

Weed counts (numbers per 0.25 m ^{2}) and dry weight (kg ha $^{-1}$) were observed maximum under the treatment weedy check (Table 36.1).

Table 36.1: Weed Counts, Dry Weed Weight (kg ha^{-1}), Weed Control Efficiency (5) and Seed Yield (Kg ha^{-1}) of Cumin and Germination Percentage of Greengram as Affected by Different Weed Control Treatments

Sl.No.	*Treatment Details*	*Weed Counts (No./m^2) at Harvest*		*Dry Weed Weight at Harvest (kg ha^{-1})*	*WCE (Per cent)*	*Seed Yield (kg ha^{-1})*	*Germination Percentage*
		Moncot	*Dicot*				
1.	T_1: Pendimethalin @ 1000g/ha as pre-emergence	3	6	485	85.2	605	84
2.	T_2: Trifluralin @ 500 g/ha as PPI	6	8	815	77.0	530	86
3.	T_3: Trifluralin @ 1000 g/ha as pre-emergence	5	6	675	81.9	553	84
4.	T_4:Oxadiazyl @ 75 g/ha as pre-emergence	10	13	1395	62.3	333	87
5.	T_5:Oxadiazyl @ 100 g/ha as pre-emergence	6	8	1410	77.0	476	85
6.	T_6:Oxadiazyl @ 125 g/ha as pre-emergence	6	10	1212	73.8	189	85
7.	T_7:Oxadiazyl @ 75 g/ha as post-emergence (25 DAS)	9	12	1324	65.6	357	87
8.	T_8:Oxadiazyl @ 100 g/ha as post-emergence (25 DAS)	8	5	625	78.7	553	85

Contd...

Table 36.1–Contd...

Sl.No.	Treatment Details	Weed Counts (No./m²) at Harvest		Dry Weed Weight at Harvest (kg ha⁻¹)	WCE (Per cent)	Seed Yield (kg ha⁻¹)	Germination Percentage
		Moncot	Dicot				
9.	T_9:Oxadiazyl @ 125 g/ha post-emergence (25 DAS)	4	2	408	90.2	616	86
10.	T_9: Hand weeding at 25 and 35 DAS	4	4	75	86.9	521	88
11.	T_{11}: Weed free	1	2	40	95.1	625	90
12.	T_{12}: Weedy check	16	45	4038		163	90
	SEm±	–	–	–	–	45	–
	CD at 5 per cent	–	–	–	–	143	–
	CV per cent	–	–	–	–	18.60	–

Weed Control Efficiency

Weed control efficiency was recorded highest under the treatment T_{11} followed by the treatments T_9, T_{10}, T_1, T_3, T_8, T_5 and T_2; while weed control efficiency was least under the treatment T_4 followed by T_7.

Seed Yield

The results indicated that T_{11} (weed free) recorded significantly higher seed yield of cumin than that recorded in the rest of the treatments, but it was at par with T_9, T_{10}, T_1, T_3, T_8, T_5, T_2 and T_6. Significantly the lowest seed yield was recorded under T_{12} (Table 36.1).The similar results were also observed by Meena and Chaudhary (2007) and Mehariya *et al.* (2007).

Phytotoxicity on Succeeding Crop

The highest germination percentage of green gram was recorded under the treatment T_{12} followed by T_{11}, T_{10}, T_4, T_7, T_9 and T_2 (Table 36.1).

References

Meena, N.L. and Chaudhary, G.R., 2007. Effect of oxidiarzyl vis-à-vis pendimethalin in controlling weeds in cumin. In: *A Paper Presented in National Seminar on Production, Development, Quality and Export of Seed Spices: Issues and Strategies*, February 2–3, pp. 22.

Mehariya, M.L., Yadav, R.S. and Jangir, R.P., 2007. Production of cumin as influenced by integrated weed control. In: *A Paper Presented in National Seminar on Production, Development, Quality and Export of Seed Spices: Issues and Strategies*, February 2–3, pp. 25.

Recent Advances in Seed Spices (2011) *Pages* **329–332**
Editors: **Y. Ravindrababu, R.K. Jaiman, K.D. Patel, N.R. Patel and S.B.S. Tikka**
Published by: **DAYA PUBLISHING HOUSE, NEW DELHI**

37

Effect of Spacing and Clove Size on Growth and Yield of Garlic Under Akola Conditions

S.M. Ghawade, S.R. Dalal and S.R. Bhople
College of Horticulture, Dr. Panjabrao Deshmukh Krishi Vidyapeeth, Akola, M.S.

ABSTRACT

The field experiment was conducted during *rabi* season of 2006 at Main Garden, University Department of Horticulture, Dr. Panjabrao Deshmukh Krishi Vidyapeeth, Akola to assess the effect of spacing and clove size on yield and quality of garlic. The result indicates that, the closer spacing (10 x 10 cm) recorded the highest yield. However, the quality bulb was produced under wider spacing (15 x 15 cm). Regarding clove size, moderate size clove (12 g 10 cloves^{-1}) gives highest yield whereas, quality bulbs were produced from large size clove (16 g 10 clove^{-1}).

Introduction

Garlic (*Alium sativum* L) is an important condiment crop grown in winter season, under irrigated condition in Vidarbha region. It is well known that, among the various inputs used for the production of garlic, optimum mother clove size and plant spacing have a great influence on the yield and quality of garlic. Owing to the absence of relevant information on different aspects of agronomy, particularly in this part of the region, the growers are reluctant to take the production of this crop on commercial scale. In view of above points, the field experiment was conducted to study the effect of spacing and clove size on yield and quality of garlic.

Materials and Methods

The present investigation was conducted at Main Garden, University Department of Horticulture, Dr. Panjabrao Deshmukh Krishi Vidyapeeth, Akola. The experiment was laid out in Factorial Randomized Block Design (FRBD). In the investigation, three levels of spacing (10x10, 15x10 and 15x15 cm) and four levels of clove size (4, 8, 12 and 16 g 10^{-1} clove) with twelve treatment combinations were tested. The planting were carried out on 18th October, 2006.

Fertilizers were applied @ 100:50:50 kg NPK per hectare in the form of urea, single super phosphate and muriate of potash. The full dose of phosphorus and potash and half dose of nitrogen were applied at the time of planting and rest of the fertilizers after one month of the first split.

Results and Discussion

The data presented in Table 37.1 indicates that, the various spacing and clove size had significant effect on the growth, yield and quality parameters.

Table 37.1: Influence of Different Spacing and Clove Size on Growth, Yield and Quality Gf garlic

Treatment	*Height of Plant (cm)*	*Leaves Plant^{-1}*	*Neck Thickness*	*Bulb Yield (q ha^{-1})*	*Wt. of Bulb (g)*	**Clove Index (g)*	*Cloves Bulb^{-1}*
Spacing (S)							
S_1 (10 x 10cm)	74.54	10.93	0.869	85.44	12.05	102.75	18.43
S_2 (15x10 cm)	79.50	11.47	0.982	73.55	14.35	107.02	20.45
S_3 (15x15 cm)	79.00	11.44	0.980	58.57	15.66	107.09	19.91
SE(m)±	0.174	0.166	0.012	0.216	0.162	0.296	0.259
CD at 5 per cent	0.510	0.488	0.035	0.635	0.477	0.868	0761
Clove size (W)							
W_1 (4 g 10 clove^{-1})	74.63	10.89	0.881	65.16	13.68	99.32	16.70
W_2 (8g 10 clove^{-1})	78.12	11.00	0.911	67.62	13.55	105.29	19.40
W_3 (12 g 10 clove^{-1})	79.69	11.88	0.986	78.61	14.25	108.02	20.78
W_4 (16 g 10 clove^{-1})	78.29	11.22	1.000	77.91	14.92	109.85	21.50
SE(m)±	0.201	0.192	0.014	0.250	0.188	0.341	0273
CD at 5 per cent	0.589	0.564	0.041	0.734	0.551	1.043	0.801
Interaction (S x W)							
SE(m)±	0.348	0.333	0.024	0.250	0.325	0.592	0.519

*: Weight of 100 cloves.

Growth

Perusal of data presented in Table 37.1 exhibited significant influence on height of plant, leaves per plant and neck thickness. Medium spacing (15 x 10 cm) and moderate cloves weight (12 g 10^{-1} clove) resulted into maximum plant height (79.50 cm and 79.69 cm, respectively), leaves per plant (11.47 and 11.88, respectively) and neck thickness of plant (0.982 and 1.00 cm, respectively). However, these were recorded minimum (74.54 and 74.63 cm, 10.93 and 10.89, 0.869 and 0.88, respectively) in closer spacing S, (10 x 10 cm) and small size cloves W1 (4 g 10^{-1} clove).

The maximum growth of the garlic plant was obtained with medium spacing and moderate clove size might be due to the fact that, the greater spacing of plant seems to have helped the individual plant to utilize more soil water, nutrition, air and light to help it to put up better growth. Similarly, initial storage of enough food material might be helpful in vigorous growth of plant. The results of the present investigation are in agreement with Purewal and Dragan (1961), Lawande *et al.* (1993), Kotgirwar *et al.* (1997) and Thakur (1997) in garlic.

Yield

The data in respect of bulb yield of garlic revealed that, the closer spacing (10 x 10 cm) and moderate clove size (12 g 10^{-1} clove) recorded the maximum (85.44 and 78.61 q ha^{-1}, respectively) bulb yield. However, it was minimum in 15 x 15 cm spacing (58.57 q ha^{-1}) and in 4 g 10^{-1} clove size (65.16 q ha^{-1}).

The higher bulb yield under maximum plant density was attributed due to significantly more number of plants per unit area. Similarly, enough storage of food material in moderate size clove helps to improve the individual bulb weight. The results are in agreement with Thakur (1997) and Raghin *et al.* (2004) in garlic.

Quality

The data in respect of weight of bulb, clove index and cloves per bulb exhibited significant differences among the different treatments of spacing and clove size. Significantly the maximum bulb weight and clove index (15.66 g and 107.09, respectively) were recorded in wider spacing (S3) while, maximum cloves per bulb (20.45) were recorded in treatment S2. The minimum (12.05 g) weight of bulb, clove index (102.75 g) and cloves per bulb (18.43) were registered in treatment S1

The planting of larger clove (16 g 10^{-1} clove) recorded maximum weight of bulb (14.92 g), clove index (109.85 g) and cloves per bulb (21.50), whereas, these were found minimum (13.68 g, 99.32 g and 16.70, respectively) under the planting of smallest cloves (4g 10^{-1} clove).

This might be due to the fact that, the plants under lower plant densities perhaps were subjected to low degree of competition for space, nutrients and for all other resources of environments and hence, instead of vertical growth, plant have grown horizontally producing more number of leaves, resulted into increased interception, absorption and utilization of radiant energy resulting in higher amount of photosynthates, which showed positive effect on quality parameters. The result of the present investigation is in line with Lawande *et al.* (1993).

The quality of garlic bulb improved by planting the larger cloves might be due to fact that, it helped in better development of root system enabling the proper absorption and utilization of water and nutrient which would results into better quality. Kotgirwar *et al.* (1997) and Thakur (1997) obtained the similar result in garlic.

References

Kotgirwar, V.V., Chaudhary, M.H., Damke, M.M. and Jadhao, B.J., 1997. Effect of planting method and clove size on growth, yield and quality of garlic cv. Godawari. *PKV Res. J.*, 21(2): 182–184.

Lawande, K.E., Pawar, R.D., Supe, V.S. and Kale, P.N., 1993. Effect of spacing and different levels of NPK on yield of garlic var. Godawari. *Maharashtra J. Hort.*, 7(2): 61–64.

Purewal, S.S. and Dragan, K.S., 1961. Effect of fertilizers and spacing on the development and yield of garlic. *Indian J. Agron.*, 17(2): 263–265.

Raghin, M.A., Siddique, M.A. and Hoosain, M.M., 1984. Effect of time of planting, mother bulb size and plant density on the yield of garlic. *Bangladesh J. Agric. Res.*, 9(2): 112–118.

Thakur, P., 1997. Propagations and variability studies in garlic. *M.Sc. Thesis*, (Unpublished), MPKV, Rahuri.

Recent Advances in Seed Spices (2011) *Pages* **333–339**
Editors: **Y. Ravindrababu, R.K. Jaiman, K.D. Patel, N.R. Patel and S.B.S. Tikka**
Published by: **DAYA PUBLISHING HOUSE, NEW DELHI**

38

Effect of Sub-surface Compaction and Moisture Regimes in Fenugreek on Highly Permeable Coarse Textured Soils

B.L. Kumawat, Arvind Kumawat, Amit Kumawat and Sangeeta Kumawat
Department of Soil Science and Agricultural Chemistry,
S.K.N. College of Agriculture, Rajasthan Agricultural University,
Campus-Jobner – 303 329, Rajasthan

ABSTRACT

A field experiment was conducted to study the effect of sub-surface (15-30 cm depth) compaction on utilization efficiency of field water by fenugreek grown on highly permeable coarse textured loamy sand soil under varying irrigation regimes. The experiment consisted of four replications of 16 treatment combinations of four levels each of compaction (0, 2, 4 and 6 passings of 500 kg iron roller) and irrigation frequencies (4, 5, 6 and 8 irrigations at different growth stages of crop) in strip plot design. Compaction by 6 passings increased the soil bulk density, soil water content and root weight density of fenugreek and decreased saturated hydraulic conductivity of soil. It was observed that root nodulation, yield attributes, seed and straw yield and total N and P uptake increased due to sub-surface compaction with 4 passings and 6 irrigations, whereas, field water use efficiency significantly improved with increasing levels of compaction up to 4 passings but decreased with increasing irrigation frequencies more so at higher irrigation frequency. The available N content in soil after harvest of crop also increased due to soil compaction. Thus, sub-surface compaction improved yield and field water use efficiency of fenugreek under highly permeable coarse textured soils of arid and semi-arid plains of Rajasthan.

Keywords: *Sub-surface compaction, Irrigation frequency, Field water use efficiency, Fenugreek, Highly permeable coarse textured soil.*

Introduction

Soil compaction refers to the packing effect of a mechanical force on the soil. This packing effect decreases the volume occupied by pores and increases the density and strength of the soil mass. Bulk density, water infiltration and penetrometer resistance are the indices of soil compaction. Pressure exerted on the soil surface depends on the static load, characteristics of the wheel or track and characteristics of the soil. The integral product of the vertical component of surface stress with the soil surface area on which it acts is equal to the total weight carried on any wheel or track. Theory suggests that the total axle load is a much more significant factor in controlling deep soil compaction than the surface contact pressure (Abu-Hamdeh and Amayreh, 2001). The loamy sand soils of Rajasthan are highly permeable and loss of water and soluble nutrients are quite high from these soils. Sub-surface compaction perhaps improves the retention of these inputs (Agrawal 1991 and Painuli *et al.* 2000). Since fenugreek is grown in *rabi* season when rainfall is excess or scanty, both the inputs assume criticality. Therefore, their proper use becomes essential for getting more economic yield, as low or high amounts of moisture supply disturb the availability of nutrients to the plant, hence, looking into the constraints associated with highly permeable soils of Rajasthan, field experiment was conducted during *rabi* season of 2006-07 to evaluate any improvement in use efficiency of field water in fenugreek crop through sub-surface compaction of highly permeable coarse textured loamy sand soils, as influenced by varying levels of irrigation regimes.

Materials and Methods

Field experiment was conducted on loamy sand soil, belonging to hyperthermic family of Typic Ustipsamment with fenugreek crop in *rabi* season. The experimental soil (0-15 cm depth) had pH 8.1, EC 0.27 dSm^{-1}, organic carbon 2.1 g kg^{-1}, available N 130 kg ha^{-1}, available P 18.1 kg ha^{-1}, available K 185 kg ha^{-1}, bulk density 1.52 Mg m^{-3} and saturated hydraulic conductivity 10.4 cm h^{-1}. The experiment was laid out in strip plot design with 16 treatment combinations of 4 levels of compaction and 4 levels of irrigation with four replications. The compaction levels were no compaction (C_0) *i.e.* two cross cultivation by tine cultivator followed by planking and comapction (C_2), (C_4) and (C_6) *i.e.* same operation as for C_0 were followed by 2, 4 and 6 passings of bullock-driven 500 kg iron roller. Four irrigation regimes were created by application of 4 (I_0), 5 (I_5), 6 (I_6) and 8 (I_8) irrigations at different growth stages of fenugreek. The depth of each irrigation was 5cm measured by parshal flume. Field was ploughed as described earlier (C_0) after 24 hours of heavy (7 cm) irrigation. The soil compaction was done with the help of 500 kg bullock driven iron roller at optimum moisture content (12 g per 100 g) for maximum compaction (Proctor 1933). Required number of shallow furrows then opened with the help of hand hoe in uncompacted and compacted plots.

A uniform recommended dose of nitrogen and phosphorus was applied @ 40 kg ha^{-1} through DAP and urea as basal. After application of fertilizers in furrows, the furrows were slightly covered with soil followed by seeding of fenugreek (cv. RMt-1). Undistured soil samples were collected with the help of cores (7 cm diameter and 8 cm in length) from 0-15, 15-30 and 30-45 cm soil depths to determine the bulk density and saturated hydraulic conductivity of soil. Soil water content was determined gravimetrically at 0-15 cm interval upto 45 cm depth at harvest. Soil cores (7 cm diameter and 8 cm in length) for determining root growth were sampled at flowering stage of crop from four locations in plots under compaction treatments by centring the auger in the middle of the row and between two adjacent rows at 0-15 cm interval down to the 0-45 cm depth. All four samples for a given depths were bulked and washed on a 1 mm screen. Root dry weight was determined after drying at 65°C. Seed and straw yields were recorded after harvest. The seed and straw samples collected from all the plots were analysed for total N and P contents (Jackson, 1973).

Results and Discussion

Soil Physical Properties and Root Density

Data (Table 38.1) reveal that compaction increased bulk density and soil water content and decreased saturated hydraulic conductivity at both the stages of sampling during field investigation. The bulk density ranged from 1.50 to 1.52 Mg m^{-3} in uncompacted control, while, it was recorded in the range of 1.58 to 1.62 Mg m^{-3} under compaction treatment involving 6 passings of 500 kg iron roller. The bulk density was maximum in sub-surface layer (15-30 cm depth) due to compaction by making 2, 4 and 6 passings of roller over control. In contract, the saturated hydraulic conductivity decreased on account of compaction under 2, 4 and 6 passings of roller and the decrease was most pronounced in 15-30 cm soil layer. Agarwal *et al.* (1987) also observed that the bulk density of lower soil layers was important for irrigation water management in sandy soils. Some workers have attached greater importance to hydraulic conductivity determination (sands *et al.* 1979) but in present investigation both bulk density and saturated hydraulic conductivity appear equally useful parameters for judging the effects of soil compaction (Table 38.1).

Table 38.1: Effect of Compaction of Bulk Density (Mg m^{-3}), Saturated Hydraulic Conductivity (cm h^{-1}) Water Content (cm^3 cm^{-3}) of Soil and Root Weight Density (mg cm^{-3}) of Fenugreek

Compaction Levels	*Depth (cm)*	*Stages of Sampling*						
		Just After Compaction			*Flowering Stage*			
		Bulk Density	*Saturated Hydraulic Conductivity*	*Water Content*	*Bulk Density*	*Saturated Hydraulic Conductivity*	*Water Content*	*Root Weight Density*
C_0	0-15	1.50	10.4	0.144	1.50	10.7	0.130	245.2
	15-30	1.51	9.5	0.156	1.52	9.9	0.144	75.6
	30-45	1.52	9.1	0.166	1.52	9.4	0.147	70.4
C_2	0-15	1.54	9.1	0.149	1.53	9.2	0.133	269.6
	15-30	1.57	8.3	0.160	1.54	8.8	0.146	96.5
	30-45	1.56	8.6	0.171	1.54	9.0	0.150	84.1
C_4	0-15	1.56	8.7	0.161	1.54	8.9	0.137	356.8
	15-30	1.59	7.4	0.169	1.56	7.8	0.150	112.5
	30-45	1.57	7.9	0.177	1.55	8.3	1.155	87.7
C_6	0-15	1.59	8.1	0.172	1.56	8.6	0.149	426.7
	15-30	1.62	6.9	0.179	1.59	7.4	0.164	121.3
	30-45	1.60	7.2	0.183	1.58	7.7	0.168	96.9

Records of root weight density (RWD) indicate that root weight increased under increasing levels of compaction treatment. The maximum enhancement of RWD due to compaction of 6 passings of roller in present study was caused in the 0-15 surface layer. The increased RWD under compaction has been attributed to increase in nutrients per unit volume (Ogunremi *et al.* 1986). Kar *et al.* (1976) observed that contact between root and soil particles also increased due to compacting of loamy sand soils.

Nodulation, Yield Attributes and Yield

Increasing levels of compaction upto 4 passings significantly increased root nodulation at 40 DAS, pods per plant, seeds per pod, seed and straw yield of crop (Table 38.2). An increase in root nodulation, pods per plant, seeds per pod, test weight, seed and straw yield by 4 passings over no-passing and 2 passings of roller was of the order of 34.86 and 10.83, 27.68 and 7.67, 10.51 and 3.42, 6.29 and 3.18, 36.65 and 11.61 and 26.50 and 9.21 per cent, respectively. Higher root nodulation, pods per plant, seeds per pod, seed and straw yields obtained under C_4 is ascribed to the favorable soil moisture regime. The beneficial effects of compaction have also been reported by Yadav and Vyas (2006) who stated that soil compaction significantly increased the growth, yield attributes and yield in light textured soil by decreasing the percolation losses of soluble nutrients and improving moisture storage in soil.

Table 38.2: Effect of Compaction and Irrigation Levels on Root Nodulation, Yield Attributes and Yield and Nutrient Uptake by Fenugreek Crop

Compaction Levels	*Root Nodules per Plant at 40 DAS*	*Yield Attributes*			*Crop Yield (q ha⁻¹)*		*Total Uptake (kg ha⁻¹)*	
		Pods/ Plant	*Seed/ Pod*	*Test Weight (g)*	*Seed*	*Straw*	*N*	*P*
				Compaction levels				
C_0	8.95	25.50	14.74	11.27	12.66	34.29	72.32	9.49
C_2	10.89	30.24	15.75	11.61	15.50	39.72	89.14	11.96
C_4	12.07	32.56	16.29	11.98	17.30	43.38	101.04	13.61
C_6	12.78	34.01	16.63	12.13	17.92	44.81	105.59	14.25
SEm±	0.25	0.54	0.16	0.09	0.26	0.67	1.61	0.33
CD (P = 0.05)	0.80	1.74	0.52	0.31	0.84	2.13	5.13	1.04
				Irrigation levels				
I_4	9.10	25.60	14.70	11.37	12.37	34.02	77.02	9.18
I_5	11.00	30.53	15.76	11.68	15.41	39.55	90.44	11.79
I_6	12.00	32.63	16.31	11.92	17.55	43.60	99.43	13.81
I_8	12.57	33.56	16.64	12.02	18.04	45.03	101.19	14.52
SEm±	0.23	0.46	0.12	0.07	0.23	0.63	1.49	0.30
CD (P = 0.05)	0.75	1.47	0.51	0.23	0.75	2.01	4.76	0.95

As the number of irrigations increased from 4 to 6 increased the root nodules per plant at 40 DAS (9.10 to 12.00), pods per plant (25.60 to 32.63) seeds per pod (14.70 to 16.31), test weight (11.37 to 11.92 g), seed yield (12.37 to 17.55 q ha^{-1}) and straw yield (34.02 to 43.60 q ha^{-1}), respectively. Higher yields with increasing levels of either compaction upto 4 passings and application of 6 irrigations may be due to better root growth (Gajri *et al.* 1989), more number of pods per plant, seeds per pod and other yield attributing components (Sarkar *et al.* 1994). Irrigation frequency also enhanced the yield by promoting synthesis and translocation of carbohydrates (Rani *et al.* 2000).

Uptake of Nutrients

The total uptake of N and P by fenugreek significantly increased due to compaction with 4 passings over lower levels (Table 38.2). This could be due to reduction in leaching losses of nitrogen and phosphorus owing to high bulk density in sub-surface soil that resulted in greater absorption of N and P by plant. Similar results were also reported by Ogunremi *et al.* (1986) and Mathan and Natesan (1993). Nitrogen and phosphorous uptake by fenugreek increased significantly with increase in irrigation frequency upto 6 irrigations. Thus could be due to increased solubility of nitrogen and phosphorus in soil solution which resulted in higher uptake of N and P (Sharma and Kumar, 1990). Yadav and Vyas (2006) and Yadav and Yadav (2008) also reported that more frequent irrigation regime caused denser rooting in 0-15 cm soil layer which is conducive for greater uptake of nutrients.

Field Water Use Efficiency

The field water use efficiency of fenugreek significantly increased with increasing levels of compaction upto 4 passings (Table 38.3). Such increases in water use efficiency following compaction could be attributed to increase in moisture availability concomitant with decreased percolation as a consequence of reduced infiltration and hydraulic conductivity of soil was also reported by Somani and Sharma (1985).

Table 38.3: Effect of Compaction and Irrigation Levels on Field Water Use Efficiency and Available N and P_2O_5 Contents in Soil at Harvest

Compaction Levels	*Field Water Use Efficiency (kg ha⁻¹ mm⁻¹)*	*Available nutrients (kg ha⁻¹)*	
		N	*P_2O_5*
	Compaction levels		
C_0	4.56	135.70	15.87
C_2	5.53	137.95	16.55
C_4	6.18	139.80	17.10
C_6	6.42	140.55	17.55
SEm±	0.13	0.53	0.39
CD (P = 0.05)	0.43	1.70	NS
	Irrigation levels		
I_4	6.19	139.70	17.44
I_5	6.16	138.97	17.06
I_6	5.85	137.77	16.49
I_8	4.51	137.56	16.08
SEm±	0.12	0.67	0.33
CD (P = 0.05)	0.40	NS	NS

The water use efficiency decreased significantly with irrigation at all the growth stages of fenugreek crop upto 6 irrigations, thereafter, it remained at par with 8 irrigations. This may probably due to increase in evaporation and infiltration losses owing to higher water application more so at higher level. This was not unexpected since every additional increment in irrigation water results in reduced yield increases (Sharma and Kumar 1990).

Available Nutrients

The available N content of soil after harvest of crop increased significantly with increasing levels of compaction upto 4 passings (Table 38.3). The higher residual N content under compaction could be expected due to more fixation of atmospheric N_2 by the crop as a result of higher root nodulation. Less movement of NO_3^--N into soil with water, brought about by an increase in bulk density of sub-surface layer and lesser residual N content under no compaction treatment could be due to more leaching of NO_3^--N with percolating waters as suggested by higher infiltration rate (Mandal *et al.* 1990). A non-significant increase in available P_2O_5 was also observed with increasing levels of compaction. However, irrigation levels could not influence available P_2O_5 content of soil after harvest of crop.

It is inferred that yield optimization was mediated by sub-surface compaction of loamy sand desertic coarse textured soil, as it provided conducive physical environment for better availability of nutrients culminating in higher water use efficiency by fenugreek.

References

Abu-Hamdeh, N.H. and Amayreh, J.A., 2001. Effect of tyre inflation pressure, wheel laod, and basic tillage systems on root density and soil compaction. *Journal of the Indian Society of Soil Science*, 49: 42–50.

Agrawal, R.P., 1991. Water and nutrient management in sandy soil by compaction. *Soil and Tillage Research*, 19: 121–130.

Agrawal, R.P., Jhorar, B.S., Dhankar, J.S. and Raj, M., 1987. Compaction of sandy soil for irrigation management. *Irrigation Science*, 8: 227–232.

Gajri, P.R., Prihar, S.S. and Arora, V.K., 1989. Effect of N and early irrigation on root development and water use by wheat on two soils. *Field Crops Research*, 21: 103–114.

Jackson, M.L., 1973. *Soil Chemical Analysis*. Prentice Hall of India Pvt. Ltd., New Delhi.

Kar, S., Varade, S.B., Subramanyam, T.K. and Ghildyal, B.P., 1976. Soil physical conditions affecting rice root growth. I. Bulk density and submerged soil temperature regime effects. *Agronomy Journal*, 68: 23–26.

Mandal, D.K., Kar, S. and Sharma, S.K., 1990. Effect of compaction on nitrogen mobility in coarse textured lateritic soil. *Journal of Indian Society of Soil Science*, 38: 145–147.

Mathan, K.K. and Natesan, R., 1993. Effect of compaction on yield of maize and nutrient uptake in sandy soils. *Journal of Indian Society of Soil Science*, 41: 765–767.

Ogunremi, L.T., Lal, R. and Babalola, O., 1986. Effects of tillage methods and water regimes on soil properties and yield of lowland rice from a sandy loam soil in south-west Nigeria. *Soil Tillage Research*, 6: 223–234.

Painuli, D.K., Majumdar, S.P., Kumawat, B.L. and Yadav, B.L., 2000. Su-surface compaction technology for coarse textured soils. *ICAR News*, 3: 1–3.

Proctor, R.R., 1933). Fundamental principles of soil compaction. Description of field and laboratory methods. *Engineering News Record*, 3: 286–289.

Rani T., Kumar, S. and Sharma, R.D., 2000. Studies on recovery and use efficiency of nitrogen in wheat crop on Typic Ustochrept of western Uttar Pradesh. *Journal of the Indian Society of Soil Science*, 48: 292–297.

Sands, R., Greacen, E.L. and Gera, C.J., 1979. Compaction of sandy soil in radiata pine forests. I. A penetrometer study. *Australian Journal of Soil Research*, 17: 101–105.

Sarkar, M.C., Uppal, K.S., Rana, D.S. and Banerjee, N.K., 1994. Fate of ^{15}N urea applied to wheat grown under upland irrigated condition on a Typic Ustochrept. *Journal of the Indian Society of Soil Science*, 42: 267–270.

Sharma, D.K. and Kumar, A., 1990. Effect of nitrogen on yield, uptake, recovery and nutrient use efficiency of mustard under different irrigation scheduling. *Journal of Indian Society of Soil Science*, 38: 229–232.

Somani, L.L. and Sharma, S.P., 1985. Improving moisture conservation and nutrient utilization by wheat through sub-surface compaction of a sandy soil. *Transactions of the Indian Society of Desert Technology*, 14: 31–36.

Yadav, B.L. and Vyas, K.K., 2006. Influence of sub-surface compaction on recovery and use efficiency of nitrogen in wheat crop on highly permeable soils. *Journal of the Indian Society of Soil Science*, 54: 158–162.

Yadav, B.L. and Yadav, C.P., 2008. Effect of nitrogen fertilization of soil–plant–water relationship under different soil water conservation practices in mustard. *Journal of the Indian Society of Soil Science*, 56: 114–117.

Recent Advances in Seed Spices (2011) *Pages* **340–343**
Editors: **Y. Ravindrababu, R.K. Jaiman, K.D. Patel, N.R. Patel and S.B.S. Tikka**
Published by: **DAYA PUBLISHING HOUSE, NEW DELHI**

39

Role of Rhizobacteria in Growth Promotion of Fenugreek

K.S. Shekhawat, Manoj Agarwal and Dhirendra Singh
Department of Plant Breeding and Genetics, Rajasthan Agricultural University, Bikaner Campus, Jobner – 303 329, Rajasthan

ABSTRACT

An experiment was conducted to assess the role of rhizobacteria on growth promotion of Fenugreek. Out of eight treatments, seed treatment and soil application with FK -14 + FL-18 was resulted in highest yield (897 kg/ha) and Biomass production (31.11 q/ha) followed by seed treatment with FK -14 + FL-18 (seed yield 790 kg/ha) and seed and soil application with FL-18 (seed yield 780 kg/ha). Maximum number of nodules per plant (18), fresh and dry weight of nodules per plant (739 and 223 mg), Root and Shoot length (10.90 and 18.33 cm) was observed in seed treatment and soil application of FK -14 + FL-18. Soil population of rhizobacteria was also observed maximum in seed treatment and soil application with FK -14 + FL-18. The control was resulted in minimum growth parameters and yield (661 kg/ha).

Introduction

Fenugreek is a self pollinated crop belongs to family fabaceae. It is widely grown in south-east Asia, Russia, Greece, China and some countries of Africa. In India, the major fenugreek growing states are Rajasthan, Gujarat, Madhya Pradesh, Tamil Nadu, Uttar Pradesh and Punjab. More than 80 per cent area and production of the country is contributed by Rajasthan state alone as it is fairly tolerant to salinity which makes it suitable for cultivation in major parts of the state. It has a high degree of medicinal value.

Bacterial strains isolated from the rhizosphere hold great promise as seed inoculants in innovative agricultural systems to promote plant growth and yield. These bacteria are termed as plant growth-promoting rhizobacteria (PGPR), or only 'rhizobacteria', to accentuate their intimate association with roots. Studies with PGPR indicate that the root microflora can profitably be manipulated through use of bacteria which live as epiphytes on roots and beneficially modulate the ecological niche. The direct application of micro-organisms to seed or other plant parts give them a competitive advantage over pathogens that must compete for nutrients and sites for attachment prior to infection.

Routine use of biological systems in controlling plant diseases has become more attractive due to the added benefit of enhanced plant growth besides disease control (Podile and Dube, 1988).

Materials and Methods

Two strains of rhizobacteria namely FK-14 and FL-18 were used for the study. There were applied as seed treatment as well as soil application. For seed treatment required quantity of fenugreek seeds were taken and shaken with rhizobacterial strain/s @ 8 gm/kg seeds on a mechanical rotary shaker for 2 hours to give a uniform coating. For soil application, rhizobacterial strain/s was taken @ 8 gm/sq.m and applied in the rows prior to sowing of seeds. The following combinations of FK-14 and FL-18 were used for the study- FK-14 (ST), FK-14 (ST + SA), FL-18 (ST), FL-18 (ST + SA), FK-14 + FL-18 (ST), FK-14 + FL-18 (ST + SA). *Trichoderma* (MTCC 5179) was used as standard control which was used @ 4 gm/sq.m as soil application. Untreated seeds were used as control. Observations were recorded for number of nodules per plant, nodules weight (fresh weight as well as dry weight), plant weight (fresh weight as well as dry weight), root and shoot length, soil population of rhizobacteria (through serial dilution technique), biomass (q/ha) and yield (kg/ha).

Results and Discussion

An experiment was conducted to assess the role of rhizobacteria on growth promotion of fenugreek. Out of eight treatments, seed treatment and soil application with FK- 14 + FL-18 resulted in highest yield (897kg/ha) and biomass production (31.11 q/ha) followed by seed treatment with FK-14 + FL-18 (seed yield 790 kg/ha) and seed and soil application with FL-18 (seed yield 780 kg/ha). Maximum number of nodules per plant (18), fresh and dry weight of nodules per plant (739 and 223 mg), root and shoot length (10.90 and 18.33 cm) was observed in seed treatment and soil application of FK-14 + FL-18. Soil population of rhizobacteria was also observed maximum in seed treatment and soil application with FK-14 + FL-18. The control was resulted in minimum growth parameters and yield (6.61 kg/ha) (Table 39.1).

Co-inoculation studies with PGPR and *Bradyrhizobium japonicum* have also demonstrated increased soybean plant root and shoot weight, seed yield, plant vigour, nodulation and nitrogen fixation (Verma *et al.*, 1986). Similarly, an increase in grain yield, nodule dry matter and nitorgenase activity were also obtained in chickpea inoculated with a mixture of *Azospirillum brasilense* and *Rhizobium* strains (Rai, 1983). Grimes and Mount (1984) found that a *Pseudomonas putida* strains (M-17), which had been selected as a potential biological control agent, markedly increased Rhizobium nodulation of bean in field soils. Polonenko *et al.* (1987) found similar effects of certain rhizobacteria (primarily fluorescent pseudomonads) on nodulation of soybean roots by *B. japonicum*.

Indirect mechanisms used by PGPR include biotic protection against pathogenic bacteria, reduction of iron available to phytopathoens in the rhizosphere, synthesis of fungal cell wall-lysing enzymes and competition with detrimental microorganisms for sites on plant roots. Direct mechanism of plant growth by PGPR include the provision of bioavailable phosphorus for plant uptake, nitrogen fixation

Table 39.1: Role of Rhizobacteria in Growth Promotion of Fenugreek (*Rabi* 2007-08)

Treatments	No. of Nodules/ Plant	Nodule Weight		Plant Weight		Root Length (cm)	Shoot Length (cm)	Soil Population (No. of rhizobacterial colonies at diff. conc.)				Biomass (q/ha)	Yield (kg/ha)
		Fresh (mg)	Dry (mg)	Fresh (mg)	Dry (mg)			10^4	10^5	10^6	10^7		
FK-14 Seed treatment	10	492.67	164.30	26.52	2.36	8.37	16.40	–	–	–	1	28.89	689.77
FK14 Seed treatment + soil application	13	498.33	165.33	26.83	2.85	8.60	16.43	–	–	–	4	29.33	715.11
FL18 Seed treatment + soil application	13	512.33	173.67	27.45	3.30	8.60	16.50	–	–	1	1	29.33	739.56
FK+14 FL18 Seed treatment	14	565.67	177.33	27.70	3.80	8.63	17.00	–	1	3	2	29.78	780.44
FK14+FL18 seed application	15	722.33	209.67	28.20	3.79	9.83	18.23	–	1	1	1	30.22	790.23
FK14+FL18 seed treatment + Soil application	18	739.00	223.67	31.55	4.17	10.90	18.33	4	1	3	3	31.11	897.33
Trichoderma (MTCC 5179) as per recommendation	11	456.33	159.00	24.36	2.09	8.17	16.07	1	1	1	1	28.00	673.77
Control	9	431.33	141.21	22.47	1.91	7.50	15.07	–	–	–	–	26.67	661.33
CD (5 per cent)	1.13	46.48	15.21	5.35	0.29	0.75	1.44	–	–	–	–	2.54	64.06
CV (per cent)	5.28	5.05	5.16	5.26	5.83	5.14	5.17	–	–	–	–	5.23	5.17

for plant use, sequestration of iron for plants by siderophores, production of plant hormones like auxins, cytokinins and gibberellins and lowering of plant ethylene levels (Lucy *et al.*, 2003).

References

Grimes, H.D. and Mount, M.S., 1984. Influence of *Pseudomonas putida* on nodulation of *Pheseclus vulgaris. Soil Biology and Biochemistry*, 6: 27–30.

Podile, A.R. and Dube, H.C., 1988. Plant growth promoting activity of *Bacillus subtilis* AF1. *Current Science*, 57(4): 183–185.

Polonenko, D.R., Scher, F.M., Kloepper, J.W., Singleton, C.A., Lalibete, M., Zaleska, I., 1987. Effect of root colonizing bacteria on nodulation of soybean roots by *Bradyrhobium japonicum. Canadian Journal of Microbiology*, 33: 498–503.

Rai, R., 1983. Efficacy of associative N_2-fixation by streptomycin resistant mutants of *Azospirillum brasilense* with genotypes of chickpea *Rhizobium* strains. *Journal of Agricultural Science*, 100: 75–80.

Verma, D.P.S., Fortin, M.G., Stanely, V.P., Mauro, S. Purohit, S. and Morrison, N., 1986. Nodulins and nodulin genes of *Glycine max*: A perspective. *Plant Molecular Biology*, 7: 51–61.

Recent Advances in Seed Spices (2011) *Pages* **344–352**
Editors: **Y. Ravindrababu, R.K. Jaiman, K.D. Patel, N.R. Patel and S.B.S. Tikka**
Published by: **DAYA PUBLISHING HOUSE, NEW DELHI**

40

Efficacy of Different Acaricides Against Chilli Yellow Mite (*Polyphagotarsonemus latus*, Banks) and its Safety to Predatory Mite

K.A. Patel, M.B. Patel, S.N. Gajjar, A.J. Patel and H.R. Desai
All India Network Project on Agricultural Acarology, Department of Entomology, Navsari Agricultural University, Navsari – 396 450, Gujarat

ABSTRACT

The experiment on effectiveness of newer product Fenpyroximate 5 EC at five different concentrations and other conventional acaricides against chilli yellow mite was carried out during 2001-02 and 2002-03 at College Farm, Gujarat Agricultural University, Navsari. The effectiveness was checked against the eggs and active stages of phytophagous mite as well as leaf curl and its intensity and safety evaluation against the predatory mites. The phytotoxicity was also studied based on 0-10 scaling. The result revealed that the newer acaricide Fenpyroximate 5 EC was comparable with the conventional acaricides like dicofol, monocrotophos, endosulfan and ethion against phytophagous mite *Polyphagotarsonemus latus* Banks on chilli. Out of five doses, two doses *viz.* 400 and 500 ml/ha can be taken advantage of from view point of efficacy and/or safety against predatory mite.

Introduction

Chilli, *Capsicum annum* L. is an important cash crop in Gujarat state, prone to considerable damage by the yellow or broad mite, *Polyphagotarsonemus latus* Banks during couple of years. These mites are microscopic, glossy and yellow in colour which remained fast moving on the under surface of leaves. On chilli, the feeding of mites cause crinkling and curling longitudinally deshaping the leaves with inverted boat shape appearance. In severe infestation, the whole plants remain stunted and show the common leaf curl symptoms and consequently the yield is affected. This yellow or broad mite can cause 27.78 per cent loss in potential yield of chilli (Anon, 2002). The control of *P. latus* is difficult due to its fast developmental rate, high reproductive potential and different diapausing forms in the population. In past studies conventional acaricide dicofol was found most effective against this mite. [Mote (1976), Dhooria and Bindra (1977), Raju and Rao (1981) and Awate *et al.* (1981)] Therefore, the newer product with conventional acaricides was evaluated under field condition.

Material and Methods

An experiment in randomized block design with three replications was conducted at College farm, GAU, Navsari Cv. G-4 spaced at 60 x 60 cm to investigate the bioefficacy of Fenpyroximate 5 EC with other conventional acaricides against *P. latus* infesting chilli during 2001-02 and 2002-03. Three sprayings were given at interval of 30 days. The observation on mite (Phytophagous and predatory) counts (both eggs and active stages) were recorded 1 day from each plot, 5 growing tips were selected from three randomly selected plants and from each growing tip, mite counts from top five leaves under stereo binocular microscope. Pre treatment (1 day before treatment) and post treatment (1, 3, 7, 10, 14 and 21 days after each spray) were recorded. The leaf curl affected plants and the visual symptoms of curling found at various growth stages of the plant (*i.e.*120, 150 and 180 days after planting) were also recorded by adopting the method described by Niles (1980) for studying spider mite injuries to cotton with slight modification as under.

Damage Score	*Extent of Damage to Plant as Evidenced by Symptoms by Curling*
0	No leaf curl incidence (healthy plant)
1	Approximately 1 to 25 per cent of leaves in plant showing curling symptoms
2	Approximately 26 to 50 per cent of leaves in plant showing curling symptoms
3	Approximately 51 to 75 per cent of leaves in plant showing curling symptoms and stunted growth
4	> 75 per cent of leaves in plant showing curling symptoms and stunted growth leaf defoliation and malformation.

Treatment Details

Treatment No.	*Name of Treatment*	*Dose/ha*
T_1	Fenpyroximate 5 EC	125ml
T_2	Fenpyroximate 5 EC	250ml
T_3	Fenpyroximate 5 EC	375ml
T_4	Fenpyroximate 5 EC	400ml
T_5	Fenpyroximate 5 EC	500ml
T_6	Dicofol 18.5 EC	1081ml
T_7	Ethion 50 EC	1000ml
T_8	Wettable sulphur 50 WP	1250 gm
T_9	Monocrotophos 36 WSC	694ml
T_{10}	Endosulfan 35 EC	1072ml
T_{11}	Methyl-o-dematon 25EC	500ml
T_{12}	Control (water spray)	–

Per cent intensity of curling was calculated by multiplying the damage score by number of plants in that category and the sum of the product was divided by the product of number of categories taken and total number of damaged plants in respective plots multiplied by hundred.

The data so obtained on per cent reduction of mite (both phytophagous and predatory), per cent leaf curl plants and per cent intensity of leaf curling were subjected to due transformation (arc-sine transformation) and analysed statistically using Randomized Block Design. The pooled analysis over periods and sprays as well as over years was carried out using SPD format. Fruit yield (green chilli)was also recorded separately for each treatment. The visual observation on phytotoxic effect of different acaricide was also made for entire crop period.

Results and Discussions

The data on per cent reduction of phytophagous mite and predatory mite, percent leaf curl affected plants and its intensity in different treatments recorded at individual periods at each sprays during both the years are presented in Tables 40.1–40.4.

The data on mean per cent reduction of phytophagous mite (mobile stages) pooled over periods and sprays during both the years as well as pooled over years (Table 40.1) indicated that all acaricidal treatments were significantly superior to the treatment of water spray (control). The treatment of Dicofol 1081 ml/ha., Fenpyroximate 500 ml/ha and monocrotophos 694 ml/ha were statistically at par recording 70.13 to 71.84 per cent reduction and were significantly superior to rest of the treatments. The minimum reduction was recorded in the treatment of Fenpyroximate 125 ml/ha (46.62 per cent) followed by Fenpyroximate 250 ml/ha (52.98 per cent) and wettable sulfur 1250 g/ha. (55.05 per cent). Among all the acaricides, Fenpyroximate 125 ml/ha and 250 ml/ha recorded minimum reduction (less harmful) and were significantly superior to rest of the treatments. The next group of moderately harmful acaricides were Fenpyroximate 375 ml/ha, wettable sulfur 1250 g/ha, Endosulfan 1072 ml/ha and Fenpyroximate 400 ml/ha (41.48 to 47.66 per cent reduction) and were comparable to each other.

Monocrotophos 694 ml/ha, Fenpyroximate 500 ml/ha and dicofol 1081 ml/ha were equally effective by recording lower leaf curl affected plants (15.71 to 16.84 per cent) and were significantly superior to rest of treatments. The next effective treatments were methyl-o-demeton 500 ml/ha, Endosulfan 1072 ml/ha and Fenpyroximate 400 ml/ha by recording 20.92 to 21.85 per cent leaf curl affected plants. In case of per cent intensity of leaf curling, the lowest intensity (with in damage plants) was found in the treatment of dicofol 1081 ml/ha(36.62 per cent) and was comparable with monocrotophos 694 ml/ha, Fenpyroximate 500 ml/ha and Endosulfan 1072 ml/ha (31.16 to 41.11 per cent intensity).The highest yield was recorded in the treatment of dicofol 1081 ml/ha (1.122 t/ha) which was statistically at par with the treatment of Fenpyroximate 500 ml/ha (1.074 t/ha) and monocrotophos 694 ml/ha (0.918 t/ha)

Based on visual observations throughout the crop period, no phytotoxic effect was observed in any of the treatment even after three successive sprayings at one month interval on chilli (cv. G-4).

The overall pooled data of each of the above parameters were considered for working out the average rank index (Table 40.4). The chronological order of effectiveness of different treatments based on average rank (comprising all parameters) was dicofol 1081 ml/ha > Fenpyroximate 500 ml/ha > monocrotophos 694 ml/ha >Endosulfan 1072 ml/ha >ethion 1000 ml/ha > Fenpyroximate 400 ml/

Table 40.1: Overall Effectiveness of Different Treatments Against Phytophagous Mite (*P. latus*) and Safety to Predatory Mite (*Amblyseius* sp.) on Chilli (Pooled of 2001-02 and 2002-03)

Tr.No	*Mean Per cent Reduction of Mite (Mobile Stage)*											
	Phytophagous Mite						*Predatory Mite*					
	2001-02		*2002-03*		*Pooled*		*2001-02*		*2002-03*		*Pooled*	
T_1	43.58^{f}	(47.52)*	42.55^{h}	(45.73)	43.06^{e}	(46.62)	18.16^{a}	(9.71)	25.96^{a}	(19.16)	22.06^{a}	(14.11)
T_2	46.62^{e}	(52.83)	46.80^{g}	(53.14)	46.71^{d}	(52.98)	21.18^{b}	(13.05)	30.09^{b}	(25.14)	25.63^{a}	(18.71)
T_3	50.43^{d}	(59.42)	51.36^{ef}	(61.01)	50.90^{c}	(60.22)	38.19^{c}	(38.23)	41.65^{cd}	(44.17)	39.92^{b}	(41.18)
T_4	52.74^{cd}	(63.35)	54.97^{c}	(67.05)	53.86^{b}	(65.22)	41.67^{d}	(44.20)	45.65^{ef}	(51.13)	43.66^{bc}	(47.66)
T_5	57.02^{a}	(70.37)	58.87^{ab}	(73.27)	57.95^{a}	(71.84)	52.54^{f}	(63.01)	51.10^{hi}	(60.57)	51.82^{cd}	(61.79)
T_6	55.59^{ab}	(68.06)	60.31^{a}	(75.47)	57.95^{a}	(71.84)	55.72^{g}	(68.28)	53.64^{i}	(64.85)	54.68^{d}	(66.58)
T_7	51.83^{cd}	(61.81)	52.15^{de}	(62.35)	51.98^{bc}	(62.06)	46.90^{e}	(53.31)	47.20^{fg}	(53.84)	47.05^{bcd}	(53.57)
T_8	46.24^{ef}	(52.16)	49.55^{f}	(57.91)	47.90^{d}	(55.05)	40.37^{d}	(41.95)	42.65^{df}	(45.90)	41.51^{b}	(43.92)
T_9	56.09^{a}	(68.88)	57.72^{b}	(71.48)	56.87^{a}	(70.13)	56.00^{g}	(68.73)	52.96^{i}	(63.71)	54.48^{d}	(66.25)
T_{10}	53.37^{bc}	(64.40)	53.89^{cd}	(65.27)	53.63^{bc}	(64.84)	47.33^{e}	(54.06)	39.24^{c}	(40.01)	43.28^{b}	(47.00)
T_{11}	51.16^{cd}	(60:67)	51.66^{def}	(61.52)	51.41^{bc}	(61.09)	53.93^{fg}	(65.33)	49.10^{gh}	(57.13)	51.52^{cd}	(61.28)
T_{12}	12.62^{g}	(4.77)	10.93^{i}	(3.60)	11.47^{f}	(4.16)	–	–	–			
GM	**48.10**	**(55.40)**	**49.23**	**(57.36)**	**48.67**	**(56.39)**	**42.91**	**(46.36)**	**43.57**	**(47.50)**	**43.24**	**(46.93)**

Contd...

Table 40.1–Contd...

Tr.No	*Mean Per cent Reduction of Mite (Mobile Stage)*											
	Phytophagous Mite						*Predatory Mite*					
	2001-02		*2002-03*		*Pooled*		*2001-02*		*2002-03*		*Pooled*	
C.D. (S.Em±)												
Y												
T	2.66	(0.91)	2.41	(0.82)	2.76	(0.89)	2.17	(0.74)	3.03	(1.03)	8.23	(2.61)
S	0.78	(0.27)	NS	(0.36)	NS	(0.63)	0.90	(0.32)	1.77	(0.62)	NS	(1.04)
P	0.79	(0.29)	1.32	(0.48)	3.50	(0.89)	1.11	(0.40)	1.06	(0.38)	5.92	(1.51)
YT	–	–	–	–	2.47	(0.87)	–	–	–	–	2.55	(0.89)
YS	–	–	–	–	0.89	(0.32)	–	–	–	–	1.41	(0.49)
YP	–	–	–	–	1.15	(0.41)	–	–	–	–	1.08	(0.39)
TS	NS	(0.94)	3.50	(1.23)	2.25	(0.80)	2.99	(1.05)	5.89	(2.06)	4.82	(1.63)
TP	2.75	(0.94)	4.57	(1.65)	4.00	(1.40)	3.68	(1.33)	3.51	(1.27)	NS	(1.45)
SP	1.38	(0.50)	NS	(0.83)	1.40	(0.51)	1.92	(0.69)	1.83	(0.66)	NS	(1.42)
YTS	–	–	–	–	NS	(1.10)	–	–	–	–	4.67	(1.64)
YTP	–	–	–	–	3.98	(1.43)	–	–	–	–	3.59	(1.30)
YSP	–	–	–	–	NS	(0.72)	–	–	–	–	1.88	(0.68)
TSP	4.77	(1.72)	NS	(2.86)	4.54	(1.64)	6.37	(2.19)	6.07	(2.19)	NS	(2.06)
YTSP	---	–	–	NS	(2.48)	–	–	–	–	6.22	(2.24)	
C.V. per cent	12.65		11.18		11.92		11.50		15.81		13.85	

* Figures in the parentheses are retransformed mean those outside are the arc sine transformed means.

Table 40.2: Overall Effectiveness of Different Treatments Against Leaf Curl and its Intensity (Pooled of 2001-02 and 2002-03)

Tr. No	*Mean Per cent Plants*			*Mean Per cent Intensity (Within damage plants)*			*Mean Per cent Intensity (Within total plants)*		
	2001-02	*2002-03*	*Pooled*	*2001-02*	*2002-03*	*Pooled*	*2001-02*	*2002-03*	*Pooled*
T_1	32.14^{de} (28.30)*	32.83^{d} (29.39)	32.48^{d} (28.84)	43.11^{f} (46.70)	46.61^{f} (52.81)	44.86^{gh} (49.76)	18.91^{e} (10.50)	2069^{e} (12.48)	19.80^{d} (11.47)
T_2	30.16^{bcd} (25.24)	33.18^{de} (29.95)	31.67^{d} (27.57)	41.80^{ef} (44.43)	44.37^{e} (48.90)	43.09^{efg} (46.67)	17.27^{de} (8.81)	20.03^{e} (11.73)	18.65^{cd} (10.23)
T_3	28.86^{bcd} (23.30)	31.78^{cd} (27.74)	30.32^{cd} (25.49)	39.36^{cd} (40.22)	44.02^{e} (48.29)	41.69^{def} (44.24)	15.92^{cd}) (7.52	19.07^{de} (10.67)	17.50^{c} (9.04)
T_4	27.75^{abc} (21.68)	27.99^{bc} (22.02)	27.87^{bc} (21.85)	37.89^{bc} (37.22)	42.95^{de} (46.43)	40.42^{cde} (42.04)	14.95^{bc} (6.66)	16.62^{cd} (8.18)	15.79^{b} (7.40)
T_5	24.53^{a} (17.24)	23.92^{ab} (16.44)	24.23^{a} (16.84)	36.10^{ab} (34.72)	40.10^{abc} (41.49)	38.10^{abc} (38.07)	12.82^{ab} (4.92)	13.64^{ab} (5.56)	13.23^{a} (5.24)
T_6	24.28^{a} (16.91)	22.56^{a} (14.72)	23.42^{a} (15.80)	35.48^{a} (33.69)	38.99^{ab} (39.59)	37.24^{a} (36.62)	12.42^{a} (4.63)	12.61^{a} (4.77)	12.51^{a} (4.69)
T_7	28.13^{abc} (22.23)	26.86^{ab} (20.41)	27.49^{bc} (21.31)	39.37^{cd} (40.24)	40.89^{c} (45.85)	40.13^{bcd} (41.54)	15.48^{cd} (7.12)	15.47^{bc} (7.11)	15.48^{b} (7.12)
T_8	31.60^{cde} (27.46)	32.69^{d} (29.17)	32.15^{d} (28.32)	43.51^{f} (47.40)	43.74^{e} (47.80)	43.63^{fg} (47.61)	18.86^{e} (10.45)	19.14^{de} (10.75)	19.00^{cd} (10.60)
T_9	24.18^{a} (16.78)	22.51^{a} (14.66)	23.35^{a} (15.71)	36.50^{ab} (35.38)	38.62^{a} (38.96)	37.56^{ab} (37.16)	12.56^{a} (4.73)	12.59^{a} (4.75)	12.57^{a} (4.74)
T_{10}	27.19^{ab} (20.88)	27.26^{bc} (20.98)	23.23^{b} (20.94)	39.37^{cd} (40.24)	40.38^{bc} (41.97)	39.88^{abc} (41.11)	15.13^{cd} (6.81)	15.63^{bc} (7.26)	15.38^{b} (7.03)
T_{11}	27.33^{ab} (21.08)	27.11^{ab} (20.77)	27.22^{b} (20.92)	40.55^{de} (42.26)	41.60^{cd} (44.08)	41.07^{def} (43.16)	15.49^{cd} (7.13)	15.69^{bc} (7.31)	15.59^{b} (7.22)
T_{12}	34.37^{e} (31.87)	37.74^{e} (37.46)	36.05^{e} (34.63)	46.76^{g} (53.07)	46.59^{f} (52.77)	46.67^{h} (52.91)	21.61^{f} (13.56)	23.66^{i} (16.10)	22.63^{e} (14.81)
GM	28.38 (22.59)	28.87 (23.31)	28.62 (22.94)	39.98 (41.28)	42.41 (45.49)	41.19 (43.37)	15.95 (7.55)	17.07 (8.62)	16.51 (8.08)

Contd...

Table 40.2–Contd...

Tr. No	Mean Per cent Plants			Mean Per cent Intensity (Within damage plants)			Mean Per cent Intensity (Within total plants)		
	2001-02	2002-03	Pooled	2001-02	2002-03	Pooled	2001-02	2002-03	Pooled
C.D. (S. Em±)									
Y									
T	3.99 (1.36)	4.64 (1.58)	2.88 (1.02)	1.81 (0.62)	1.68 (0.57)	2.70 (0.87)	2.23 (0.76)	2.67 (0.94)	1.69 (0.60)
P	0.94 (0.32)	1.11 (0.39)	7.50 (1.23)	0.75 (0.62)	0.84 (0.29)	3.95 (0.65)	0.55 (0.19)	0.60 (0.21)	2.72 (0.45)
YT	–	–	NS	–	–	1.70 (0.59)	–	–	NS (0.86)
YP	–	–	1.01 (0.36)	–	–	0.78 (0.28)	–	–	0.57 (0.20)
TP	3.25 (1.14)	–	NS (1.21)	2.58 (0.90)	2.90 (1.02)	NS (1.15)	NS (0.67)	NS (0.73)	NS (0.52)
YTP			3.51 (1.25)	–	–	2.71 (0.96)	–	–	NS (0.70)
C.V. per cent	14.37	16.43	15.45	4.62	4.05	4.33	14.27	16.55	15.54

* Figures in the parentheses are retransformed mean those outside are the arc sine transformed means.

ha >methyl-o-demeton 500 ml/ha > Fenpyroximate 375 ml/ha > fenpyroxiamte 250 ml/ha > wettable sulphur 1250 g/ha >Fenpyroximate 125 ml.ha > water spray (control)

Table 40.3: Yield of Green Chilli Fruits in Different Treatments (Pooled 2001-02 and 2002-03)

Tr.No	*Yield of Green Chilli Fruit (t/ha)*		
	2001-02	*2002-03*	*Pooled*
T_1	0.661	0.679	0.670[de]
T_2	0.687	0.750	0.718[cd]
T_3	0.787	0.826	0.806[cd]
T_4	0.822	0.901	0.861[bcd]
T_5	1.168	0.980	1.074[ab]
T_6	1.237	1.008	1.122[a]
T_7	0.854	0.865	0.807[bcd]
T_8	0.647	0.797	0.722[cd]
T_9	0.865	0.970	0.918[abc]
T_{10}	0.797	0.891	0.844[bcd]
T_{11}	0.852	0.858	0.855[bcd]
T_{12}	0.636	0.289	0.462[e]
GM	0.835	0.818	0.826
SEm			
Y	–	–	–
T	0.066	0.075	0.077
YT	-	-	0.071
C.D.			
Y	–	–	–
T	0.196	0.219	0.239
YT	–	–	0.202
C.V. per cent			
	13.89	15.79	14.85

Conclusion

From the foregoing results it can be concluded that the newer acaricide Fenpyroximate 5 EC was comparable with the conventional acaricides like dicofol, monocrotophos, endosulfan and ethion against phytophagous mite *Polyphagotarsonemus latus* Banks on chilli. Out of five doses, two doses *viz.* 400 and 500 ml/ha can be taken advantage of from view point of efficacy and/or safety against predatory mite.

Table 40.4: Overall Performance of Different Treatment Based on Rank Index Methods

Tr.No	Rank Indices for Different Parameters Based on Pooled Results						Total Rank	Average	Rank No.
	Phytophagous Mite	Predatory Mites	LC Plants and its Intensity			Yield (t/ha)			
			Per cent LC Plants	Per cent Intensity (Damage)	Per cent Intensity (Total)				
1	2	3	4	5	6	7	8	9	10
T_1	5	1	4	7.5	4	4.5	26	4.33	X
T_2	4	1	4	6	3.5	3.5	22	3.67	VIII
T_3	3	2	3.5	5	3	3.5	20	3.33	VII
T_4	2	2.5	2.5	4	2	3	16	2.67	V
T_5	1	3.5	1	2	1	1.5	10	1.67	II
T_6	1	4	1	1	1	1	9	1.50	I
T_7	2.5	3	2.5	3	2	3	16	2.67	V
T_8	4	2	4	6.5	3.5	3.5	23.5	3.92	IX
T_9	1	4	1	1.5	1	2	10.50	1.75	III
T_{10}	2.5	2	2	2.5	2	3	14	2.33	IV
T_{11}	2.5	3.5	2	5	2	3	18	3	VI
T_{12}	6	–	5	8	5	5	29	5.8	XI

References

Anonymous, 2002. *Annual Research Report*, All India Network Project on Agricultural Acarology, Department of Entomology, G.A.U., Navsari

Awate, B.G., Gandhale, D.N., Naik, L.M. and Patil, A.S., 1981. Chemical control of mite, *Polyphagotarsonemus latus* Banks on chilli, *Capsicum annum* L. *Indian J. Pl. Prot.*, 9(2): 193–194.

Dhooria, M.S. and Bindra, O.S., 1977. Polyphagotarsonemus latus (Banks): A mite pest of chilli and potato in Punjab. *Acar. Newsl.*, 14: 7–9.

Mote, U.N., 1976. Seasonal fluctuation in population and chemical control of chilli mites (*Polyphagotarsonemus latus*, Banks.). *Veg. Sci.*, 3(1): 54–60.

Niles, G.A., 1980. Breeding of cotton for resistance to insect pests. In: *Breeding Plant Resistance of Insects* (Eds.) F.G. Maxwell and P.R. Jennings, pp. 367–370.

Raju, K.S. and Rao, G.S., 1981. New finding on the control of chilli mites (*Hemitarsonemus latus*, Banks). *Pestology*, 5(5): 19–20.

Recent Advances in Seed Spices (2011) *Pages* 353–357
Editors: Y. Ravindrababu, R.K. Jaiman, K.D. Patel, N.R. Patel and S.B.S. Tikka
Published by: DAYA PUBLISHING HOUSE, NEW DELHI

41

Ecofriendly Management of Dill Seed Aphid, *Hydaphis coriandari* (Das) through Seed Treatments

M.N. Kapadia, P.G. Butani, V.R. Virani and D.M. Jethva
Department of Entomology, College of Agriculture,
Junagadh Agricultural University, Junagadh – 362 001, Gujarat

ABSTRACT

The experiment was carried out to study the effectiveness of two insecticides (imidacloprid and thiamethoxam) as seed dresser against aphid, *Hydaphis coriandari* (Das) on dillseed at Junagadh Agricultural University, Junagadh during 2004-05 and 2005-06. The results clearly indicated that the insecticidal seed treatments of thiamethoxam 70 WS@ 4.2g/kg seed and Imidacloprid 70WS @ 10g/kg seed proved most effective in controlling the aphid as compared to lower doses of both the insecticides. The highest seed yield and ICBR (1323 kg/ha, ICBR 1: 5.29) was recorded in the treatment with thiamethoxam 70 WS@ 4.2g/kg seed followed by thiamethoxam 70 WS@ 2.8g/kg seed (1174 kg/ha, ICBR 1: 4.99) and Imidacloprid 70WS @ 10g/kg seed (1246 kg/ha, ICBR 1: 1.69). Thus, seed treatment of the thiamethoxam 70 WS@ 4.2g/kg seed and can be recommended for effective and economical important of aphid of dill seed.

Keywords: *Hydaphis coriandari (Das), Insecticidal seed treatments, Dill seed.*

Introduction

The seed spices crop dill seed, is being cultivated in Saurashtra under limited water irrigation sources. The aphid, *Hydaphis coriandari* (Das) is the key pest of dill seed in this area. It causes the considerable damage in late growth of the crop and ultimately the yield is affected quantitatively and qualitatively. The use of newer insecticides as seed dressing treatment is highly necessary to get more effective, economical and ecofriendly management strategy in comparison to the recommended foliar application of insecticides. The present investigation was carried out with view to find out the effectiveness of neonicotinoid insecticides as seed dressers against aphids of dill seed.

Materials and Methods

The field experiment on dill seed (var. dill seed local) was conducted at the Junagadh Agricultural University campus, Junagadh during *rabi*, 2004-05 and 2005-06. The dill seeds were treated with different doses of imidacloprid 70 WS and thiamethoxam 70 WS. Both insecticides were diluted in water (1:1 ratio) and starch solution was added. Seeds were treated in polythene bag and stirred well for uniform coating. The experiment was conducted in a randomized block design replicated thrice with plot size of 5.0 X 2.7m.The conventional insecticide; dimethoate 0.03 per cent was compared with seed treatments. The observations on population of aphid as aphid index at weekly interval starting from crop germination from five randomly selected plants. The data thus, obtained were then transformed and analyzed. The aphid index was followed below (Butani and Bharodia, 1984).

Index

0.	Plant free from aphid
1.	Aphid present but colonies not built up. No injury due to apparent on plant.
2.	Small colonies of aphids present on tender parts of the plants. Such tender parts of plant exhibit slight curling due to aphid feeding.
3.	Large colonies of aphids present on tender plant parts counts are possible and the tender parts show damage symptoms due to aphid feeding.
4.	The tender plant parts completely covered with aphid colonies. Plant growth hindered due to pest feeding.

Results and Discussion

Efficacy of Seed Treatments

Reduction in aphid index per plant was observed at 7 WAG (week after germination) reflected in significant differences among various treatments (Table 41.1). The low aphid index per plant (0.17) was observed in thiamethoxam 70 WS@ 4.2 g/kg seed which was at par with imidacloprid 70 WS @ 10g/kg seed (0.27). The next best treatments were thiamethoxam 70 WS @2.8g/kg seed (0.50) and imidacloprid 70 WS @ 7.5 g/kg seed (0.67). Similar trends were observed at 10 WAG also where in minimum aphid index per plant was observed in thiamethoxam 70 WS@ 4.2g/kg seed (0.37) followed by imidacloprid 70WS @ 10g/kg seed (0.67) and thiamethoxam 70 WS@ 2.8g/kg seed (1.13). More or less similar trend was also observed at 12 WAG and mean pooled also.

Yield and Economics

The highest seed yield 1323 kg/ha was obtained from the seed treatment with thiamethoxam 70 WS@ 4.2g/kg seed and it was at par with imidacloprid 70WS @ 10g/kg seed (1246 kg/ha) (Table 41.2). The next best treatment was thiamethoxam 70 WS@ 2.8g/kg seed (1174 kg/ha). Economic returns from increased yield was highest in thiamethoxam 70 WS@ 4.2g/kg seed (Rs.3517/ha) followed by imidacloprid 70WS @ 10g/kg seed (Rs. 2844/ha) and thiamethoxam 70 WS@ 2.8g/kg

Table 41.1: Effect of Different Insecticidal Seed Treatments on Aphid Population of Dill Seed

Sl.No.	Treatment	Aphid Index/Plant/Week											
		7 WAG			10 WAG			12 WAG			Mean Pooled		
		2004-05	2005-06	Pooled	2004-05	2005-06	Pooled	2004-05	2005-06	Pooled	2004-05	2005-06	Pooled
1.	Imidacloprid 70WS @ 5g/kg seed	0.93	0.73	0.83	1.53	2.20	1.87	1.40	1.80	1.60	1.33	1.71	1.52
2.	Imidaclorpid 70WS @ 7.5g/kg seed	0.80	0.53	0.67	1.20	1.87	1.53	1.27	1.47	1.37	1.10	1.43	1.27
3.	Imidacloprid 70WS @ 10g/kg seed	0.33	0.20	0.27	0.60	0.73	0.67	0.80	0.60	0.70	0.50	0.68	0.59
4.	Thiamethoxam 70 WS @ 2.8g/kg seed	0.60	0.40	0.50	1.00	1.27	1.13	1.13	1.13	1.13	0.83	1.12	0.98
5.	Thiamethoxam70WS @ 4.2g/kg seed	0.20	0.13	0.17	0.20	0.53	0.37	0.53	0.40	0.47	0.43	0.44	0.44
6.	Dimethoate 30 EC 0.03 per cent on need base	1.13	0.87	1.00	1.27	0.87	1.07	1.13	0.27	0.70	0.78	0.86	0.82
7.	Control (water spray)	1.07	0.93	1.00	1.93	2.93	2.43	1.67	2.33	2.00	1.53	2.21	1.87
8.	Control (Untreated)	1.13	0.87	1.00	2.07	3.00	2.53	1.80	2.40	2.10	1.82	2.29	2.06
	S.Em.±for Treat	0.13	0.14	0.09	0.14	0.19	0.23	0.16	0.18	0.25	0.038	0.068	0..11
	C.D. at 5 per cent for Treat	0.39	0.41	0.27	0.42	0.57	0.77	0.48	0.56	0.84	0.114	0.208	0.36
	S.Em.±for year			0.05			0.12			0.13			0.02
	C.D. at 5 per cent for year			0.14			0.39			NS			0.056
	S.Em.±for YXT			0.13			0.17			0.17			0.055
	C.D. at 5 per cent for YXT			NS			0.48			0.50			0.16
	C.V. per cent	17.33	21.74	19.37	14.00	14.95	14.69	16.10	17.80	17.02	10.34	15.75	13.84

WAG: Week after germination.

seed (Rs.2214/ha). The treatments *viz.*, thiamethoxam 70 WS@ 4.2g/kg seed, thiamethoxam 70 WS@ 2.8g/kg seed and imidacloprid 70WS @ 10g/kg seed recorded the higher incremental cost benefit ratio of 5.29, 4.99 and 1.69, respectively. Imidacloprid @ 10 g and thiamethoxam @ 4.2 g/kg seed has been reported most effective seed dressers against sucking pests of cotton (Vadodaria *et al.*, 2001 and Bheemanna *et al.*, 2003), and okra (Kapadia and Butani, 2005). Prasanna *et al.* (2004) reported that seed treatment with thiamethoxam at higher dost (10 g/kg seed) was highly effective against leafhopper of cotton up to 50 days. However, at lower doses (2.85 and 4.28 g), it gave good control of sucking pests up to 40 days after treatment. Sreelatha and Divakar (1997) observed that seed treatment with imidacloprid (7.5 g/kg) significantly reduced sucking pest of okra. According to Rana *et al.* (2006), imidacloprid @ 2 ml, thiamethoxam @ 2 g and carbosulfan @ 4 g/kg seed were effective in controlling the sucking pest of okra. Whereas, Mathirajan and Regupathy (2001) observed that seed treatment with thiamethoxam @3 g and 6 g/kg seed and imidacloprid @5 g/kg significantly reduced sucking pests of cotton. Thus, present findings colloborate the results reported by earlier workers.

Table 41.2: Economics of Insecticidal Seed Treatments Against Dill Seed Aphids

Treatment	*Grain Yield kg/ha*			*Net Realization Rs/ha*	*ICBR*
	2004-05	*2005-06*	*Pooled*		
Imidacloprid 70WS @ 5g/kg seed	1117	946	1032	971	1:1.16
Imidaclorpid 70WS @ 7.5g/kg seed	1197	1042	1120	1741	1:1.38
Imidacloprid 70WS @ 10g/kg seed	1326	1166	1246	2844	1:1.69
Thiamethoxam 70 WS @ 2.8g/kg seed	1253	1095	1174	2214	1:4.99
Thiamethoxam70WS @ 4.2g/kg seed	1411	1236	1323	3517	1:5.29
Dimethoate 30 EC 0.03 per cent on need base	1155	975	1065	1260	1:1.23
Control (water spray)	1026	826	926	44	1:0.06
Control (Untreated)	1038	805	921	–	–
S.Em.±for Treat	53	67	43	–	–
C.D. at 5 per cent for Treat	161	202	123	–	–
S.Em.±for year	–	–	21	–	–
C.D. at 5 per cent for year	–	–	62	–	–
S.Em.±for YXT	–	–	60	–	–
C.D. at 5 per cent for YXT	–	–	NS	–	–
C.V. per cent	8.00	11.00	9.00	–	–

WAG: Week after germination.

*: Labour charges was calculated Rs. 15. 00/hrs; **: Market price of dilseed grain Rs. 8.75/kg.

Spray solution was used. 617 liter per hectare during first and second sprays each. Seed rate 12.00 kg/ha.

The study concluded that the seed treatment of thiamethoxam 70 WS@ 4.2g/kg seed was found most effective and economical for ecofriendly management of aphids in dill seed crop.

References

Bheemanna, M., Patil, B.V., Anand Kumar, V. and Hanchinal, S.G., 2003. Bioefficacy of thiamethoxam seed treatments against cotton early sucking insect pests. In: *Proc. National Symposium on Frontier Areas of Entomological Research*, 5–7 November, New Delhi, p. 169.

Butani, P.G. and Bharodia, R.K., 1984. Relation of groundnut aphid population with its natural predator, lady bird beetles. *Gujarat Agricultural University Research Journal*, 9(2): 72–74.

Kapadia, M.N. and Butani, P.G., 2005. Ecofriendly management of sucking pests in summer okra through new seed dressers. In: *Proc. National Seminar on Applied Entomology: Current Status, Challenges and Opportunities*, 26–28 September, Udaipur, pp. 109–111.

Mathirajan, V.G. and Regupathy, A., 2001. Seed treatment with thiamethoxam (CruiserReg.): An ecologically selective method for the management of sucking pests of cotton. *Pest Management and Economic Zoology*, 9(2): 177–186.

Prasanna, A.R., Bheemanna, M. and Patil, B.V., 2004. Evaluation of thiamethoxam 70 WS as seed treatment against leaf miner and early sucking pests on hybrid cotton. *Karnataka Journal of Agricultural Sciences*, 17(2): 238–241.

Rana, S.C., Singh, P.B., Pandita, V.K. and Sinha, S.N., 2006. Evaluation of insecticides as seed treatment for control of early sucking pests in seed crop of okra. *Annals of Plant Protection Sciences*, 14(2): 364–367.

Sreelatha and Divakar, B.J., 1997. Impact of imidacloprid seed treatment on insect pest incidence in okra. *Indian Journal of Plant Protection*, 25(1): 52–55.

Vadodaria, M.P., Patel, C.J., Maisuria, I.M. and Patel, U.G., 2001. Imidacloprid (Gaucho) a new seed dresser against early sucking pests of cotton. *Gujarat Agricultural University Research Journal*, 26(2): 32–38.

Recent Advances in Seed Spices (2011) *Pages* **358–362**
Editors: **Y. Ravindrababu, R.K. Jaiman, K.D. Patel, N.R. Patel and S.B.S. Tikka**
Published by: **DAYA PUBLISHING HOUSE, NEW DELHI**

42

Ecofriendly Management of Cumin Thrips, *Thrips tabaci* Lindeman through Seed Treatments

M.N. Kapadia, P.G. Butani, V.R. Virani and D.M. Jethva
Department of Entomology, College of Agriculture, Junagadh Agricultural University, Junagadh – 362 001, Gujarat

ABSTRACT

The experiment was carried out to study the effectiveness of two insecticides (imidacloprid and thiamethoxam) as seed dresser against thrips, *Thrips tabaci* Lindeman on cumin at Junagadh Agricultural University, Junagadh during 2004-05 and 2005-06. The results clearly indicated that the insecticidal seed treatments of thiamethoxam 70 WS@ 4.2g/kg seed and imidacloprid 70WS @ 10g/kg seed proved most effective in controlling the thrips as compared to lower doses of both the insecticides. The highest seed yield and ICBR (553 kg/ha, ICBR 1: 9.22) was recorded in the treatment with thiamethoxam 70 WS@ 4.2g/kg seed followed by thiamethoxam 70 WS@ 2.8g/kg seed (446 kg/ha, ICBR 1: 7.72) and imidacloprid 70WS @ 10g/kg seed (482 kg/ha, ICBR 1:2.58). Thus, seed treatment of the thiamethoxam 70 WS@ 4.2g/kg seed can be recommended for effective and economical control of thrips in cumin.

Keywords: Thrips tabaci Lindeman, Insecticidal Seed treatments, Cumin.

Introduction

The seed spices crop cumin, is being cultivated in Saurashtra under limited water irrigation sources. The thrips, *Thrips tabaci* Lindeman is the key pest of cumin in this area. It causes the considerable damage in late growth of the crop and ultimately the yield is affected quantitatively and qualitatively. The use of newer insecticides as seed dressing treatment is highly necessary to get more effective, economical and ecofriendly management strategy in comparison to the recommended foliar application of insecticides. The present investigation was carried out with view to find out the effectiveness of neonicotinoid insecticides as seed dressers against thrips of cumin.

Materials and Methods

The field experiment on cumin (var. cumin G-2) was conducted at the Junagadh Agricultural University campus, Junagadh during *rabi*, 2004-05 and 2005-06. The cumin seeds were treated with different doses of imidacloprid 70 WS and thiamethoxam 70 WS. Both insecticides were diluted in water (1:1 ratio) and starch solution was added. Seeds were treated in polythene bag and stirred well for uniform coating. The experiment was conducted in a randomized block design replicated thrice with plot size of 5.0 X 2.7m.The conventional insecticide; dimethoate 0.03 per cent was compared with seed treatments. The observations on population of thrips were recorded at weekly interval starting from crop germination from five randomly selected plants. The data thus, obtained were then transformed and analyzed.

Results and Discussion

Efficacy of Seed Treatments

Reduction in thrips per plant was observed at 3 WAG (week after germination) reflected in significant differences among various treatments (Table 42.1). The low thrips per plant (2.41) was observed in imidacloprid 70 WS @ 10g/kg seed which was at par with thiamethoxam 70 WS@ 4.2 g/kg seed (2.76). The next best treatments were thiamethoxam 70 WS @2.8 g/kg seed (5.66) and imidacloprid 70 WS @ 7.5 g/kg seed (6.26). Similar trends were observed at 6 WAG also where in minimum thrips per plant was observed in thiamethoxam 70 WS@ 4.2g/kg seed (12.85) followed by imidacloprid 70WS @ 10g/kg seed (14.29) and thiamethoxam 70 WS@ 2.8g/kg seed (15.74). More or less similar trend was also observed at 9 WAG and mean pooled also.

Yield and Economics

The highest seed yield 553 kg/ha was obtained from the seed treatment with thiamethoxam 70 WS@ 4.2g/kg seed and it was at par with imidacloprid 70WS @ 10g/kg seed (482 kg/ha) (Table 42.2). The next best treatment was thiamethoxam 70 WS@ 2.8g/kg seed (446 kg/ha). Economic returns from increased yield was highest in thiamethoxam 70 WS@ 4.2g/kg seed (Rs.10224/ha) followed by imidacloprid 70WS @ 10g/kg seed (Rs.7225/ha) and thiamethoxam 70 WS@ 2.8g/kg seed (Rs.5704/ha). The treatments *viz.*, thiamethoxam 70 WS@ 4.2g/kg seed and thiamethoxam 70 WS@ 2.8g/kg seed recorded the higher incremental cost benefit ratio of 9.22 and 7.72, respectively compared to imidacloprid @ 10 g/kg seed (2.58 ICBR),. Imidacloprid @ 10 g and thiamethoxam @ 4.2 g/kg seed has been reported most effective seed dressers against sucking pests of cotton (Vadodaria *et al.*, 2001 and Bheemanna *et al.*, 2003), and okra (Kapadia and Butani, 2005). Prasanna *et al.* (2004) reported that seed treatment with thiamethoxam at higher dose (10 g/kg seed) was highly effective against leafhopper of cotton up to 50 days. However, at lower doses (2.85 and 4.28 g), it gave good control of sucking pests up to 40 days after treatment. Sreelatha and Divakar (1997) observed that seed treatment with imidacloprid (7.5 g/kg) significantly reduced sucking pest of okra. According to Rana *et al.* (2006), imidacloprid @ 2 ml, thiamethoxam @ 2 g and carbosulfan @ 4 g/kg seed were effective in controlling

Table 42.1: Effect of Different Insecticidal Seed Treatments on Thrips Population of Cumin

Sl.No.	Treatment	Average Thrips/5 Plants/Week											
		3 WAG			6 WAG			9 WAG			Mean Pooled		
		2004-05	2005-06	Pooled	2004-05	2005-06	Pooled	2004-05	2005-06	Pooled	2004-05	2005-06	Pooled
1.	Imidacloprid 70WS @ 5g/kg seed	3.18 (10.10)	2.34 (5.47)	2.76 (7.61)	3.84 (14.76)	6.44 (41.44)	5.14 (26.42)	5.55 (30.78)	8.74 (76.36)	7.14 (51.03)	4.20 (17.66)	6.09 (37.09)	5.15 (26.52)
2.	Imidaclorpid 70WS 7.5g/kg seed	2.96 (8.79)	2.04 (4.15)	2.50 (6.26)	3.67 (13.45)	5.48 (30.08)	4.58 (20.94)	5.40 (29.11)	8.51 (72.37)	6.95 (48.32)	3.94 (15.52)	5.66 (32.08)	4.80 (23.06)
3.	Imidacloprid 70WS @ 10g/kg seed	1.76 (3.10)	1.34 (1.81)	1.55 (2.41)	3.38 (11.44)	4.18 (17.46)	3.78 (14.29)	5.46 (29.78)	8.60 (74.01)	7.03 (49.42)	3.45 (11.87)	4.91 (24.13)	4.18 (17.46)
4.	Thiamethoxam 70 WS @ 2.8g/kg seed	2.90 (8.42)	1.86 (3.45)	2.38 (5.66)	3.53 (12.45)	4.41 (19.42)	3.97 (15.74)	5.43 (29.45)	8.57 (73.43)	7.00 (48.97)	3.73 (13.93)	5.35 (28.58)	4.54 (20.60)
5.	Thiamethoxam70WS @ 4.2g/kg seed	1.86 (3.45)	1.46 (2.14)	1.66 (2.76)	3.24 (10.48)	3.93 (15.46)	3.58 (12.85)	5.33 (28.45)	8.41 (70.73)	6.87 (47.22)	3.33 (11.10)	4.77 (22.75)	4.05 (16.41)
6.	Dimethoate 30EC 0.03 per cent on need base	3.71 (13.76)	2.79 (7.78)	3.25 (10.56)	3.97 (15.77)	6.59 (43.45)	5.28 (27.89)	5.57 (31.07)	5.71 (32.61)	5.64 (31.83)	4.40 (19.35)	5.96 (35.53)	5.18 (26.83)
7.	Control (water spray)	3.66 (13.41)	2.60 (6.78)	3.13 (9.81)	3.84 (14.76)	6.52 (42.46)	5.18 (26.83)	5.49 (30.10)	8.67 (75.08)	7.08 (50.06)	4.35 (18.94)	6.35 (40.29)	5.35 (28.62)
8.	Control (Untreated)	3.80 (14.42)	2.73 (7.48)	3.27 (10.67)	3.93 (15.46)	6.77 (45.77)	5.35 (28.61)	5.58 (31.12)	8.81 (77.67)	7.20 (51.78)	4.44 (19.73)	6.48 (42.02)	5.46 (29.83)
	S.Em.±for Treat	0.20	0.14	0.12	0.1715	0.1785	0.47	0.18	0.26	0.53	0.05	0.47	0.24
	C.D. at 5 per cent for Treat	0.60	0.43	0.35	NS	0.5415	1.57	NS	0.80	NS	0.16	NS	0.69
	S.Em.±for year			0.06			0.24			0.27			0.12
	C.D. at 5 per cent for year			0.18			0.79			0.89			0.34
	S.Em.±for YXT			0.17			0.18			0.23			0.34
	C.D. at 5 per cent for YXT			NS			0.51			0.66			NS
	C.V. per cent	11.50	11.32	11.59	8.08	5.58	6.58	5.82	5.57	5.76	2.29	14.71	12.19

$\sqrt{x}$ transformation used. Data in parentheses are retransformed values.

the sucking pest of okra. Whereas, Mathirajan and Regupathy (2001) observed that seed treatment with thiamethoxam @3 g and 6 g/kg seed and imidacloprid @5 g/kg significantly reduced sucking pests of cotton. Thus, present findings colloborate the results reported by earlier workers.

Table 42.2: Economics of Different Insecticidal Seed Treatments Against Cumin Thrips

Treatment	*Grain Yield kg/ha*			*Net Realization Rs/ha*	*ICBR*
	2004-05	*2005-06*	*Pooled*		
Imidacloprid 70WS @ 5g/kg seed	420	254	337	1098	1:0.78
Imidaclorpid 70WS @ 7.5g/kg seed	469	301	385	3126	1:1.49
Imidacloprid 70WS @ 10g/kg seed	586	377	482	7225	1:2.58
Thiamethoxam 70 WS @ 2.8g/kg seed	543	349	446	5704	1:7.72
Thiamethoxam70WS @ 4.2g/kg seed	673	433	553	10224	1:9.22
Dimethoate 30 EC 0.03 per cent on need base	395	270	332	887	1:1.73
Control (water spray)	370	246	308	–	–
Control (Untreated)	383	239	311	–	–
S.Em.±for Treat	45	41	30		
C.D. at 5 per cent for Treat	135	124	88		
S.Em.±for year			15		
C.D. at 5 per cent for year			44		
S.Em.±for YXT			43		
C.D. at 5 per cent for YXT			NS		
C.V. per cent	16.00	23.00	19.00		

*: Labour charges was calculated Rs. 15. 00/hrs; **: Market price of cumin grain Rs. 42.25/kg.

Spray solution was used 617 liter per hectare during first and second sprays each Seed rate 20 kg/ha.

The study concluded that the seed treatment of thiamethoxam 70 WS@ 4.2g/kg seed was found most effective and economical for ecofriendly management of thrips in cumin crop.

References

Bheemanna, M., Patil, B.V., Anand Kumar, V. and Hanchinal, S.G., 2003. Bioefficacy of thiamethoxam seed treatments against cotton early sucking insect pests. In: *Proc. National Symposium on Frontier Areas of Entomological Research*, 5–7 November, New Delhi, p. 169.

Kapadia, M.N. and Butani, P.G., 2005. Ecofriendly management of sucking pests in summer okra through new seed dressers. In: *Proc. National Seminar on Applied Entomology: Current Status, Challenges and Opportunities*, 26–28 September, Udaipur, pp. 109–111.

Mathirajan, V.G. and Regupathy, A., 2001. Seed treatment with thiamethoxam (CruiserReg.): An ecologically selective method for the management of sucking pests of cotton. *Pest Management and Economic Zoology*, 9(2): 177–186.

Prasanna, A.R., Bheemanna, M. and Patil, B.V., 2004. Evaluation of thiamethoxam 70 WS as seed treatment against leaf miner and early sucking pests on hybrid cotton. *Karnataka Journal of Agricultural Sciences*, 17(2): 238–241.

Rana, S.C., Singh, P.B., Pandita, V.K. and Sinha, S.N., 2006. Evaluation of insecticides as seed treatment for control of early sucking pests in seed crop of okra. *Annals of Plant Protection Sciences*, 14(2): 364–367.

Sreelatha and Divakar, B.J., 1997. Impact of imidacloprid seed treatment on insect pest incidence in okra. *Indian Journal of Plant Protection*, 25(1): 52–55.

Vadodaria, M.P., Patel, C.J., Maisuria, I.M. and Patel, U.G., 2001. Imidacloprid (Gaucho) a new seed dresser against early sucking pests of cotton. *Gujarat Agricultural University Research Journal*, 26(2): 32–38.

Recent Advances in Seed Spices (2011) *Pages* **363–367**
Editors: **Y. Ravindrababu, R.K. Jaiman, K.D. Patel, N.R. Patel and S.B.S. Tikka**
Published by: **DAYA PUBLISHING HOUSE, NEW DELHI**

43

Ecofriendly Management of Fenugreek Jassid, *Empoasca spinosa* Dworowaska and Sohi through Seed Treatments

M.N. Kapadia, P.G. Butani, V.R. Virani and D.M. Jethva
Department of Entomology, College of Agriculture,
Junagadh Agricultural University, Junagadh – 362 001, Gujarat

ABSTRACT

The experiment was carried out to study the effectiveness of two insecticides (imidacloprid and thiamethoxam) as seed dresser against jassid, *Empoasca spinosa* Dworowaska and Sohi on fenugreek at Junagadh Agricultural University, Junagadh during 2004-05 and 2005-06. The results clearly indicated that the insecticidal seed treatment of thiamethoxam @ 4.2 g/kg seed gave significantly highest seed yield (1827 kg/ha) and net profit (Rs. 3300), followed by imidacloprid @ 10 g/kg seed (1798 kg/ha) and Rs. 2952. However, incremental cost benefit ratio was obtained highest (ICBR, 1:2.12) in thiamethoxam @ 2.8 g/kg seed, followed by thiamethoxam @ 10 g/kg seed (1:1.98) and imidacloprid @ 10 g/kg seed (1:0.70). Thus, thiamethoxam @ 4.2 g/kg seed and imidacloprid @ 10 g/kg seed were found the most effective and economical seed dressers for the management of jassids in fenugreek.

Keywords: Empoasca spinosa Dworowaska and Sohi, Insecticidal seed treatments, Fenugreek.

Introduction

The seed spices crop fenugreek, is being cultivated in Saurashtra under limited water irrigation sources. The jassids, *Empoasca spinosa* Dworowaska and Sohi is the key pest of fenugreek in this area.

It causes the considerable damage in late growth of the crop and ultimately the yield is affected quantitatively and qualitatively. The use of newer insecticides as seed dressing treatment is highly necessary to get more effective, economical and ecofriendly management strategy in comparison to the recommended foliar application of insecticides. The present investigation was carried out with view to find out the effectiveness of neonicotinoid insecticides as seed dressers against jassids of fenugreek.

Materials and Methods

The field experiment on fenugreek (var. Local Fenugreek) was conducted at the Junagadh Agricultural University campus, Junagadh during *rabi*, 2004-05 and 2005-06. The Fenugreek seeds were treated with different doses of imidacloprid 70 WS and thiamethoxam 70 WS. Both insecticides were diluted in water (1:1 ratio) and starch solution was added. Seeds were treated in polythene bag and stirred well for uniform coating. The experiment was conducted in a randomized block design replicated thrice with plot size of 5.0 X 2.7m.The conventional insecticide; dimethoate 0.03 per cent was compared with seed treatments. The observations on population of jassid were recorded at weekly interval starting from crop germination from five randomly selected plants. The data thus, obtained were then transformed and analyzed.

Results and Discussion

Efficacy of Seed Treatments

Reduction in jassid per plant was observed at 8 WAG (week after germination) reflected in significant differences among various treatments. The low jassid per plant (0.15) was observed in thiamethoxam 70 WS @2.8 g/kg seed which was at par with thiamethoxam 70 WS@ 4.2 g/kg seed (0.16) and imidacloprid 70 WS @ 10g/kg seed (0.23). The next best treatment was imidacloprid 70 WS @7.5 g/kg seed (0.36). Similar trends were observed at 9 WAG also where in minimum jassid per plant was observed in thiamethoxam 70 WS@ 4.2g/kg seed (0.43) followed by imidacloprid 70WS @ 10g/kg seed (0.66) and thiamethoxam 70 WS@ 2.8g/kg seed (0.79). More or less similar trend was also observed at 10 WAG, 11 WAG and mean pooled also.

Yield and Economics

The highest seed yield 1827 kg/ha was obtained from the seed treatment with thiamethoxam 70 WS@ 4.2g/kg seed and it was at par with imidacloprid 70WS @ 10g/kg seed (1798 kg/ha) and thiamethoxam 70 WS@ 2.8g/kg seed (1748 kg/ha) (Table 43.2). Economic returns from increased yield was highest in thiamethoxam 70 WS@ 4.2g/kg seed (Rs.3300/ha) followed by imidacloprid 70WS @ 10g/kg seed (Rs. 2952/ha) and thiamethoxam 70 WS@ 2.8g/kg seed (Rs.2352/ha). Whereas, highest ICBR was obtained in thiamethoxam @ 2.8 g/kg seed (1:2.12), followed by thiamethoxam @ 4.2 g/kg seed (1:1.98) and imidacloprid @ 10 g/kg seed (1:0.70). Imidacloprid @ 10 g and thiamethoxam 4.2 g/kg seed has been reported most effective seed dressers against sucking pests of cotton (Vadodaria *et al.*, 2001 and Bheemanna *et al.*, 2003) and okra (Kapadia and Butani, 2005). Prasanna *et al.* (2004) reported that seed treatment with thiamethoxam at higher dost (10 g/kg seed) was highly effective against leafhopper of cotton up to 50 days. However, at lower doses (2.85 and 4.28 g), it gave good control of sucking pests up to 40 days after treatment. Sreelatha and Divakar (1997) observed that seed treatment with imidacloprid (7.5 g/kg) significantly reduced sucking pest of okra. According to Rana *et al.* (2006), Imidacloprid @ 2 ml, thiamethoxam @ 2 g and carbosulfan @ 4 g/kg seed were effective in controlling the sucking pest of okra. Whereas, Mathirajan and Regupathy (2001) observed that seed treatment with thiamethoxam @3 g and 6 g/kg seed and imidacloprid @5 g/kg significantly reduced sucking pests of cotton. Thus, present findings collaborate the results reported by earlier workers.

Table 43.1: Effect of Different Insecticidal Seed Treatments on Jassid Population of Fenugreek

Sl.No.	*Treatment*	*Jassids/3 Leaves*												*Mean Pooled*
		8 WAG			*9 WAG*			*10 WAG*			*11 WAG*			
		2004-05	*2005-06*	*Pooled*	*2004-05*	*2005-06*	*Pooled*	*2004-05*	*2005-06*	*Pooled*	*2004-05*	*2005-06*	*Pooled*	
1.	Imidacloprid 70WS @ 5g/kg seed	0.94 (0.39)	1.05 (0.59)	1.00 (0.49)	1.40 (1.46)	1.47 (1.66)	1.43 (1.56)	1.70 (2.39)	1.60 (2.05)	1.65 (2.22)	1.66 (2.25)	1.72 (2.45)	1.69 (2.35)	1.44 (1.58)
2.	Imidaclorpid 70WS @ 7.5g/kg seed	0.87 (0.26)	0.98 (0.46)	0.93 (0.36)	1.28 (1.13)	1.35 (1.32)	1.31 (1.22)	1.44 (1.59)	1.40 (1.46)	1.42 (1.52)	1.47 (1.66)	1.53 (1.85)	1.50 (1.76)	1.29 (1.17)
3.	Imidacloprid 70WS @ 10g/kg seed	0.91 (0.33)	0.79 (0.13)	0.85 (0.23)	1.05 (0.59)	1.10 (0.72)	1.08 (0.66)	1.07 (0.65)	1.05 (0.59)	1.06 (0.62)	1.14 (0.79)	1.16 (0.86)	1.15 (0.83)	1.03 (0.57)
4.	Thiamethoxam 70 WS @ 2.8g kg seed	0.71 (0.00)	0.91 (0.33)	0.81 (0.15)	1.10 (0.72)	1.17 (0.86)	1.14 (0.79)	1.25 (1.06)	1.19 (0.92)	1.22 (0.99)	1.30 (1.19)	1.35 (1.32)	1.33 (1.26)	1.12 (0.76)
5.	Thiamethoxam70WS @ 4.2g kg seed	0.79 (0.13)	0.83 (0.19)	0.81 (0.16)	0.94 (0.39)	0.98 (0.46)	0.96 (0.43)	0.87 (0.26)	0.87 (0.26)	0.87 (0.26)	1.07 (0.65)	1.11 (0.73)	1.09 (0.69)	0.94 (0.37)
6.	Dimethoate 30 EC 0.03 per cent on needbase	1.01 (0.53)	1.14 (0.79)	1.08 (0.66)	1.72 (2.46)	1.85 (2.92)	1.78 (2.68)	2.29 (4.73)	2.27 (4.66)	2.28 (4.69)	2.45 (5.52)	2.56 (6.05)	2.51 (5.78)	1.91 (3.15)
7.	Control (water spray)	1.07 (0.65)	1.22 (1.00)	1.15 (0.82)	1.70 (2.39)	1.81 (2.79)	1.76 (2.59)	2.37 (5.13)	2.20 (4.32)	2.28 (4.72)	2.40 (5.26)	2.51 (5.79)	2.45 (5.52)	1.91 (3.15)
8.	Control(Untreated)	1.05 (0.59)	1.19 (0.93)	1.12 (0.76)	1.76 (2.59)	1.89 (3.05)	1.82 (2.82)	2.33 (4.93)	2.23 (4.45)	2.28 (4.69)	2.47 (5.59)	2.59 (6.19)	2.53 (5.89)	1.94 (3.25)
	S.Em.±for Treat	0.05	0.05	0.04	0.07	0.07	0.05	0.07	0.07	0.05	0.08	0.08	0.05	0.12
	C.D. at 5 per cent for Treat	0.16	0.16	0.11	0.21	0.21	0.14	0.21	0.21	0.14	0.24	0.23	0.16	0.34
	S.Em.±for year	–	–	0.02	–	–	0.02	–	–	0.02	–	–	0.03	–
	C.D. at 5 per cent for year	–	–	0.05	–	–	0.07	–	–	0.07	–	–	0.08	–
	S.Em.±for YXT	–	–	0.05	–	–	0.07	–	–	0.07	–	–	0.08	–
	C.D. at 5 per cent for YXT	–	–	NS	–	–	NS	–	–	NS	–	–	NS	–
	C.V. per cent	9.96	9.03	9.47	8.81	8.41	8.60	7.35	7.57	7.46	7.72	7.19	7.45	15.88

$\sqrt{x} + 0.5$ transformations used. Data in parentheses are retransformed values. WAG: Week after germination.

Table 43.2: Economics of Different Insecticidal Seed Treatments Against Jassid in Fenugreek

Treatment	*Grain Yield kg/ha*			*Net Realization Rs/ha*	*CBR*
	2004-05	*2005-06*	*Pooled*		
Imidacloprid 70WS @ 5g/kg seed	1802	1417	1610	696	1:0.33
Imidaclorpid 70WS @ 7.5g/kg seed	1889	1485	1687	1620	1:0.51
Imidacloprid 70WS @ 10g/kg seed	2000	1597	1798	2952	1:0.70
Thiamethoxam 70 WS @ 2.8g kg seed	1951	1546	1748	2352	1:2.12
Thiamethoxam70WS @ 4.2g kg seed	2025	1629	1827	3300	1:1.98
Dimethoate 30 EC 0.03 per cent on need base	1778	1363	1570	216	1:0.42
9. Control (water spray)	1765	1375	1570	216	–
Control (Untreated)	1753	1351	1552	–	–
S.Em.±for Treat	65.26	64.07	45.73		
C.D. at 5 per cent for Treat	198.0	194.4	132.5		
S.Em.±for year	–	–	22.87		
C.D. at 5 per cent for year	–	–	66.23		
S.Em.±for YXT	–	–	64.67		
C.D. at 5 per cent for YXT	–	–	NS		
C.V. per cent	6.04	7.55	6.71		

Labour charges was calculated Rs. 15.00/hrs; Market price of fenugreek grain Rs. 12.00/kg.

Spray solution was used 617 liter per hectare during first spray.

The study concluded that thiamethoxam 4.2 g/kg seed and imidacloprid 10 g/kg seed were found the most effective and economical for management of jassid in fenugreek.

References

Bheemanna, M., Patil, B.V., Anand Kumar, V. and Hanchinal, S.G., 2003. Bioefficacy of thiamethoxam seed treatments against cotton early sucking insect pests. In: *Proc. National Symposium on Frontier Areas of Entomological Research*, 5–7 November, New Delhi, p. 169.

Kapadia, M.N. and Butani, P.G., 2005. Ecofriendly management of sucking pests in summer okra through new seed dressers. In: *Proc. National Seminar on Applied Entomology: Current Status, Challenges and Opportunities*, 26–28 September, Udaipur, pp. 109–111.

Mathirajan, V.G. and Regupathy, A., 2001. Seed treatment with thiamethoxam (CruiserReg.): An ecologically selective method for the management of sucking pests of cotton. *Pest Management and Economic Zoology*, 9(2): 177–186.

Prasanna, A.R., Bheemanna, M. and Patil, B.V., 2004. Evaluation of thiamethoxam 70 WS as seed treatment against leaf miner and early sucking pests on hybrid cotton. *Karnataka Journal of Agricultural Sciences*, 17(2): 238–241.

Rana, S.C., Singh, P.B., Pandita, V.K. and Sinha, S.N., 2006. Evaluation of insecticides as seed treatment for control of early sucking pests in seed crop of okra. *Annals of Plant Protection Sciences*, 14(2): 364–367.

Sreelatha and Divakar, B.J., 1997. Impact of imidacloprid seed treatment on insect pest incidence in okra. *Indian Journal of Plant Protection*, 25(1): 52–55.

Vadodaria, M.P., Patel, C.J., Maisuria, I.M. and Patel, U.G., 2001. Imidacloprid (Gaucho) a new seed dresser against early sucking pests of cotton. *Gujarat Agricultural University Research Journal*, 26(2): 32–38.

Recent Advances in Seed Spices (2011) *Pages* **368–370**
Editors: **Y. Ravindrababu, R.K. Jaiman, K.D. Patel, N.R. Patel and S.B.S. Tikka**
Published by: **DAYA PUBLISHING HOUSE, NEW DELHI**

44

Coriander Seed Midge (*Systole albipennis* Walker) Emerging as Important Pest of Coriander in Rajasthan

V.K. Agrawal and M.G. Patil
S.K.N. College of Agriculture, Rajasthan Agriculture University, Jobner, Rajasthan

ABSTRACT

The infestation of seed midge, *Systole albipennis* Walker on coriander was observed in storage condition and at different localities of Rajasthan and this may be the first report from Rajasthan on coriander. The pest with it's pointed ovipositor oviposit through the fruit wall between pericarp and ovule in field. After hatching from egg larvae feed upon embryo and/or endosperm. The damage starts at field and continued in storage.

Introduction

Coriander, *Coriandrum sativum* L. is an annual herb and an important seed spice crop, commonly known as *Dhania*. It is mainly cultivated for its fruits as well as for tender leaves. Fruits have fragrant odour and pleasant aromatic taste due to essential oil content which varies from 0.1 – 1.0 per cent in dry seeds. The essential oil is used for flavouring liquors, cocoa preparations in confectionary and also to mask the offensive odour in pharmaceutical preparations. The dried ground fruits are major ingredient of the curry powder and the whole fruits are used to flavour foods like pickles, sauces and confectionary. The young plants as well as the leaves are used in the preparation of *chutney* and green leaves are also used for seasoning the curries, soups, sauces and *chutneys*. Coriander possesses medicinal properties too. Fruits are said to have carminative, diuretic, tonic, stomachic and aphrodisiac properties.

Coriander, a winter season crop and is cultivated throughout the country with a total area of 3,52,000 hectares and seed production of 2,45,000 tonnes with average productivity of 696 kg ha^{-1} during 2005-06 (Malhotra and Vashistha, 2008). Rajasthan occupies the highest position in production and acreage of coriander with 60 per cent of total production in India. In Rajasthan, it is cultivated in an area of 1,48,390 hectares with production of 1,69,830 tonnes and productivity of 1145 kg ha^{-1} during 2004-05 (Rathore, 2007). In the state it is mainly cultivated in Kota, Baran, Bundi, Jhalawad, Sikar, Jaipur, Tonk and Alwar districts.

There are many constraints in the production of coriander. Abiotic constraints may be drought, soil toxicity, etc. On the other hand technological constraints includes are lack of improved techniques to develop superior varieties and most important constraints could be biotic factors which includes the infestation of pest and diseases.

Insect-pests are one of the major limiting factors for higher production of good quality coriander leaves as well as seeds. The insect pests of coriander are, coriander aphid, *Hydaphis coriandari* (Das), white fly, *Bemisia tabaci* (Genn.), *Spodoptera exigua* (Hub.), *Myzus persicae* (Sulzer), *Chrotogonus trachyperus* (Blanch), and *Thrips tabaci* (Linn.) beside mite, *Petrobia latens* (Muller) and seed midges which feed on this crop.

Seed midges cause both qualitative and quantitative losses in seed yield of coriander. It attacks in field as well as in storage. Seed midges reported on coriander are *Systole albipennis* Walker and *Systole coriandri* Nikol, amongst these two seed midges, *Systole albipennis* was emerging as important pest in Rajasthan. This insect is also observed damaging fennel (Agrawal, 2004). The pest makes small holes and feeds on inner content, leaving the papery coat intact. Hence studies were undertaken to assess the damage intensity of *S. albipennis* in coriander during storage condition at farmers as well as market.

Material and Methodology

Studies on insect damage intensity in coriander was conducted by observing seed samples procured, just before sowing time from farmers as well as from open markets (10 from each level) weighing 10 gm each. For studying the damage intensity of the pest, the seeds were observed for characteristic damage caused by seed midge and numbers of bored as well as total numbers of seed were recorded and per cent seed damage was computed.

Results and Discussion

From the observations taken on seeds procured from farmers revealed that the damage varies from 23 to 38 per cent while the damage on market procured seeds was 5 to 18 per cent (Table 44.1). Agrawal *et al.* (2004) observed epidemic attack of *S. albipennis* during field survey in year 2003 in fennel crop causing more than 50 per cent damage.

The observations are indicating that damage of seed midge was comparatively higher at farmer's level in comparison to market. Enquiry was also made regarding precautions taken against the pests at both the levels and studies revealed that farmers are not adopting any precaution against the pest and hence multiplies during storage, where as, at market level the concerned persons were not aware regarding the precautions against the pest taken by their procedures.

Table 44.1: Samples Obtained from Market

Sl.No.	Name of Store	Damage (Per cent)
1.	Moti Kirana Store	5
2.	Ramesh Kirana Store	11
3.	Choudhary Provisional	8
4.	Aurvedic Medicinal Store	18
5.	Gurukripa Provisional	10
6.	Jyoti Kirana Store	13

Certainly, the damaged seeds are not fit for next crop as the pest feed on embryo of the fruit/seed. These studies have been supported by Kalra (2006), reported that the eggs of seed midge, *S. albipennis* were laid in the developing fruits in field condition. The larvae, on hatching, feds on embryo/endosperm before entering for pupation within the fruit/seed.

References

Agrawal, V.K., Sharma, S.L., Rajput, S.S. and Sharma, B.M., 2004. New record of seed midge fly (*Systole albipennis* Walker) on fennel from Rajasthan. In: *National Seminar on New Perspectives in Commercial Cultivation, Processing and Marketing of Seed Spices and Medicinal Plants*, Rajasthan Agriculture University, Campus Jobner, pp. 74.

Kalra, V.K., 2006. Insect pests of seed spice crops and their management. In: *Emerging Trend in Economic Entomology*, (Eds.) B.S. Chhillar, R.K. Saini and Roshan Lal. Center of Advance Studies, Department of Entomology, HAU, Hissar, pp. 136–142.

Malhotra, S.K. and Vashishtha, B.B., 2008. *Package of Practices for Production of Seed Spices.* National Research Center on Seed Spices, Ajmer, Rajasthan, p. 3–19.

Rathore, P.S., 2007. Seed spices: scope for crop diversification in Rajasthan. In: *Production, Development, Quality and Export of Seed Spices*, (Eds.) S.K. Malhotra and B.B. Vashishtha. National Research Center on Seed Spices, Ajmer, Rajasthan, pp. 162–168.

Recent Advances in Seed Spices (2011) *Pages* **371–376**
Editors: **Y. Ravindrababu, R.K. Jaiman, K.D. Patel, N.R. Patel and S.B.S. Tikka**
Published by: **DAYA PUBLISHING HOUSE, NEW DELHI**

45

Antagonistic Potentiality of *Trichoderma* spp. against *Fusarium oxysporum* var *cumini* Causing Wilt of Cumin and Effect of Temperature and pH on Mycelial Growth of *Trichoderma* spp.

R.K. Jaiman[1], S. Gangopadhyay[2], S.K. Yadav[2] and V.S. Yadav[2]
[1]Centre for Research on Seed Spices, S.D. Agricultural University, Jagudan – 382 710. Dist. Mehsana, Gujarat
[2]College of Agriculture, Rajasthan Agricultural University, Bikaner, Rajasthan

ABSTRACT

Wilt of cumin, caused by *Fusarium oxysporum* var cumini is one of the most important and destructive disease which causes substantial yield loss. Present study was done to control the pathogen by *Trichoderma spp* and effect of temperature and pH on mycelial growth of antagonists at College of Agriculture, Rajasthan Agricultural University, Bikaner (Rajasthan) during 2003 - 04, 2004-05. All the eight different isolates of *Trichoderma* spp. (*Trichoderma viride* isolates TV-1,TV-2, TV-3, TV-4, and TV-5, *Trichoderma harzianum* isolates TH-1 and TH-8 and *Trichoderma virens*) inhibited the growth of wilt pathogen. Among them *T. viride* (TV-2) and *T. harzianum* (TH-8) proved most potential antagonists to cumin wilt pathogen. These two isolates further tested under field condition as seed treatment and soil application with and without FYM. Minimum disease incident (10.20 per cent) with maximum yield (5.10 q/ha.) observed in *T. harzianum* (TH-8) as seed treatment (6gm/kg.) and soil application (10 kg/ha.) with FYM (10 t/

laboratory experiment was conducted to study the effect of temperature on mycelial growth of fungal antagonists. Maximum mycelial growth of all the eight Trichoderma isolates were recorded at 30°C followed by 25°C. TH-8 and *T. virens* did not grow at 35°C and 40°C. The effect of pH on mycelial growth of eight fungal antagonists were also studied. Mycelial growth of most Trichoderma isolates was higher at pH 5.0 to 6.0. Mycelial growth was slightly less in alkaline pH *i.e.* pH 8.0 to 9.0.

***Keywords:** Fusarium oxysporum f.sp. cumini, Trichoderma, Bavistin, Cumin wilt.*

Introduction

Cumin wilt caused by *Fusarium oxysporum* f.sp. cumini is the most important disease of Rajasthan. It causes 40 per cent reduction in yield specially arid districts of Rajasthan where cumin is extensively cultivated as a cash crop (Lodha *el. al.*, 1986). The wilt pathogen being soil-borne in nature is difficult to control chemically and by other means of disease control. Different species of Trichoderma and their strains when incorporated in the soil and applied to seeds were reported to be more effective against *F. oxysporum* causing wilt of different crops including cumin (Mukhopadhyay, 1987; Mukhopadhyay *et al.*, 1989; Dwivedi *et al.*, 1993; Israel, 2002). An attempt has been made to isolate *Trichoderma spp.* locally were screened for the efficacy in the control of wilt of cumin under different field condition.

Materials and Methods

Isolation of Pathogen

Roots of cumin crop sowing apparent infection of wilt disease was collected from the farmers field during rabi season 2003-04. Small bits of the infected roots were surface sterilized with 0.2 $HgCl_2$ per cent solution for one minutes and repeatedly washed with sterilized distilled water to remove the traces of $HgCl_2$ and inoculated on potato dextrose agar(PDA) at 25±2 °C. Subsequently, pure culture was established by hypel tip isolation method. Fungus was identified by comparing its morphological characters.

Isolation of the Antagonists

Two hundred five soil samples were collected from different crop fields *viz.*, cumin, chickpea, groundnut, cotton, mothbean, mungbean, clusterbean etc. covering seven agroclimatic zones and more than thirteen districts of Rajasthan. The collected soil samples were air dried in shade for 12 hours before analysis. 10 grams of soil was placed in 90 ml sterile distilled water in Erlenmeyer flask(250 ml capacity) and serial dilutions were made from stock soil suspensions. A 0.5 ml soil suspension having 10^3 dilution was added on surface of media [Martins Rose Bangal Agar Medium (MRBA) and Trichoderna Selective Medium(TSM) (Elad *et al.*, 1981)] in petri dishes and spread uniformly. The inoculated petridishes were incubated at 25±2 °C. The fungus cultures were purified and observed under microscope for their identification. The antagonists were identified on the basis of morphological characters.

Zone of Inhibition Technique

The antagonistic potential of five isolates of *Trichoderma viride*(TV1, TV2, TV3, TV4 and TV5), two isolates of *T. Harzianum*(TH1 and TH8) and one isolate of *T. Virins*(TVi 1) were screened against *F.*

oxysporum f.sp. cumini by dual culture technique (Dennis and Webster, 1971). Required quantity of potato dextrose agar(20 ml) was poured in each of the sterilized petri plates. Actively growing 5 mm disc of both the organisms were placed in plates at the opposite of each other. The inoculated petridishes were kept at 25±2°C in BOD incubator. In case of control the mycelial disc was inoculated at the center of the petridish and incubated. Three replications were kept for each antagonists. Mycelial growth of antagonist and pathogen was recorded and inhibition percentage of mycelial growth was calculated by using following formula:

$$\text{Per cent inhibition of mycelial growth} = C - T/C \times 100$$

where,

C: Mycelial growth in control

T: Mycelial growth in treatment

Field Evaluation

A field trial using Trichoderma as seed treatment, soil application with and without FYM and bavistin was conducted in field at research farm College of Agriculture, RAU, Bikaner. The experiment was laid out in randomized block design with three replications. The cumin cv. RZ-19 was sown in a plot size of 3x2 sqm with a row spacing of 30 cm in during crop season 2003-04 and 2004-2005. Seed treatment with Trichoderma(6gm/kg), soil application(5kg/ha), seed treatment bavistin(2gm/kg) and soil application with FYM(10t/ha) was done. Test pathogen was raised in sand: miaze meal media. Fifty gram inocula of pathogen was applied in each plot for disease development. The fungal antagonists were raised on molasses yeast extract broth. The mycelial mat was homogenized and mixed with talc powder at the ratio 1; 2 and dried in shade. Carboxy methyl cellulose was added 6gm/kg formulation and mixed thoroughly. The talc based formulation was used for seed treatment and soil application in field experiment. The disease incidence was recorded periodically. Infected plants were brought to laboratory and isolations were made in orded to determine the involvement of the pathogen.

Effect of Temperature and pH on Mycelial Growth of *Trichoderma* Isolates

One set of laboratory experiments was conducted to study the effect of temperature and pH on mycelial growth of Trichoderma isolates. Eight Trichoderma isolates showed promising antagonistic potential towards *F. oxysporum* f.sp. cumini were included in this experiment.The mycelial growth on PDA was recorded at five different temperature *i.e.* 20, 25, 30, 35 and 40°C and five different pH levels *i.e.* 5.0, 6.0, 7.0, 8.0 and 9.0. Different temperatures were maintained in BOD and pH was adjusted by adding NaoH or Hcl before pouring the media. Three replications were kept for each treatment. The linear mycelial growth was recorded at 72 hours of incubation.

Results and Discussion

Zone of Inhibition Technique

Results of in vitro study are present in Table 45.1. The results revealed that all eight Trichoderma isolates were inhibitory to the mycelial growth of *F. oxysporum* f.sp. cumini. *T. viride* isolate TV2 and *T. harzianum* isolate TH8 showed maximum inhibition in radial growth of *F. oxysporum* f.sp. cumini. Inhibition was minimum in TV5. Patel and Patel (1998a) reported effective inhibition of *F. oxysporum* f.sp. cumini by *T. Harzianum*, *Aspergillus flavus*, and *A. niger*. Sangle and Bambawale(2004) have also reported the inhibition of *F. oxysporum* f.sp. sesami by *T. viride* and *T. harzianum*.

Table 45.1: Antagonistic Potentiality of *Trichoderma* spp. Isolated from Soil Against *F. oxysporum* f.sp. *cumini in vitro*

Sl.No.	Fungal Species	Per cent Inhibition of Mycelial Growth
1.	*Trichoderma viride* (TV1)	62.21(52.10)*
2.	*T. viride* (TV2)	69.62(56.58)
3.	*T. viride* (TV3)	63.70952.56)
4.	*T. viride* (TV4)	68.88(56.12)
5.	*T. viride* (TV5)	59.62(50.56)
6.	*T. harzianum* (TH1)	67.40(55.21)
7.	*T. harzianum* (TH8)	69.25(51.35)
8.	*T. virins* (TVi1)	68.51(55.90)
	SEm(±)	0.94
	CD(P = 0.05)	2.73

*: Figures in parentheses are angular transformed value.

Table 45.2: Effect of Bioagents in Controlling Wilt of Cumin Under Field Condition

Sl.No.	Treatments	Disease Incidence (Per cent)			Grain Yield (kg/ha)		
		2003-04	2004-05	Pooled	2003-04	2004-05	Pooled
1.	*T. viridi* ST+SD	16.00 (23.57)*	21.50 (27.61)	18.75 (25.59)	313	253	283
2.	*T. viridi* ST+SD + FYM	14.80 (22.62)	20.30 (26.77)	17.55 (24.69)	386	326	356
3.	*T. harzianum* ST+SD	13.70 (21.71)	19.10 (25.90)	16.40 (23.81)	445	355	400
4.	*T. harzianum* ST+SD+ FYM	10.20 (18.62)	15.70 (23.33)	12.95 (20.98)	510	450	480
5.	Carbendazim ST	23.20 (28.78)	28.50 (32.25)	25.85 (30.52)	285	225	255
6.	Carbendazim ST+ FYM	25.30 (30.19)	30.70 (33.63)	28.00 (31.91)	290	230	260
7.	Control	38.50 (38.35)	43.90 (41.48)	41.20 (39.91)	238	178	208
	SEm(±)	0.21	0.09	0.10	3.22	5.60	5.67
	CD (P = 0.05)	0.65	0.27	0.30	9.92	17.26	19.62

Figures in parentheses are angular transformed value.

Field Evaluation

All the treatments significantly reduced the incidence of wilt over untreated check. The minimum disease incidence (12.95 per cent) was recorded in *T. harzianum* used as seed treatment and soil application with FYM followed by *T. harzianum* used as seed treatment and soil application without FYM (16.40 per cent). Grain yield increased in response to bio-agents treatments (Table 45.2). In pooled analysis maximum grain yield (480kg/ha) was recorded in *T. harzianum* used as seed treatment and soil application along with FYM. Akhtar *et al.* (1982) have also reported *T. harzianum* to be most effective against *F. oxysporum* f.sp. lini causing linseed wilt. Mukhopadhyay (1995) found seed treatment with *T. harzianum* coupled with chemical fungicides to be effective in controlling wilt disease caused by *Fusarium* spp.

Effect of Temperature on Mycelial Growth of Fungal Antagonists

Maximum mycelial growth of all the eight Trichoderma isolate listed was recorded at 30°C followed by 25°C. Six Trichoderma isolates grew well at 35°C also. But two isolates *i.e. T. harzianum* (TH8) and *T. virins* (Tvi1) did not grow at 35 and 40°C (Table 45.3). Umamaheswari *et al.* (2003) also observed maximum growth and sporulation in CIAH-186 at 25°C. Even 35°C growth upto 33 mm of this isolate was observed. This variation is due to strains, it is vary to strain to strain.

Table 45.3: Effect of Temperature on Mycelial Growth of Trichoderma Isolates Isolated from Soil

Temperature (°C)	*Mycelial Growth (in mm)*							
	T. viridi (TV1)	*T. viridi (TV2)*	*T. viridi (TV3)*	*T. viridi (TV4)*	*T. viridi (TV5)*	*T. harzianum (TH1)*	*T. harzianum (TH8)*	*T. virens (TVi1)*
20	32.25	34.25	27.25	28.50	30.50	27.25	25.75	16.25
25	73.00	85.50	84.25	78.25	73.50	84.50	72.75	52.25
30	87.75	87.50	88.00	81.75	81.00	89.00	78.25	60.50
35	76.00	77.75	87.00	75.50	69.75	83.25	1.50	1.50
40	24.60	16.30	21.00	26.60	27.30	21.60	0.00	0.00
SEm (±)	1.54	1.27	0.40	0.27	0.97	1.00	0.26	0.78
CD (P = 0.05)	4.73	3.92	1.22	0.82	2.98	3.09	0.80	2.41

Table 45.4: Effect of pH on Mycelial Growth of Trichoderma Isolates Isolated from Soil

Temperature (°C)	*Mycelial Growth (in mm)*							
	T. viridi (TV1)	*T. viridi (TV2)*	*T. viridi (TV3)*	*T. viridi (TV4)*	*T. viridi (TV5)*	*T. harzianum (TH1)*	*T. harzianum (TH8)*	*T. virens (TVi1)*
5.0	90.00	90.00	81.66	83.33	83.33	89.33	88.66	80.60
6.0	87.00	81.00	80.33	83.66	82.66	81.00	89.00	81.00
7.0	87.00	81.33	82.00	79.00	79.33	83.66	89.33	76.66
8.0	80.00	80.33	77.66	80.33	80.00	87.66	82.00	65.33
9.0	85.00	81.00	82.66	80.66	79.33	85.66	89.33	66.66
SEm (±)	0.26	0.60	0.62	0.70	0.47	0.60	1.44	0.58
CD (P = 0.05)	0.81	1.87	1.96	2.20	1.48	1.88	4.52	1.82

Effect of pH on Mycelial Growth of Fungal Antagonists

The results given in Table 45.4 showed that mycelial growth of most of the Trichoderma isolates was higher at pH 5.0 to 6.0. Mycelial growth was slightly less in alkaline pH *i.e.* 8.0 to 9.0. One trichoderma isolate *T. harzianum* (TH8) grew equally well at pH 6.0, 7.0 and 9.0. Another Trichoderma isolate *i.e. T. virins* (TVi1) was sensitive to alkaline pH as its growth was less at pH 8.0 and 9.0. Umamaheswari *et al.* (2003) found that most of Trichoderma isolates preferred alkali range of pH(8-9) for better growth and conidiogensis. The pH variation due to strain to strain.

References

Akhtar, C.M., Amin, C.M. and Salim, C.M., 1982. Investigation of the biological control of *Rhizoctonia solani* and *Fusarium oxysporum* f.sp. lini causing root rot and wilt diseases in cotton and linseed crop, respectively. *Avub. Agric. Res. Inst.*, Faisalabad, Pakistan, 2: 63.

Dennis, C. and Webster, J., 1971. Antagonistic properties of species group of *Trichoderma*. I. Production of non-volatile antibiotics. *Trans. Brit. Mycol. Soc.*, 57: 47–48.

Dwivedi, S.K., Ambasht, R.S. and Dwivedi, R.S., 1993. Toxicity of some antagonistic fungi on pathogenic fungi of economic crops. *National Academy Science Letters*, 16(7–8): 209–210.

Elad, Y., Chet, I. and Heniz, Y., 1981. A selective medium for improving quantitative isolation of *Trichoderma spp.* from soil. *Phytoparasitica*, 9: 59–67.

Israel, S., 2002. Effect of amendments on wilt of cumin caused by *Fusarium oxysporum* f.sp.*cumini* Prasad and Patel in Aridsoils. *Ph.D. Thesis*, JNV University, Jodhpur.

Lodha, S., Gupta, G.K. and Singh, S., 1986. Crop disease situation and new records in India arid zone. *Ann. Arid. Zone*, 25: 311–320.

Mukhopadhyay, A.N., 1987. Biological control of soil-borne diseases of chickpea and lentil. *Indian Phytopath.*, 42: 275.

Mukhopadhyay, A.N., Kaur, N. and Saxena, H.C., 1989. Biological control of soil-borne diseases of vegetables and pulses by *Trichoderma* spp. *Indian Phytopath.*, 49: 275.

Mukhopadhyay, A.N., 1995. Exploitation of *Gliocladium virens* and *Trichoderma harzianum* for biological seed treatment against soil borne diseases. *J. Mycol. Pl. Pathol.*, 25(1 and 2): 124.

Patel, S.M. and Patel, B.K., 1998a. Inhibitory effect of different fungal bio-agents on *Fusarium oxysporum* f.sp. *cumini* causing cumin wilt. *Annals of Plant Protection Sciences*, 6: 25–27.

Sangle, U.R. and Bambawale, O.M., 2004. New strains of *Trichoderma* spp. strongly antagonistic against *Fusarium oxysporum* f.sp. *sesami*. *J. Mycol. Pl. Pathol.*, 34(1): 107–109.

Umamaheswari, C., Nallthambi, P. and Dhandar, D.G., 2003. Influence of growth factors on performance of desert isolates of *Trichoderma*. *J. Mycol. Pl. Pathol.*, (Abstract), 33(3): 481.

Recent Advances in Seed Spices (2011) *Pages* **377–383**
Editors: **Y. Ravindrababu, R.K. Jaiman, K.D. Patel, N.R. Patel and S.B.S. Tikka**
Published by: **DAYA PUBLISHING HOUSE, NEW DELHI**

46

A Video Conference Service: New Ways to Create More Immediate, Interactive Links Between the Scientists and Seed Spices Growers

L.D. Parmar[1], A.P. Prajapati[2], V.P. Prajapati[1] and P.K. Patel[2]
[1]**Centre for AICT, S.D. Agricultural University, Sardarkrushinagar, Dist. Banaskantha, Gujarat**
[2]**Centre for Research on Seed Spices, S.D. Agricultural University, Jagudan – 382 710. Dist. Mehsana, Gujarat**

Introduction

A video conference (also known as a *videoteleconference*) is a set of interactive telecommunication technologies which allow two or more locations to interact via two-way video and audio transmissions simultaneously. It has also been called visual collaboration and is a type of groupware. It differs from videophone in that it is designed to serve a conference rather than individuals. Video conferencing uses telecommunications of audio and video to bring people at different sites together for a meeting. This can be as simple as a conversation between two people in private offices (point-to-point) or involve several sites (multi-point) with more than one person in large rooms at different sites. Besides the audio and visual transmission of meeting activities, video conferencing can be used to share documents, computer-displayed information, and whiteboards. High speed Internet connectivity has become more widely available at a reasonable cost and the cost of video capture and display technology has decreased. Consequently personal video teleconference systems based on a webcam, personal computer system, software compression and broadband Internet connectivity have become affordable for the farmers through state government projects such as 'e-Gram Vishwagram'. Government of Gujarat dedicate his ambitious 'e-Gram Vishwagram Project' by offering broadband connectivity to all 13,693 Gram Panchayats. Effective Wide Area Network was established in SDAU. Sardarkrushinagar, the head quarter of SDAU facilitate with BSNL Leased line Copper link from Telephone Exchange, Dantiwada (2.0 MBPS), GSWAN Radio link from Mamlatdar Office, Dantiwada (0.5 MBPS) and ERNET VSAT Link from ERNET, New Delhi (0.5 MBPS). All sub-canters of SDAU facilitate with GSWAN Radio link and/or ERNET VSAT Link. Looking to this IT infrastructure of Gujarat State and SDAU, SDAU planned to start Video Conferencing Services for farmers. The video conferencing facilities are made available on newly established local area network by installing core equipments for video conferencing. SDAU is also planning for the establishment of Video Conferencing Studio which will make it possible to develop new ways to create more immediate, interactive links between the scientists and seed spices growers in Gujarat.

Technology

Multipoint Video-conferencing

Simultaneous video conferencing among three or more remote points is possible by means of a Multipoint Control Unit (MCU). This is a bridge that interconnects calls from several sources (in a similar way to the audio conference call). All parties call the MCU unit, or the MCU unit can also call the parties which are going to participate, in sequence. There are MCU bridges for IP and ISDN-based video conferencing. There are MCUs which are pure software, and others which are a combination of hardware and software. An MCU is characterised according to the number of simultaneous calls it can handle, its ability to conduct transposing of data rates and protocols, and features such as Continuous presence, in which multiple parties can be seen onscreen at once.

MCUs can be stand-alone hardware devices, or they can be embedded into dedicated VTC units.

Some systems are capable of multipoint conferencing with no MCU, stand-alone, embedded or otherwise. These use a standards-based H.323 technique known as "decentralized multipoint", where each station in a multipoint call exchanges video and audio directly with the other stations with no central "manager" or other bottleneck. The advantages of this technique are that the video and audio will generally be of higher quality because they don't have to be relayed through a central point. Also, users can make ad-hoc multipoint calls without any concern for the availability or control of an MCU. This added convenience and quality comes at the expense of some increased network bandwidth, because every station must transmit to every other station directly.

Standards

H.264/AVC/MPEG-4 Part 10 contains a number of new features that allow it to compress video much more effectively than older standards and to provide more flexibility for application to a wide variety of network environments. In particular, some such key features include:

- ☆ Multi-picture inter-picture prediction
- ☆ Spatial prediction from the edges of neighboring blocks for intra coding, rather than the "DC"-only prediction found in MPEG-2 Part 2 and the transform coefficient prediction found in H.263v2 and MPEG-4 Part 2. This includes luma prediction block sizes of 16×16, 8×8, and 4×4 (of which only one type can be used within each macroblock).
- ☆ Lossless macroblock coding features including:

Figure 46.1: Inaugration of SDAU Video Conferencing Service by Hon'ble C.M. wiith Farmers

- Flexible interlaced-scan video coding features, including:
- New transform design features, including:
- A quantization design including:
- An in-loop deblocking filter which helps prevent the blocking artifacts common to other DCT-based image compression techniques, resulting in better visual appearance and compression efficiency.
- An entropy coding design including:
- Loss resilience features including:
- Switching slices, called SP and SI slices, allowing an encoder to direct a decoder to jump into an ongoing video stream for such purposes as video streaming bit rate switching and "trick mode" operation. When a decoder jumps into the middle of a video stream using the SP/SI feature, it can get an exact match to the decoded pictures at that location in the video stream despite using different pictures, or no pictures at all, as references prior to the switch.
- A simple automatic process for preventing the accidental emulation of start codes, which are special sequences of bits in the coded data that allow random access into the bitstream and recovery of byte alignment in systems that can lose byte synchronization.

- Supplemental enhancement information (SEI) and video usability information (VUI), which are extra information that can be inserted into the bitstream to enhance the use of the video for a wide variety of purposes.
- Auxiliary pictures, which can be used for such purposes as alpha compositing.
- Support of monochrome, 4:2:0, 4:2:2, and 4:4:4 chroma subsampling (depending on the selected profile).
- Support of sample bit depth precision ranging from 8 to 14 bits per sample (depending on the selected profile).
- The ability to encode individual color planes as distinct pictures with their own slice structures, macro block modes, motion vectors, etc., allowing encoders to be designed with a simple parallelization structure (supported only in the three 4:4:4-capable profiles).
- Picture order count, a feature that serves to keep the ordering of the pictures and the values of samples in the decoded pictures isolated from timing information, allowing timing information to be carried and controlled/changed separately by a system without affecting decoded picture content.

These techniques, along with several others, help H.264 to perform significantly better than any prior standard under a wide variety of circumstances in a wide variety of application environments. H.264 can often perform radically better than MPEG-2 video—typically obtaining the same quality at half of the bit rate or less, especially on high bit rate and high resolution situations.

Networks to be Used for Video Conferencing Facility for Farmers

Through the Local Area Network (LAN) 51 buildings of SDAU are planned to be connected which include offices, colleges, hostels, public service centers, research stations and library. This network is unique of its kind. The design of this LAN is prepared by IT Centre of SDAU with technical expertise from various network experts. The networking will be helpful to scientists, professors, employees, students and farmers in communicating with each other with good reliability and speed and for atomization of University activities (Figure 46.2).

Effective Wide Area Network was established in SDAU, Sardarkrushinagar, the head quarter of SDAU facilitate with BSNL Leased line Copper link from Telephone Exchange, Dantiwada (2.0 MBPS), GSWAN Radio link from Mamlatdar Office, Dantiwada (0.5 MBPS) and ERNET VSAT Link from ERNET, New Delhi (0.5 MBPS). All sub-centers of SDAU facilitate with GSWAN Radio link and/or ERNET VSAT Link (SDAU News Volume-5 No. 3-4 July – December, 2008, Page-3).

In Gujarat, all Districts/Taluka Panchayats, different Government/Semi Government offices, Universities etc. are connected with GSWAN Radio link. Government of Gujarat also dedicate his ambitious 'e-Gram Vishwagram Project' by offering V-set connectivity to all 13,693 Gram Panchayats.

Progressive Farmers/processors must have computer and information technology savvy in order to be competent and productive in the workforce. To participate and make informed decisions in the agricultural industry, a global person must possess technological and information literacy skills that include the ability to gather, process, and manipulate data.

As per Information note to the Press (Press Release No 11/2009) of Telecom Regulatory Authority of India (TRAI), Total Broadband subscribers base has reached 5.45 million by the end of December 2008 and the annual growth rate of Broadband subscribers is 74.12 during the year 2008. These data indicate that Indian households and businesses are finally getting up to speed with broadband. With

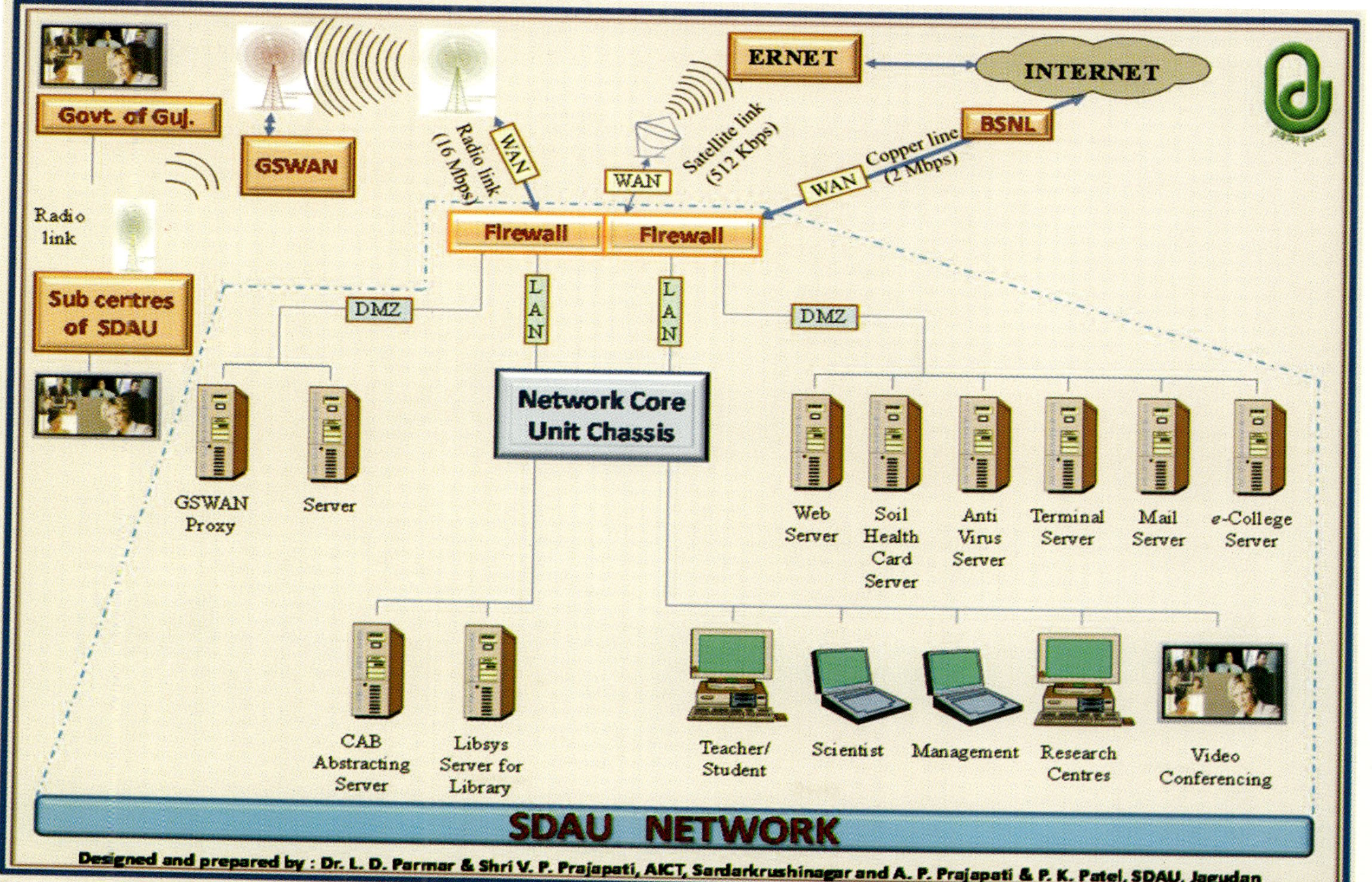
Govt. of Guj.
GSWAN
Radio link
Sub centres of SDAU
WAN
Radio link (16 Mbps)
WAN
Satellite link (512 Kbps)
ERNET
INTERNET
BSNL
WAN
Copper line (2 Mbps)
Firewall
Firewall
DMZ
LAN
LAN
DMZ
Network Core Unit Chassis
GSWAN Proxy
Server
Web Server
Soil Health Card Server
Anti Virus Server
Terminal Server
Mail Server
e-College Server
CAB Abstracting Server
Libsys Server for Library
Teacher/ Student
Scientist
Management
Research Centres
Video Conferencing
SDAU NETWORK
Designed and prepared by : Dr. L. D. Parmar & Shri V. P. Prajapati, AICT, Sardarkrushinagar and A. P. Prajapati & P. K. Patel, SDAU, Jagudan

Figure 45.2

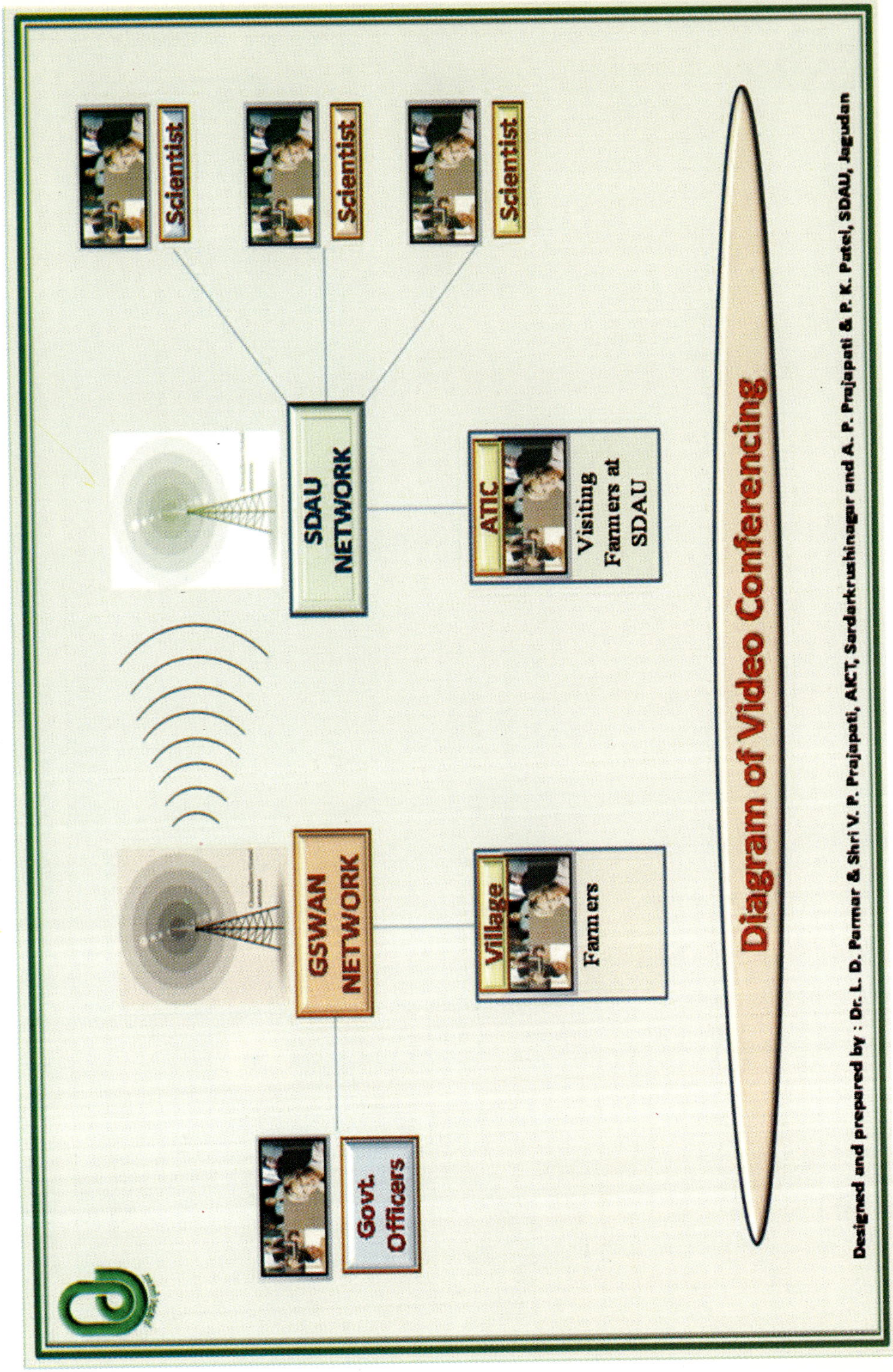

Govt. Officers
GSWAN NETWORK
Village
Farmers
SDAU NETWORK
ATIC
Visiting Farmers at SDAU
Scientist
Scientist
Scientist
Diagram of Video Conferencing
Designed and prepared by : Dr. L. D. Parmar & Shri V. P. Prajapati, AICT, Sardarkrushinagar and A. P. Prajapati & P. K. Patel, SDAU, Jagudan

Figure 45.3

the infrastructure in place and cheaper options from which farmers can choose, we are poised to take advantage of the benefits of the ICT revolution in our homes and workplaces.

SDAU Starts Video Conferencing Facility for Farmers

Consequently personal video teleconference systems based on a webcam, personal computer system, software compression and broadband Internet connectivity have become affordable for the farmers through SDAU's Networks and Government of Gujarat's Wide Area Networks and Broadband connections available in homes and workplaces of progressive farmers and Agriculture Businessmen (Figure 46.3).

The video conferencing facilities will also be made available on newly established local area network of this campus by installing core equipments for video conferencing. It will provide seamless interaction among officers, scientists and farmers for their problems and conferencing to the residential farmers who are connected with GSWAN. The financial outlay of this project is Rs.217.00 lakh. (SDAU News Volume-5 No. 3-4 July – December, 2008, Page-3)

Impact on the Seed Spices Growers

High speed Internet connectivity has become more widely available at a reasonable cost and the cost of video capture and display technology has decreased. Consequently personal video teleconference systems based on a webcam, personal computer system, software compression and broadband Internet connectivity have become affordable for the seed spices growers through proposed "Video Conferencing Service for farmers". Also, the hardware used for this technology has continued to improve in quality, and prices have dropped dramatically. The availability of freeware (often as part of chat programs) has made software based video conferencing accessible to many.

Sometimes it is simply not possible or practical to have a face-to-face meeting among farmers and scientists. Video conferencing adds another possible alternative, and can be considered when:

- ☆ A live conversation is needed;
- ☆ Visual information is an important component of the conversation;
- ☆ The parties of the conversation can't physically come to the same location; or
- ☆ The expense or time of travel is a consideration.

Video conferencing can enable individuals in faraway places to have meetings on short notice. Time and money that used to be spent in traveling can be used for other works. Through this facility, Seed Spices growers can obtain valuable information and guidelines immediately from the Scientists of SDAU and solve their problems without spending more money and time.. Thus, SDAU's "Video Conferencing Services for farmers" make it possible to develop new ways to create more immediate, interactive links between the scientists and seed spices growers in Gujarat.

References

Ferran, Carlos and Watts, Stephanie, 2008. Video conferencing in the field: A heuristic processing model. *Management Science*, 54(9).

SDAU, 2008. News Volume 5(3)4: 3.

Wolfe, Mark, 2007. Broadband video conferencing as knowledge management tool. *Journal of Knowledge Management*, 11(2).

Recent Advances in Seed Spices (2011) *Pages* **384–391**
Editors: **Y. Ravindrababu, R.K. Jaiman, K.D. Patel, N.R. Patel and S.B.S. Tikka**
Published by: **DAYA PUBLISHING HOUSE, NEW DELHI**

47

E-commerce: A New Market Channel in Seed Spices for Domestic and Export Market Promotion

A.P. Prajapati and Y. Ravindrababu
Centre for Research on Seed Spices, S.D. Agricultural University
Jagudan – 382 710, Dist. Mehsana, Gujarat

Introduction

People must have computer and information technology savvy in order to be competent and productive in the workforce. To participate and make informed decisions in the agricultural industry, a global person must possess technological and information literacy skills that include the ability to gather, process, and manipulate data.

Within a decade of its introduction, Internet has become an extremely important marketing tool for small and medium enterprises. Today, it is a major source for acquiring customers. Understanding behavior pattern in the web has become a crucial element in attracting potential customers and transact business. As E-commerce becomes adopted by a growing majority of Indian homes and businesses, according to domestic and export market promotion in seed spices, we would expect to use greater amounts of E-commerce taking place. One of the key requirements of E-commerce is the availability of internet connectivity. This Activity looks at the take up of broadband in the India compared to some of our international competitors. As per Information note to the Press (Press Release No 11/2009) of Telecom Regulatory Authority of India (TRAI), Total Broadband subscribers base has reached 5.45 million by the end of December 2008 and the annual growth rate of Broadband subscribers is 74.12

during the year 2008. These data indicate that Indian households and businesses are finally getting up to speed with broadband. With the infrastructure in place and cheaper options from which consumers can choose, we are poised to take advantage of the benefits of the ICT revolution in our homes and workplaces. Seed Spices Organization can use their web site and E-mail as a low cost way of extending the geographical reach of business marketing activities, and to develop a global export market for their seed spices value added products. Text messages, documents, graphics, photographs, music, video, CAD/CAM files, and much more, can be converted into data and sent as E-mail attachments or presented on a web site. There are no postage, printing and packaging costs with e-Commerce communications. So, use this new communications method to work quicker, and in new ways, with customers and suppliers outside of your locality or country.

By Juhan Samuel Kochi, May 26 (ANI): India's spice exports have reached another milestone by crossing the one billion US dollar mark in 2007-08. During the financial year 2007-08, a total of 4,44,250 tonnes of spices and spice products valued Rs.4435.50 crores were exported from the country as against 3,73,750 tonnes valued Rs.3575.75 crores in the previous year. The spices export in 2007-08 has also far exceeded the target fixed for the year, both in term of volume and value. Against the target of 3,80,000 tonnes valued Rs.3600 crores fixed for the year, the achievement has been 117 per cent in volume, 123 per cent in rupee value and 126 per cent in dollar value. "The export has reached a new peak in volume as well. Against last year's performance, the export has shown an increase of 19 per cent in volume, 24 per cent in rupee value and 39 per cent in dollar terms of value", said V.J. Kurian, Chairman, Spices Board, Kochi. During the year 2007-08, the export of pepper, chilli, seed spices like coriander, cumin, fennel, fenugreek and other seeds has shown substantial growth both in volume and value as compared to last year. The export of value added products like curry powder, mint products and spice oils and oleoresins have also shown impressive growth in 2007-08.

Thus, we can say that tremendous development in IT Sector in India is also beneficial to promote domestic and export market of spices. In future, seed spices organizations in India will boosting their performance through e-commerce and we can expect that E-commerce will be proven an important market channel in seed spices for domestic and export market promotion in near future.

Benefits of E-Commerce in Seed Spices Industry

E-commerce introduce an another channel to promote domestic and export market of seed spices. Business web site of any seed spices processing unit can take orders 24 hours a day, every day of the year, and is readily accessible to customers all over the world. A successful business web site can produce a highly profitable revenue stream.

Lower the Cost of Order Processing

The direct cost-of-sale for an order taken from a web site should be lower than through traditional means (retail, paper based), as there is no human interaction during the on-line electronic purchase order process. Also, electronic selling should virtually eliminate processing errors, as well as being faster and more convenient for the visitor. The domestic and export market of seed spices will be speed up through this feature of E-commerce.

No Physical Space Limitations

Seed Spices industries can offer more value added products of seed spices with an eCatalogue, than you can physically stock on shelves.

Buy and Sell Quicker

In respect to seed spices processing industries, the requirements, brochures, quotations, purchase orders, delivery schedules, etc., can be eMailed within seconds to your suppliers and customers all over the world. The whole process of buying and selling can be made much more efficient and responsive using eMails. Therefore, it is necessary to have fast and efficient internet connectivity and trained staff in seed spices processing industry.

Boost Sales with Exports

Seed spices industries can use their web sites and eMails as a low cost way of extending the geographical reach of your business marketing activities, and to develop a global export market for your products and services. Thus, seed spices export market can be enhanced through E-commerce.

Exchange and Share Data Across the World

The useful and important data of seed spices processing units in the form of text messages, documents, graphics, photographs, music, video, CAD/CAM files, and much more, can be sent as eMail attachments or presented on a web site. There are no postage, printing and packaging costs with eCommerce communications. So, use this new communications' method to work quicker, and in new ways, with customers and suppliers outside of your locality or country.

Update Employees Instantly with New Policies or Procedures

The employees of the seed spices processing unit can update the knowledge regarding new policies and procedures through web site and eMail. They can works well when communicating with colleagues who are off-site, out of the country, or tele-working.

Hold Web Meetings (Data Conferencing)

A great way to have a virtual meeting is to upload a document (spreadsheets, project plans, etc) on to a web site and enable it to be viewed and edited in real-time through the Internet. Holding virtual meetings can save significantly on travel expenses and time.

Use the Internet to Improve Business Administration of Seed Processing Unit

With a mobile phone connected to a portable computer, employees working off-site can access and up-date internal information, such as customer records, price lists, time sheets, schedules, and job reports. Providing remote Internet access for staff makes it possible for the latest information to be at hand, for administrative tasks such as invoicing, employee whereabouts, and expense claims.

Introduce Collaborative Working

With fast, reliable, and cheap eCommerce communications, it is possible for colleagues, customers, suppliers and partners, to use collaborative working practices to manage, share, and enhance project work, regardless of location.

Some Facts about E-commerce

I have compiled here some interesting facts and statistics about suffers behavior in the Net. Let us understand these and examine their implication on your marketing strategy, specifically how this knowledge can contribute to your success in Internet.

Fact No. 1: Website is Essential for International Business

First thing first–to be successful in Internet, you must have an Internet presence. The 9th Annual Middleburg/Ross Survey of Media Professionals found that over 70 per cent of working journalists list the company website as the first stop when researching a breaking story or a feature. Same is the case with other professionals whose first port of call on any subject is the company web-site. Having no web-site mostly mean no customer and hence no business. There are other more compelling reasons like building trust, 24-hour storefront, business E-mail etc. The days of doing business over Internet using free or paid e-mail are truly over–if you are serious about your business and expect serious customers from Internet–a professional web-site is essential.

Fact No. 2: You have Barely 5 Seconds to Impress and Engage a Visitor

Research shows that web surfers tend to scan through pages before they read anything. What it means in reality is–a surfer may leave your website even if you have loads of useful information. You must be able to engage him/her before he/she decides to move out. In the dial-up days--it used to be called 'Battle for 20 seconds' when surfers decided to read-on or move out of a web-page within maximum 20 seconds. Today, with broadband connection and faster servers–a web-site has barely 5-seconds to impress a visitor.

What is the implication of this fact ? In addition to fast loading, attractive design and lay-out–you must provide the surfer easy and convenient click options. As soon as the scan is over and the surfer feels comfortable about the web-site, he/she should be able to locate interesting links to click. Select right keywords, highlight and bold at least 7 per fold–and you are telling the customer what to do within those 7 highlighted or bolded words... Your website has more of a chance of converting a visitor into a customer.

Fact No. 3: The Three-Click Rule

The three-click rule states that any page in a web-site should be accessible in three clicks. Call your visitor lazy, idle or bore–he/she is unlikely to click more than three times to reach the destination page.

The implication for this fact is simple–make your web-site really easy to find, easy to remember, easy to understand and most important–easy to navigate.

Do you use navigation bar in each page ? How many clicks are required to reach home page from any other page ? Do you use an internal search engine ? How easy is it to find 'Order Form' ? How many clicks required to reach 'Send Enquiry' or 'Contact Us' page ?

Fact No. 4: The 12-Click Deadline

An average user clicks 12 times before leaving a website. If any page is accessible within 3-clicks, one may safely assume that the surfer will visit 3 or 4 pages of your website. Implication of this fact–you must present your visitor with something to act on (purchase/enquire/e-mail/fill order-form) in each page, so that chances of a sale is high within 12 clicks. Without your site being easy, attractive, memorable, and clear... you'll just lose that visitor.

Fact No. 5: Importance of 7th Visit

Most buyers complete their purchase-related research within two or more weeks before they buy. During this period–he/she makes, on an average, 7 visits to the web-site. During this 2-weeks/7-visit duration–your web-site must be able to make a lasting impression on the customer. Some of the key success factors during these pre-sale days are

- ☆ Your web-site must be up and running. If the buyer finds your web-site down on crucial purchase-day–the implication needs no explanation.
- ☆ Prompt customer service. Nothing is more disastrous than delayed and incomplete response to pre-sale queries
- ☆ Apart from timely and precise communication, the tone and style of communication should be able to inspire confidence. For example, while paying to any PayPal merchant–PayPal displays the number of people who have paid this merchant before you. Such reference of successful sale significantly impresses a semi-decided mind to go for a deal.

Fact No. 6: Importance of Wednesday and Saturday

All days in a week are same–right ? Obviously wrong, as most of us will not equate a weekly-off day (*e.g.* Sunday) with any other working day.

So is the case with web surfers. Research has shown that Saturdays and Wednesdays are the most popular days of the week to surf the Net. So, these are the two days when you want your email to land in your customers' inbox or its the two days when you want to test market a new sales item. All in all, its the two days you should do whatever it takes to double your daily e-commerce income.

This is not to suggest that you slack off other days of the week but only to highlight the importance of these two days when customers are out there–perhaps in larger number. It's more like prime time TV–when more viewers can be found before the Idiot box.

Website Development Steps for Seed Spices Processing Unit

Every seed spices processing unit has its own unique set of requirements and mission to determine and achieve the goal. The integral part of the story is well understood and taken on behalf of several assortments in real time.

E-commerce development process can be defined and amalgamated with different shopping tools and shopping cart with gateway interfaces. This would allow you to retain in global business and makes your company inevitable on online.

The first interaction will be the gathering of all the information about the seed processing units and the audience whom a client is targeting and project the objectives to determine the requirements of unit. The web site designer ask many questions and also look over existing web site of the Seed Spices Processing Unit, if exist. He also asks for your ideas.

1. Discovery–determining the requirements of seed spices processing unit
2. Formalizing the project
3. Acquiring necessary resources

Database driven ecommerce sites require the following subtle as a part of the project. The account details and all necessary fees are covered by the website owner and even some company provides you the domain name, hosting as per SEO and best keyword research.

- ☆ Web Hosting Account is host must meet database, SSL, and email requirements.
- ☆ SSL Certificate is needed for secure transactions and account pages.
- ☆ Merchant Bank Account is needed for credit card transaction deposits.
- ☆ Payment Gateway is gateway needed for real-time credit card transactions.

Following are the different phases of web site development.

- ✰ Design Phase
- ✰ Development Phase
- ✰ Training
- ✰ Launch
- ✰ Support

Table 47.1: Twenty Most Popular Business Websites as on 1st February, 2009 [Ranked by a combination of Inbound Links, Google Page Rank, Alexa Rank, and U.S. traffic data from Compete and Quantcast (*eBusiness Data*)]

Sl.No.	Name of Website	Inbound Links	Complete Monthly Visitors	Quantcast Monthly Visitors	Alexa Ranking	Page Ranking
1.	Yahoo! Finance	11636557	NA	4500000	NA	8
2.	Forbs.com	4413504	14322102	4900000	444	5
3.	CNNMoney.com	3708493	NA	15200000	NA	8
4.	Wall Street Journal.com	16626219	5809566	3600000	386	7
5.	Bloomberg.com	4582399	3745050	6600000	330	6
6.	MarketWatch.com	6832813	4663916	2000000	463	7
7.	Business.com	1683562	5887646	7400000	1622	7
8.	BusinessWeek.com	4230771	3308732	4700000	900	8
9.	FT.com	5103227	2270543	3400000	909	7
10.	CNBC.com	1837781	2249228	3700000	791	7
11.	MSN Money	3001990	NA	4100000	NA	7
12.	Fool.com	2233128	3146147	4000000	1357	7
13.	BizJournals.com	2499706	2773222	6100000	1838	5
14.	BankRate.com	2015505	4799170	1600000	2191	7
15.	TheStreet.com	1650336	2377914	1900000	2002	7
16.	Hoovers.com	4822552	1458040	1000000	3600	8
17.	Google Finance.com	1563773	NA	458000	NA	8
18.	SeekingAlpha.com	1296909	1322130	679000	2780	7
19.	AllBusiness.com	501773	1359549	1000000	2136	6
20.	Entrepreneur.com	887057	571146	1000000	3574	7

Important Tips for E-commerce Sites of Seed Spices Processing Unit

For any kind of business, we need customers. To get customer, we need a market range and how we can get them to be interested in our products or services. Business use to have the boundaries of location but with internet today, globalization is just around the corner. It is only the matter of taking it or not. The internet has the potential to help you promoting your business to the entire world. Therefore, you need a reliable web hosting company as your business partner. Web hosting companies will be the one to put your website in the international market through the internet.

Web hosting which worth the partnership are those who can provide you with great reliability. This is because they play an important role to your business. The more reliable they are the better chance of your business succeeding. To know about the reliability of a web hosting company, you can search their existing clients to look at their hosting up time, infrastructure and their experience in providing web host service.

Next, you should also pay attention to the amount of storage space and bandwidth that they provide. You don't want to have insufficient space to promote you business. For any medium sized business, a normal amount of storage and bandwidth will be good enough. The thing to look for is companies that allow you to upgrade. You never know when your business is getting notice from the crowd and you want to make sure your web hosting could cope with the heavy traffic when the day comes.

To have a successful business, you must have a web hosting that possesses all those features. For a business website, an e-commerce website design is always an advantage. This is because e-commerce application will allow you to sell products directly from your website. Prices, items and other details are all listed clearly in your website so that people can make decision on the spot to purchase the product. With the e-commerce application like shopping cart and SSL encryption, transactions can be done under a safe and secure system.

Thus, for a business to succeed in the internet, a business hosting with e-commerce features is necessary. The features will not only provide a better user experience, it will also give a more professional image onto your business. For example like HostGator, they provide a toll number to their business hosting partner at no additional cost. Imagine the positive feedback you will get from your potential customers. Isn't it professional and reliable looking? Thus it is important to have a good business hosting provider.

If you are a business owner and you are new to e-commerce Internet solution products, here are some tips that you need to remember when looking for a service provider:

Be Specific About the Features that you Need

For the most part, e-commerce Internet solution service providers offer almost the same set of features including shopping carts, payment processing, website design, content management system and website marketing.

However, there are e-commerce Internet solution providers which offer a more sophisticated set of features so that you can manage your business more efficiently.

Make Sure that the E-commerce Internet Solution Provider Offers a Multitude of Payment Options for your Online Customers

A major factor which will make customers turn away from your website is if your E-commerce Internet solution does not offer the mode of payment that they have. Look for an e-commerce Internet solution provider with a payment system for all the major credit and debit card companies, as well as other popular online payment options.

Security of the E-commerce Internet Solution should be your Number One Priority

When looking for an e-commerce Internet solution service provider, security should be your number one priority. Remember that online customers are often hesitant to provide their personal and financial information online.

From your end, you should make sure that the payment processes conducted on your site are secure and encrypted. Ask about these security features when deciding which e-commerce Internet solution provider it is that you will choose.

To sum it all up–security, affordability, ease of the order-taking process, a wide array of payment options for your customers and good product support for you as a business owner are the features that you should look for from an e-commerce Internet solution provider.

Conclusion

After detailed analyzing and studying of the effects and benefits of incorporating an e-commerce solution to an existing seed spices business, it is clear that an e-commerce solution will benefit the seed spices business in every aspect. The implementation of an e-commerce solution will generate a brand new revenue stream, expand the market exposure, and decrease the operation cost. Thus E-commerce will be a new market channel in seed spices for domestic and export market promotion.

Recommendation

Based on the conclusion of this study, I recommend the following:

- ☆ Every seed spices business unit should implement an e-commerce solution into the business operation.

References

E-Commerce Times, 1999. *Everything You Wanted to.Know About Doing Business Online*. E-Commerce Times, 1999 (9 August 1999).

Web site http://www.all-about-e-commerce.com

Web site http://www.ebizmba.com

Following articles published on the Web site http://ezinearticles.com

- ☆ These Are What an E-Commerce Site Should Possess by Justin Job
- ☆ Five Tips For Building Your Own Website and Shopping Cart by Dan Boudreau
- ☆ Ecommerce Shopping Cart Design by Quinlan Murray
- ☆ All About E-commerce Internet Solutions by Mabelle Sese
- ☆ Ecommerce Website Development Steps by John Enderson

http://www.ebizmba.com/articles/business

Recent Advances in Seed Spices (2011) *Pages* **392–394**
Editors: **Y. Ravindrababu, R.K. Jaiman, K.D. Patel, N.R. Patel and S.B.S. Tikka**
Published by: **DAYA PUBLISHING HOUSE, NEW DELHI**

48

Technology Adoption Pattern of Coriander Production in Humid South Eastern Plain of Rajasthan

Rajkumar*[1] and Pratap Singh[2]
[1]Agricultural Research Station, Ummedganj, P.B. No. 7, GPO Nayapura, Kota – 324 001, Rajasthan
[2]Maharana Pratap University of Agriculture and Technology, Udaipur

ABSTRACT

The present study was conducted in south-eastern zone of Rajasthan during 2006-07, with an object to know the technology adoption pattern of coriander. The productivity of coriander may be boost up by adaption of selected potential technologies, organizing intensive extension activities and to insure the supply of critical inputs at proper time.

Introduction

Coriander (*Coriander sativum* L.) is an important crop of the south east zone of Rajasthan. It covers (2.1 lakh ha.) 97 per cent of the total area under coriander and also stands in first production in the state. Coriander is considered as an important spice of Indian diet. It is rich in Vitamin A (carotene) and is used for making curry powder and medicine. The seeds are used as carminative, refrigerant, diuretic and aphrodisiac. Coriander is valuable cash crop for cultivators, as it brings ready cash and

* Corresponding Author: E-mail: rajkumar_57@hotmail.com.

contributes to the exports earning to the country. Recently Government of Rajasthan in consultation with APEDA announced to establish the Agri. Export Zone. (AEZ) at Kota to promote the export of coriander from this zone. But, from the last two years the crop is facing various challenges and is loosing its attraction amongst. The cultivators and lower productivity is the one of important reasons. Though several modern technologies have been introduced for increasing the production of coriander crop, farmers are reluctant to adopt all the recommended technologies. They adopt them selectively and do not adopt them absolutely. This results in low yield levels. Thus, there is a wide gap between potential and real yield obtained by the farmers. In order to understand the reasons for low yields of coriander study was conducted in heavy textured, non-calcareous soil micro farming situation in humid south eastern plain of Rajasthan.

Methodology

The study is confined to south-east zone of Rajasthan. Administratively it covers four district of the state namely; Kota, Bundi, Baran and Jhalawar. A survey of 100 coriander growers from 10 villages of zone was conducted to determine the extent of adoption of recommended technology by coriander growers of the study area multi stage stratified raindom sampling was adopted for collecting data. The farmers having land holding with in 2 ha. placed under small category, those having 2-4 ha. under medium category and the rest having lald holding above 4 ha. under large category. The primary data related to coriander crop of the crop calendar year 2006-07 were collected by survey method with the help of pre tested schedules to find out the extent of adoption rate of improved technology for coriander. The data were subjected to tabular analysis to examine the adoption of production technology of coriander.

Table 48.1: Extent of Adoption of Improved Technology in Coriander Cultivation

Sl.No.	*Major Practices*	*Recommended (Per ha.)*	*Adopted (Per ha)*	*Percentage of Adoption*
1.	Variety	RCr-41, 436, 684	RCr-41, 436, 684	62
2.	Seed rate (kg/ha)	15-20	20-25	35
	(i) Certified seed	–	–	35
	(ii) Home produce (own seed)	–	–	65
3.	Fertilizer (kg/ha)			
	N	60	14.2	25
	P	30	28.2	40.5
	K	20	–	–
4.	No. of irrigation	4	2	50
5.	Plant protection measures	100	72	–
6.	Row to row distance	30	22.5	25

Results

The results of the study revealed that 62 per cent of the respondents adopted recommended varieties of coriander and the most preferred variety were RCr 41, 436, 684 majority of the farmers (65 per cent) are using more than recommended seed ratė. Only 35 per cent farmers are using certified seed due to non availability of certified seed in the area at proper time. Only 25 per cent respondents followed row to row distance 30 cm because sowing of the crop by nine tine seed drills is a common practice in the area. Maximum coriander growers are using less then recommended dose of fertilizer resulting poor vegetative growth, 72 per cent farmers are using recommended dose of plant protection chemicals after appearance of insect/pest and disease majority of farmers are not using any weedicides in their fields.

Conclusion

From the study it is inferred that, yield of coriander may be increased by adoption of selected potential technologies any by organizing intensive extension activities, declaration of support price, availability of certified seed, use of limited irrigation water at critical stage and developing storage facilities in the zone.

References

Executive Summary, 1999. Economics of Pulses production and identification of constraints in raising their production in Bihar. *Agricultural Situation in India*, 56(2): 75–77.

Pant, D.C., Singh, C.P. and Singh, H., 2000. Adoption of improved cultivation practices of pigonpea in tribal area of Rajasthan. *Indian Journal of Agricultural Economics*, 55 (3): 538.

Vasantkumar, J. and Selvi, G. Tamil, 1999. Yield gap in paddy and extent of adoption of improved paddy practices. *Agricultural Extension Review*, p. 16–21.

Recent Advances in Seed Spices (2011) *Pages* **395–399**
Editors: **Y. Ravindrababu, R.K. Jaiman, K.D. Patel, N.R. Patel and S.B.S. Tikka**
Published by: **DAYA PUBLISHING HOUSE, NEW DELHI**

49

Adoption of Kharif Fennel Cultivation Practices by Fennel Growers of Banaskantha District of Gujarat State

S.V. Kaid[1], M.R. Prajapati[2] and R.R. Prajapati[1]
[1]Agricultural Officer, Directorate of Extension Education, S.D. Agricultural University, Sardarkrushinagar, Gujarat
[2]Professor and Head, Department of Extension Education, C.P. College of Agriculture, S.D. Agricultural University, Sardarkrushinagar, Gujarat

Introduction

Fennel is an important spice crop of Palanpur taluka of Banaskantha district of Gujarat state. Palanpur taluka has been considered as productively potential region of fennel crop due to assured irrigation facilities and favourable soil and climatic conditions. However, there is still a wide gap between the production potential and the actual production realized by the fennel growers. This may be due to partial adoption of recommended package of practices by the fennel growers. So far, no systematic efforts were made to study the adoption of recommended cultivation practices of kharif fennel cultivation, the present study was therefore undertaken to find out the adoption of recommended *kharif* fennel production technology constrains experienced by them in adoption of cultivation practices.

Objectives of the Study

The present study was conceived with a general objective of adoption of recommended fennel cultivation technology by the fennel growers.

The specific objectives of the study were:

1. To ascertain the knowledge level of the fennel growers pertaining to recommended fennel production technology.
2. To know the extent of adoption of recommended fennel production technology by the fennel growers.
3. To study the constraints experienced by the farmers in adopting recommended fennel production technology.

Methodology

The present study was undertaken in Palanpur taluka of Banaskantha district of Gujarat state. There are 40 villages in the taluka where fennel crop is being grown by majority farmers. Ten villages were randomly selected for the present study. From each of the selected village, twelve fennel growers were randomly selected, making a total random sample of 120 fennel growers.

Knowledge level of the respondents about fennel production technology was determined by using the teacher made test based on the scale developed by Jha and Singh (1970). Extent of adoption of recommended fennel production technology was measured by using adoption quotient the formula developed by Sengupta (1967).

A simple ranking technique was applied to measure the constraints in adoption of fennel production technology. The data were collected with the help of structured interview schedule. The statistical tools used to analyze the data were percentage, mean score, standard deviation and coefficient of correlation.

Knowledge Level of the Fennel Growers Regarding Recommended Fennel Production Technology

Knowledge is the cognitive behaviour of an individual. Knowledge plays an important role in covert as well as overt behaviour of an individual. Knowledge was measured with the help of teacher made test developed for the purpose. As discussed earlier, the knowledge index for each farmer was calculated and the respondents were grouped into three categories *viz.*, low, medium and high. Data regarding this aspect are presented in Table 49.1.

Table 49.1: Distribution of the Respondents According to their Level of Knowledge Regarding Fennel Production Technology

Level of Knowledge	*Frequency*	*Percentage*
Low (0 to 33.33 index)	008	006.66
Medium (33.34 to 66.66 index)	110	091.67
High (Above 66.67 to 100.00 index)	002	001.67
Total	120	100.00

It is evident from the Table 49.1 that majority (91.67 per cent) of the respondents had medium level of knowledge, whereas, 6.66 per cent and 1.67 per cent of the respondents had low and high level of knowledge about fennel production technology.

From this finding, it can be concluded that majority of the respondents had medium level of knowledge. The probable reason might be that most of the respondents were educated so they could easily understand and acquire skills about fennel production technology.

This finding is in conformity with findings of Singh (1998) and Sai (2002).

Extent of Adoption of Recommended Fennel Production Technology by the Fennel Growers

The "adoption process" is the mental process through which an individual passes from first hearing about an innovation to its final adoption. An 'adoption' is a decision to continue full use of an innovation. With a view to find out the extent of adoption of *kharif* fennel production technology, the farmers were asked to give information about package of practices adopted by them. Data regarding this aspect are presented in Table 49.2.

Table 49.2: Distribution of the Respondents According to Extent of Adoption of Recommended Fennel Production Technology

Extent of Adoption	*Frequency*	*Percentage*
Low (0 to 33.33 index)	22	18.33
Medium (33.34 to 66.66 index)	95	79.17
High (Above 66.67 to 100.00 index)	03	02.50
Total	120	100.00

The data presented in Table 49.2 reveal that majority (79.17 per cent) of the respondents were in medium level of adoption followed by only 18.33 per cent in low and only 2.50 per cent respondents were observed having high level of adoption.

The probable reason might be their medium to low level of knowledge.

These findings are in line with the findings of Patel (2001).

Practice-wise Adoption of Kharif Fennel by the Farmers

It could be seen from Table 49.3 that the adoption of for different recommended practices was varied. The average adoption among the fennel growers was 63.20 per cent.

Table 49.3: Practice-wise Adoption of Kharif Fennel Practices by the Fennel Growers (N = 120)

Recommended Practices	*Adoption (per cent)*	*Rank*
Recommended variety	39.17	III
Nursery raising	31.35	II
Transplanting	75.69	VII
Farm yard manure	90.83	IX
Chemical fertilizer	50.42	V
Irrigation schedule	40.94	IV
Gap filling	97.50	XI
Interculturing and earthing up	93.33	X
Hand weeding	85.83	VIII
Plant protection measures	23.97	I
Harvesting	66.11	VI
Average	63.20	

It could be further inferred that about one fifth of the farmers had adopted plant protection measures and about one third of the farmers had adopted nursery raising.

While majority of the farmers had adopted transplanting, hand weeding, farm yard manure, interculturing, earthing up and gap filling.

The overall adoption of all the listed 11 recommendation practices of fennel crop together by the fennel growers was 63.20 per cent

These findings are supported by the earlier findings of Prajapati (1995).

Constraints Experience by the Fennel Growers in Adoption of Recommended Fennel Production Technology

The respondents were requested to express the constraints faced by them in adopting recommended fennel production technology. According to the number of respondents rank was given and data in this regard are presented in Table 49.4.

Table 49.4: Constraints Experience by Fennel Growers in Adoption of Recommended Fennel Production Technology (N = 120)

Constraints	*No. of Respondents*	*Per cent*	*Rank*
Lack of technical guidance	59	49.17	I
More problem of disease and insect	58	48.33	II
Non-getting remunerative price of the produce	57	47.50	III
Long duration crop	52	43.33	IV
High charge and irregular supply of electricity	47	39.17	V
Unavailability of certified seeds	23	19.17	VI

As seen from Table 49.4 that major constraints faced by the farmers were "Lack of Technical Guidance" (49.17 per cent). More problems of disease and insects (48.33 per cent), non-getting remunerative price of the produce (47.50 per cent), Long duration crop (43.33 per cent), high charge and irregular supply of electricity (39.7 per cent).

Conclusion

Majority of the fennel growers had medium level of knowledge regarding the recommended fennel production technology. Majority of the fennel growers were found to have medium extent of adoption. About one fifth of the fennel growers had adopted plant protection measures and about one third of the fennel growers had adopted nursery raising. The overall adoption for 11 different practices together by the fennel growers was 63.20 per cent. Major constraints perceived by the fennel growers were lack of technical guidance, more problem of disease and insect, non-getting remunerative price of the produce and long duration crop.

References

Jha, P.N. and Singh, K.N., 1970. A test to measure farmers' knowledge about high yielding varieties programme. *Interdiscipline*, 7(1): 67–77.

Patel, J.N., 2001. Stated on knowledge and adoption of vegetable (Tomato and Brinjal) production technology among the vegetable growers of Navsari district. *M.Sc. (Agri.) Thesis*, Gujarat Agricultural University, Navsari (Unpublished).

Prajapati, R., 1995. A study on technological gap in *kharif* fennel cultivations in Vijapur taluka of Mehsana district of Gujarat state. *M.Sc. (Agri.) Thesis*, Gujarat Agricultural University, Anand (Unpublished).

Sai, K.A., 2002. A study on gain in knowledge and attitude of trained women towards fruits and vegetable preservation training of Anand district. *M.Sc. (Agri.) Thesis*, Gujarat Agricultural University, Anand (Unpublished).

Sengupta, T., 1967. A simple adoption scale for selection of farmers for higher yielding varieties programme on rice. *Indian J. Ext. Edn.*, 3(3): 107–115.

Singh, A.K., 1998. Attitude towards vocational bakery training and gain in knowledge about bakery items with reference to ex-trainers of school of bakery, Anand of Gujarat state. *M.Sc. (Agri.) Thesis*, Gujarat Agricultural University, Anand (Unpublished).

Recent Advances in Seed Spices (2011) *Pages* **400–404**
Editors: **Y. Ravindrababu, R.K. Jaiman, K.D. Patel, N.R. Patel and S.B.S. Tikka**
Published by: **DAYA PUBLISHING HOUSE, NEW DELHI**

50

Cyber Extension: A Useful ICT Concept for Transfer of Technologies in Seed Spices Crops

A.P. Prajapati, Y. Ravindrababu and D.B. Patel
Centre for Research on Seed Spices, S.D. Agricultural University, Jagudan – 382 710, Dist. Mehsana, Gujarat

Introduction

During last decade, the tremendous progress in information technology (IT) is affecting all spheres of our life. For seed spices, this progress in IT provides new opportunities to improve the speed and performance of communication among scientists, extension officers/workers, seed spices growers. The research challenge here is to build cost-effective IT based systems to improve the living standards of seed spices growers by providing them information about modern technologies developed by the scientists for various seed spices crops. ICTs are proving their value in helping to deliver information to and from intermediary information providers such as SAUs, government departments, NGOs etc. Some of the most successful ICT for development projects are focused on supporting the work of intermediaries who are relaying information to and from farmers and others at the grassroots level who do not themselves have access to the technology. Looking to this aspect, Government of Gujarat dedicate his ambitious 'e-Gram Vishwagram Project' by offering broadband connectivity to all 13,693 Gram Panchayats. This development will offer strong potential to improve our existing extension system.

India and Connectivity

- ☆ 107 Million Telephones

- 15 Million PCs (Proposed 75 Million by 2010)
- 5 Million Internet Connections (45 M. by 2010)
- Connectivity is highly skewed
- Internet connectivity more skewed
- Technology for Rural Connectivity available
- Efforts required to promote Rural Connectivity
- 5.45 million Total Broadband subscribers by the end of December 2008
- 74.12 per cent annual growth rate of Broadband subscribers during the year 2008.

[Press (Press Release No 11/2009) of Telecom Regulatory Authority of India (TRAI)]

Extension Communication: The Most Critical Gap

- Lack of feedback mechanism in the Research Extension Linkages identified as one of the weaknesses in existing Extension Systems (by many studies, evaluation reports)
- Communication is the most critical element in Research-Extension-Farmer-Marketing loop
- Media and ICTs offer strong potential to improve this mechanism.

Defining Cyber Extension

Cyber

According to Oxford Dictionary the word Cyber means, "relating to Information Technology, the Internet, and virtual reality, the Cyber Space"

The Cyber Space

Cyber Space is the imaginary or Virtual space of computers connected with each other on Networks, across the globe. These computers can access information in form of Text, Graphics, audio, video and animation files. Software tools on networks provide facilities to interactively access the information from connected servers. The cyber space thus can be defined as the imaginary space behind the interconnected telecommunications and computer networks, the virtual world.

Extension

Extension stands for "the action or process of enlarging or extending something". It could be extension of area, time or space.

Agricultural Extension

Agricultural Extension, according to Dr. D. Benor, " relates to the process of carrying the technology of scientific agriculture to the farmer in order to enable him to utilize the knowledge and a better economy. Agriculture extension service seeks to impart the necessary skills to the farmers for undertaking improved agricultural operations, to make available to them timely information improved practices in an easily understandable form suited to their level of literacy and awareness, and to create in them a favourable attitude for innovation and change" (Benor *et al.*, 1984). The Extension is an on-going process of getting useful information to people (the communication dimension) and then in assisting those people to acquire the necessary knowledge, skills and attitudes to utilize effectively this information or technology (the educational dimension). Thus Extension is central mechanism in the agricultural development process, both in terms of technology transfer and human resources development (Samanta, 1993).

Cyber Extension

Cyber Extension thus can be defined as the "Extension over Cyber Space". As the word Extension is subject-neutral, so is Cyber Extension. But in the applied context of Agriculture, Cyber Extension means "using the power of online networks, computer communications and digital interactive multimedia to facilitate dissemination of agricultural technology". Cyber Extension includes effective use of Information and Communication technology, national and international information Networks, Internet, Expert Systems, Multimedia Learning Systems and Computer based training systems to improve information access to the Farmers, Extension Workers, Research Scientists and Extension Managers.

Cyber extension is an agricultural information exchange mechanism over cyber space, the imaginary space behind the interconnected computer networks through telecommunication means. It utilizes the power of networks, computer communications and interactive multimedia to facilitate information sharing mechanism (Wijekoon, 2003).

Elements of Cyber Extension

- ✰ Use of CDROMs as Information tools in seed spices crops
- ✰ Use of Interactive Multimedia to Develop low cost Audio Visual Aids
- ✰ E-Mail;
- ✰ Web-based Extension Information Services;
- ✰ Interactive Expert Systems on Pests and diseases;
- ✰ INTERNET browsing for Extension Information;
- ✰ Video Conferencing- Static, Mobile, village level.

Now, we discuss that how can cyber extension be useful to the seed spices growers and processors.

Use of Audio-Video CDs

The Audio-Video CDs of various modern production technologies in seed spices should be developed under the guideline of concerned senior scientists and with the assistance of ICT Experts (Multimedia Designers). Objective of using these CDs as ICT strategy was to provide relevant agricultural content to the seed spices growers in an aesthetically pleasing and entertaining environment. Extension workers can use these CDs as a teaching tool (audio visual aid). Farmers also can use the CDs as a self-learning package. Farmers trained on use of information technologies in agriculture, will become 'e-farmer'. The extension workers will become 'e-extensionist'. In addition to technical information on each crop, organized under several chapters a separate chapter was devoted for compiling database on research papers and articles published by local authors in local and foreign journals. Each CD has a separate link for video films pertaining to each seed spices crop, which were produced and telecast through *TV Programme*.

Use of Interactive Multimedia to Develop low cost Audio Visual Aids

Extension and training are inevitable in transferring technical knowhow to seed spices growers and most of these trainings are not supported with better quality audio visual media (Wijekoon *et al.*, 1999a). Therefore a CDROM to develop presentation skills (especially flipchart skills) of Agriculture

Instructors was developed and given to all Cyber units of seed spices. As a consequence, extension workers were able to produce low cost audio visual aids at their door step with locally available material (Hi-tech to improve low cost instructional media).

Use of E-book/E-materials

An E-book is a book in electronic format. It can be stored in a CDs or downloaded to a computer, PC, Mac, laptop, PDA or any other kind of computer, and is read on the screen. It can have numbered pages, table of contents, pictures and graphics, exactly like a printed book. Page turn is facilitated by two icons; 'Next' (to go next page) and "Previous' (to go previous page). The contents were organized into chapters, topics and sub-topics. Most of the pages contain variety of multimedia presentations (video, sound/voiceovers, animations, graphics and text). All media are interactive and users may review and/or skip section, as they desire. Each page is printable and stakeholders will be able to get a print. Therefore, E-book/E-materials should be prepared by the experts of various fields in each seed spices crop.

New Dimension for Web Based (Internet)

The SDAU has setup the web of www.sdau.edu.in to serve farmers and processors as well as the general public. A link of Centre for Research on Seed Spices, SDAU, Jagudan is also available at this web site for seed spices growers and processors. The website, presents most of the seed spices crops related information as well as statistics, news, recently published books etc. The Cyber extension project has established a mechanism to distribute web to *Cyber units*. Furthermore, seed spices growers and other stakeholders can download the useful information and solve their problems by using various web sites such as Soil Health Card, Govt. Departments, NGOs, Panchayat Institutes etc. The problem of updating the websites is very serious and it is very essential to update the information regularly on related web sites. Therefore, there should be a powerful Content Management System for each web site.

Develop Digital Training Material (Audio Visual Aids) for Local Extension and Training

The most of the seed spices growers and processor's problems, reported in the recent past were location specific. In certain circumstances other elements of Cyber Extension produced by the national centre or SAUs would not give answers for their location specific problems. Therefore, extensionists were able to produce their own 'Power Point' presentations and desktop publications by using the facility of Cyber extension unit for their extension training at local level. They were also asked to compile a visual database of local problems in each season and there by researchers would be able to investigate the history of local problems.

Special Email Service for Seed Spices Growers and Processors

In present days, Electronic Mail is very efficient, fast and easy way for communication among the stakeholders. Therefore, there is a need to start a Special Email Service for seed spices growers and processors. Through this service, seed spices growers and processors can send their queries for the solution of their burning problem and the immediate response should be given by the experts of particular fields in seed spices crops periodically.

Video Conferencing Service Seed Spices Growers and Processors

Consequently personal video teleconference systems based on a webcam, personal computer system, software compression and broadband Internet connectivity have become affordable for the farmers through SDAU's Networks and Government of Gujarat's Wide Area Networks and Broadband connections available in homes and workplaces of progressive farmers and Agriculture Businessmen.

The video conferencing facilities will also be made available on newly established local area network of Sardarkrushinagar campus of SDAU by installing core equipments for video conferencing. It will provide seamless interaction among officers, scientists and farmers for their problems and conferencing to the residential farmers who are connected with GSWAN. The financial outlay of this project is Rs.217.00 lakh. The Seed Spices Growers and processors can take advantage of video conferencing service to be started by SDAU (SDAU News Volume-5 No. 3-4 July – December, 2008, Page-3).

Scope of Cyber Extension in Gujarat and SDAU

Government of Gujarat dedicate his ambitious 'e-Gram Vishwagram Project' by offering V-sat connectivity to all 13,693 Gram Panchayats of Gujarat. In Gujarat, All Panchayati Raj Institutes (PRI), District Agriculture Departments, Agriculture Research Stations etc. have Radio Link connectivity of GSWAN. The cost of PCs and High Speed Internet connectivity is decreased dramatically and therefore PCs and High speed Internet connectivity has become more widely available at a reasonable cost in Rural Area of Gujarat. Now, BSNL also provide Broadband connection in many rural areas of Gujarat. Wireless Internet Connectivity is also offered by various private cell phone networks. Progressive farmers and agricultural businessmen have their own PC/Lap top and internet connectivity at home as well as at work place. Now a days, the facility of basic Computer and IT Education is also available in Rural Area of Gujarat. Therefore, the trained manpower needed for the implementation of Cyber Extension will be available in all rural areas of Gujarat. This development will offer strong potential to improve our existing extension system and implement cyber extension system in Gujarat

Strong and Effective IT Infrastructure, Local Area Network and Wide Area Network were already established in SDAU. Sardarkrushinagar, the head quarter of SDAU facilitate with BSNL Leased line Copper link from Telephone Exchange, Dantiwada (2.0 MBPS), GSWAN Radio link from Mamlatdar Office, Dantiwada (0.5 MBPS) and ERNET VSAT Link from ERNET, New Delhi (0.5 MBPS). All sub-canters of SDAU facilitate with GSWAN Radio link and/or ERNET VSAT Link. In current financial year, computer and IT training was given to most of the scientists and other technical staff of SDAU under CCC/CCC+ training programme. Looking to this IT infrastructure and trained human resources, it is proper time to implement newly cyber extension system in SDAU for transfer of technologies.

References

Cyber-extension: A new era in agricultural extension in India. *MANAGE Bulletin*, April-May, 1999 Issue.

Cyber extension: Innovative uses of ICTs for agriculture lessons from case studies: A presentation by Dr. V.P. Sharma, Director (Information Technology, Documentation and Publications), National Institute of Agricultural Extension Management, Hyderabad.

Wijekoon, R.R.A. and Emitiyagoda, S., 2006. Rice knowledge bank for cyber extension in Sri Lanka. In: *Paper Presented to Workshop on Linking Extension and Research Needs through Information Technology*, Bankok, p. 6-8.

Recent Advances in Seed Spices (2011) *Pages* **405–420**
Editors: **Y. Ravindrababu, R.K. Jaiman, K.D. Patel, N.R. Patel and S.B.S. Tikka**
Published by: **DAYA PUBLISHING HOUSE, NEW DELHI**

51

Evaluation of Different Transfer of Technology of KVKS by Using the Developed Instrument

A.S. Sheikh[1] and S.S. Shekhawat[2]
[1]Assistant Professor and Head of the Department
Department of Veterinary and Animal Husbandry Extension,
S.D.Agricultural University, Sardarkrushinagar – 385 506, Gujarat
[2]Assistant Professor, C.P. College of Agriculture, S.D. Agricultural University,
Sardarkrushinagar, Gujarat

ABSTRACT

The present study was conducted on three KVKs which were working under different host institutions in Rajasthan *viz.*, KVK Udaipur under NGO, KVK, Jodhpur under ICAR and KVK, Banswara under SAU. These KVKs were selected purposively. Total 45 contact farmers were selected by randomly (15 from each KVK). The effectiveness of KVKs in regard to transfer of technology was measured by developed scientific instrument. This instrument consisted of five major activities *viz.*, Agronomical, Horticultural, Animal Husbandry and allied field, Agricultural Engineering and Home Science. There were three major techniques in KVKs *viz.*, training, demonstration and advisory services for transferring knowledge and skill to the farmers. The important summarized findings of training on transfer of technology of five major activities. On agronomical aspects maximum training was organized on "Methods of using insecticides and pesticides" (KVK Udaipur), "Seed treatment" (KVK Jodhpur and Banswara). Similarly on Horticultural aspects maximum training was organized on "Nursery raising and transplanting" (KVK Udaipur and Jodhpur), "Vegetable cultivation" (KVK, Banswara). On Animal Husbandry and allied field maximum training was organized on "Increasing milk production" (KVK Udaipur and Jodhpur), "Cross breeding" (KVK, Banswara).All KVKs organized training on "Smokeless Chula" in Agricultural Engineering and in Home Science maximum training was organized on "Tailoring" (KVK Udaipur) and "Cleanness of Home" (KVK, Jodhpur and Banswara).

Overall results regarding transfer of technology bring to the fore that trainings were conducted frequently on 'Agronomy' followed by 'Home Science,' 'Horticulture,' 'Animal Husbandry and allied field' and 'Agricultural Engineering' in descending order.

Keywords: Krishi Vigyan Kendra, Transfer of technology.

Introduction

The KVK is being meticulously developed as a practical centre for rural training. The major purpose of evaluation is to assist in programmed decision. Formal evaluation is worth doing only, if they have a chance of affecting such decision.

An additional consideration is the significance of the need. A major extension effort involving large numbers of farmers and substantial resources will be (with other things being equal) more significant than smaller programmed centered around a new farmer. Significance can also be gauged by looking at the issues with which the evaluation is concerned and assessing the likely impact of the evaluation on the resolution of these issues. Thus, looking to huge budgetary expenditure and the deployment of large numbers of agriculture extension staff in the KVKs, it is essential that the impact and extension approach must be critically evaluated time to time and discrepancies be removed.Productivity and accountability of extension can be greatly advanced through evaluation of it programmed. The effectiveness of programmes of KVKs should be evaluated by standardized scale. In view of these facts, the present study was conducted with the objective of evaluation of different transfer of technology of KVKs of Rajasthan by using the developed instrument. Thus, it is needless to high- light the importance of KVK in quick and regular transfer of technology among the farmers.

Research Methodology

The present study was conducted on three KVKs, which were working under different host institution in Rajasthan *viz.*, KVK, Udaipur under NGOs, KVK, Jodhpur under ICAR institutes (CAZRI) and KVK, Banswara under Rajasthan Agricultural University. These KVKs were established during the year 1983-1984. These KVKs selected purposively because they together yield as nearly as possible the same yield or proportion as the totality with respect to those characteristics, which are already a matter of statistical knowledge. Thus, for the study purpose Western and Southern region of Rajasthan were under taken. Keeping in view the high concentration of extension work performed by selected KVKs in this zone.

Checklist of all the activities which could be carried under the programmes was prepared with the help of literature and discussion held with the experts. A final scale was developed for measuring the effectiveness of KVK on scientific line for this purpose. Fifteen contact farmers from each KVK were selected randomly. Thus, a sample of 45 contact farmers was drawn for the study purpose. An interview schedule was constructed for study purpose keeping in view the objective of the study. The scale was pre tested before administrating into actual respondents. Personal interview method was adopted by the investigator to record the responses. Data so collected were tabulated and analyzed in light of the objective.The mean score of each item was calculated by multiplying the responses to the weight given to each items/statement and then by dividing it with the total number of respondents. Interferences were drawn after subjecting the data in statistical analysis.

Results and Discussion

Instrument consisted of five major activities *viz.*, Agronomical, Horticultural, Animal husbandry and allied fields, Agricultural Engineering and Home Science. There were three major techniques in KVKs *viz.*, Training, Demonstration and Advisory services for transferring knowledge and skill to the farmers.

The important findings of trainings on five major heads have been given in Table 51.1.

Organizing training is the main objective of the KVK. A KVK can be said effective if it is organising trainings frequently on different aspects of related to agriculture. Table 51.1 expresses the frequency of trainings organized by different KVKs as per views of the respondents.

KVK, Udaipur

Agronomical Activities

Looking to the mean score in Table 51.1 revealed that in KVK, Udaipur amongst "Agronomical activities" – the training were conducted frequently in 'Methods of using insecticide and pesticide' (2.67) as well as 'Irrigation and drainage' (2.67) and both the activities were ranked combined at the top followed by 'Method of fertilizer application' (2.47), 'Preparation of FYM and Compost' (2.40) and so on.

However, the training was not conducted at all on 'Graded bunding,' 'Development of shelterbelts' and 'Soil survey and mapping' activities as per the opinion of all the respondents and as such these were ranked at the bottom.

Horticultural Activities

Training were conducted more frequently in 'Nursery raising and transplanting' (2.67) as well as 'Planning and layout of garden' (2.67) and as such these were combined ranked first place. 'Vegetable cultivation' (2.20) as well as 'Kitchen gardening' (2.20) were considered equally important and were ranked second.

'Pruning and Training' (1.87), 'Fruit production technology of Aonla' (1.87), 'Farm forestry' (1.87) and 'Plantation of fodder tree' (1.87). These four activities were ranked combined at the third place followed by 'Fruit production technology of lemon' (1.80); 'Fruit, vegetable and flower production' (1.67) and so on.

However, training were not held on 'Seed production technology,' 'Fruit and vegetable preservation,' 'Plantation of new tree,' 'Beekeeping' activities as per the opinion of the respondents and as such these were ranked at the bottom.

Animal Husbandry and Allied Field Activities

'Increasing milk production' (3.0) and 'Hygienic milking production' (3.0) were ranked at the top places followed by 'Preparation of feed mixture', 'Fodder crop and varieties recommended for *rabi* and *kharif*' and 'Sheep and goat rearing.'

However, the trainings were not conducted on 'Artificial insemination',' Rabbit rearing,' 'Piggery development,' 'Sericulture,' 'Fisheries' and 'Mushroom cultivation' and as such these were ranked at the bottom.

Table 51.1: Frequency of Training Organized by KVKs as Perceived by Farmers

Sl.No.	Activity/Item	Number of Farmers Under Each Response Category																	
		KVK, Udaipur						KVK, Jodhpur						KVK, Banswara					
		Training Once in a						Training Once in a						Training Once in a					
		Month	Six Month	Year	Never	M.S.	Rank	Month	Six Month	Year	Never	M.S.	Rank	Month	Six Month	Year	Never	M.S.	Rank
[I]	**AGRONOMY**																		
	(a) Soil																		
1.	Soil testing	–	–	10	5	0.67	21.0	–	1	10	4	0.80	10.5	–	–	–	15	0.00	20.5
2.	Soil reclamation	–	–	–	15	0.00	23.5	–	4	2	9	0.67	13.0	–	–	–	15	0.00	20.5
	(b) Soil conservation																		
3.	Land conservation	–	11	4	–	1.73	14.0	–	9	3	3	1.40	4.0	–	–	–	15	0.00	20.5
4.	Land leveling	–	5	10	–	1.33	19.0	–	–	4	11	0.27	19.0	–	–	–	15	0.00	20.5
5.	Graded bunding	–	–	–	15	0.00	23.5	–	–	–	15	0.00	23.0	–	–	–	15	0.00	20.5
6.	Development of shelter belts	–	–	–	15	0.00	23.5	–	–	9	6	0.60	14.5	–	–	–	15	0.00	20.5
7.	Bench terracing	–	2	13	–	1.13	20.0	–	–	–	15	0.00	23.0	–	–	–	15	0.00	20.5
8.	Soil survey and mapping	–	–	–	15	0.00	23.5	–	–	–	15	0.00	23.0	–	2	13	–	1.13	15.0
	(c) Seed																		
9.	Seed treatment	4	11	–	–	2.27	6.0	–	14	–	1	1.87	1.0	–	15	–	–	2.00	9.5
10.	Use of seed culture	2	13	–	–	2.13	7.0	–	10	–	5	1.33	1.5	–	15	–	–	2.00	9.5
11.	Storage of imp. Seed	2	10	3	–	1.93	13.0	–	10	–	5	1.33	1.5	–	13	2	–	1.87	13.5
	(d) Fertilizer																		
12.	Method of fertilizer application	7	8	–	–	2.47	3.0	–	4	11	–	1.27	7.0	5	10	–	–	2.33	4.5
13.	Preparation of FYM and compost	6	9	–	–	2.40	4.5	–	–	6	9	0.40	17.5	3	12	–	–	2.20	6.0
14.	Green manuring	3	12	–	–	2.20	10.0	–	–	–	15	0.00	23.0	–	–	–	15	0.00	20.5
15.	Soil fertility management	3	12	–	–	2.20	10.0	–	3	5	7	0.73	12.0	–	15	–	–	2.00	9.5

Contd...

Table 51.1–Contd...

Sl.No.	Activity/Item	Number of Farmers Under Each Response Category																	
		KVK, Udaipur						KVK, Jodhpur						KVK, Banswara					
		Training Once in a						Training Once in a						Training Once in a					
		Month	Six Month	Year	Never	M.S.	Rank	Month	Six Month	Year	Never	M.S.	Rank	Month	Six Month	Year	Never	M.S.	Rank
	(e) Crop production																		
16.	Dryland farming	–	12	–	3	1.60	17.3	–	–	13	2	0.87	9.0	–	15	–	–	2.00	9.5
17.	Improved practices for irrigated areas	3	12	–	–	2.20	10.0	–	–	3	12	0.20	20.0	8	7	–	–	2.53	2.0
18.	Improved practices of un-irrigated area	3	12	–	–	2.20	10.0	–	–	15	–	1.00	8.0	–	15	–	–	2.00	9.5
	(f) Plant protection																		
19.	Control of rats	–	12	–	3	1.60	17..3	–	3	6	6	0.80	10.5	–	13	2	–	1.87	13.5
20.	Handling of plant protection equipment	6	9	–	–	2.40	4.5	–	11	4	–	1.73	2.0	10	3	–	2	2.40	3.0
21.	Control of whitegrub	3	12	–	–	2.20	10.0	–	2	3	10	0.47	16.0	–	–	–	15	0.00	20.5
22.	Methods of using insecticide and pesticide	10	5	–	–	2.67	1.5	–	10	5	–	1.67	3.0	10	5	–	–	2.67	1.0
23.	Control of katra	–	12	–	3	1.60	17.3	–	–	–	15	0.00	23.0	–	–	–	15	0.00	20.5
	(g) Water																		
24.	Water conservation	2	8	3	2	1.67	15.0	–	–	9	6	0.60	14. 5	–	15	–	–	2.00	9.5
25.	Irrigation and drainage	10	5	–	–	2.67	1.5	–	–	6	9	0.40	17. 5	5	10	–	–	2.33	4.5
	Average (I)	**3**	**8**	**2**	**3**	**1.70**		**0**	**3**	**5**	**7**	**0.70**		**2**	**7**	**1**	**6**	**1.33**	
[II]	**HORTICULTURE**																		
	(a) Fruit, vegetable and flower production																		
1.	Fruit production technology of beer	–	10	5	–	1.67	10.0	12	–	3	–	2.60	3.0	–	–	15	–	1.00	10.0
2.	Vegetable cultivation	–	15	–	–	2.20	3.5	–	10	5	–	1.67	6.5	15	–	–	–	3.00	1.5
3.	Pruning and training	–	13	2	–	1.87	6.5	–	–	15	–	1.00	12.0	–	10	5	–	1.67	5.0

Contd...

Table 51.1–Contd...

Sl.No.	Activity/Item	Number of Farmers Under Each Response Category																	
		KVK, Udaipur						KVK, Jodhpur						KVK, Banswara					
		Training Once in a						Training Once in a						Training Once in a					
		Month	Six Month	Year	Never	M.S.	Rank	Month	Six Month	Year	Never	M.S.	Rank	Month	Six Month	Year	Never	M.S.	Rank
4.	Plant propagation by cutting, budding, grafting etc.	–	–	15	–	1.00	13.5	–	12	3	–	1.80	5.0	–	10	5	–	1.67	5.0
5.	Fruit production technology of pomegranate	–	–	13	2	0.87	15.0	11	4	–	–	2.73	2.0	–	–	10	5	0.67	13.0
6.	Nursery raising and transplanting	10	5	–	–	2.67	1.5	14	1	–	–	2.93	1.0	10	5	–	–	2.67	3.0
7.	Planning and layout of garden	10	5	–	–	2.67	1.5	–	–	15	–	1.00	12.0	–	–	8	7	0.53	14.0
8.	Fruit production technology of Aolna	–	13	2	–	1.87	6.5	4	–	11	–	1.53	8.0	–	–	5	10	0.33	15.0
9.	Kitchen gardening	3	12	–	–	2.20	3.5	10	–	2	3	2.13	4.0	15	–	–	–	3.00	1.5
10.	Fruit production technology of lemon	–	12	3	–	1.80	9.0	–	–	–	15	0.00	17.5	–	10	5	–	1.67	5.0
11.	Seed production technology	–	–	–	15	0.00	17.5	–	–	–	15	0.00	17.5	–	–	10	5	1.00	10.0
12.	Flower cultivation	–	3	12	–	1.20	12.0	–	–	–	15	0.00	17.5	–	5	10	–	1.33	7.0
	(b) Preservation and processing of fruit and vegetable																		
13.	Fruit and vegetable preservation	–	–	–	15	0.00	17.5	–	–	15	–	1.00	12.0	–	–	–	15	0.00	17.5
14.	Food preservation	–	–	15	–	1.00	13.5	–	10	5	–	1.67	6.5	–	–	–	15	0.00	17.5
	(c) Forestry																		
15.	Farm forestry	–	13	2	–	1.87	6.5	–	–	15	–	1.00	12.0	–	–	15	–	1.00	10.0
16.	Plantation of new tree	–	–	–	15	0.00	17.5	–	–	15	–	1.00	12.0	–	–	15	–	1.00	10.0
17.	Plantation of fodder tree	–	13	2	–	1.87	6.5	–	–	15	–	1.00	12.0	–	–	15	–	1.00	10.0

Contd...

Table 51.1–Contd...

Sl.No.	Activity/Item	Number of Farmers Under Each Response Category																	
		KVK, Udaipur						KVK, Jodhpur						KVK, Banswara					
		Training Once in a						Training Once in a						Training Once in a					
		Month	Six Month	Year	Never	M.S.	Rank	Month	Six Month	Year	Never	M.S.	Rank	Month	Six Month	Year	Never	M.S.	Rank
18.	Energy plantation	–	10	3	2	1.53	11.0	–	–	15	–	1.00	12.0	–	–	–	15	0.00	17.5
19.	Bee keeping	–	–	–	15	0.00	17.5	–	–	–	15	0.00	17.5	–	–	–	15	0.00	17.5
	Average (II)	**1**	**7**	**4**	**3**	**1.40**		**3**	**2**	**7**	**3**	**1.42**		**2**	**2**	**6**	**5**	**1.15**	
[III]	**ANIMAL HUSBANDRY AND ALLIED FIELD**																		
	(a) Breeding																		
1.	Artificial insemination	–	–	–	15	0.00	21.0	–	–	2	13	0.13	18.0	–	–	–	15	0.00	20.0
2.	Cross breeding	–	12	3	–	1.80	13.0	–	12	3	–	1.80	4.5	15	–	–	–	3.00	1.0
3.	Castration of scrub animals	–	–	15	–	1.00	5.7	–	7	4	4	1.20	13.0	–	–	10	5	0.67	13.0
	(b) Feeding																		
4.	Preparation of feed mixture	10	5	–	–	2.67	3.0	15	–	–	–	3.00	3.0	–	15	–	–	2.00	6.0
5.	Fodder crop and varieties recommen-ded for Kharif and rabi	–	15	–	–	2.00	8.0	–	–	15	–	1.00	14.5	–	15	–	–	2.00	6.0
6.	Preparation of nutrition receipt for balance nut	–	15	–	–	2.00	8.0	15	–	–	–	3.00	3.0	–	15	–	–	2.00	6.0
7.	Urea feeding to animals	–	15	–	–	2.00	8.0	–	10	–	5	1.33	11.0	–	15	–	–	2.00	6.0
8.	Silage making	–	–	–	15	0.00	21.0	–	5	10	–	1.33	11.0	–	15	–	–	2.00	6.0
	(c) Animal production																		
9.	Effective poultry production	–	–	15	–	1.0	5.7	–	–	–	15	0.00	21.5	–	–	9	6	0.60	15.0
10.	Sheep and goat rearing	–	15	–	–	2.0	8.0	–	12	3	–	1.80	4.5	–	–	10	5	0.67	13.0

Contd...

Table 51.1–Contd...

Sl.No.	Activity/Item	Number of Farmers Under Each Response Category																	
		KVK, Udaipur						KVK, Jodhpur						KVK, Banswara					
		Training Once in a						Training Once in a						Training Once in a					
		Month	Six Month	Year	Never	M.S.	Rank	Month	Six Month	Year	Never	M.S.	Rank	Month	Six Month	Year	Never	M.S.	Rank
11.	Preparation of economic shelter belts for animals	–	15	–	–	2.0	8.0	–	5	10	–	1.33	11.0	–	–	–	15	0.00	20.0
12.	How to purchase and select better animals	–	15	–	–	2.0	8.0	–	5	5	5	1.00	14.5	–	–	12	3	0.80	11.0
13.	Shearing and grading of wool	–	10	–	5	1.33	14.0	–	5	–	10	0.67	17.0	–	–	–	15	0.00	20.0
14.	Rabbit rearing	–	–	–	15	0.00	21.0	–	–	–	15	0.00	21.5	–	–	–	15	0.00	20.0
15.	Piggery development	–	–	–	15	0.00	21.0	–	–	–	15	0.00	21.5	–	–	–	15	0.00	20.0
16.	Seri-culture	–	–	–	15	0.00	21.0	–	–	–	15	0.00	21.5	–	–	–	15	0.00	20.0
17.	Fisheries	–	–	–	15	0.00	21.0	–	–	–	15	0.00	21.5	–	–	–	15	0.00	20.0
18.	Mushroom cultivation	–	–	–	15	0.00	21.0	–	–	–	15	0.00	21.5	–	–	–	15	0.00	20.0
	(d) Animal health																		
19.	First aid of animals	–	15	–	–	2.00	8.0	–	10	5	–	1.67	7.5	–	15	–	–	2.00	6.0
20.	Animal vaccination	–	15	–	–	2.00	8.0	–	10	5	–	1.67	7.5	–	15	–	–	2.00	6.0
21.	Animal hygiene	–	15	–	–	2.00	8.0	–	10	5	–	1.67	7.5	–	15	–	–	2.00	6.0
	(e) Milk production																		
22.	Increasing milk production	15	–	–	–	3.00	1.5	15	–	–	–	3.00	3.0	–	15	–	–	2.00	6.0
23.	Preparation of byproducts of milk	–	–	15	–	1.00	5.7	–	–	12	3	0.80	16.0	–	–	–	15	0.00	20.0
24.	Hygienic milking production	15	–	–	–	3.00	1.5	–	10	5	–	1.67	7.5	–	10	–	5	0.67	13.0
	Average (III)	**2**	**7**	**2**	**5**	**1.37**		**2**	**4**	**4**	**6**	**1.14**		**1**	**6**	**2**	**7**	**1.01**	

Contd...

Table 51.1–Contd...

Sl.No.	Activity/Item	Number of Farmers Under Each Response Category																	
		KVK, Udaipur						KVK, Jodhpur						KVK, Banswara					
		Training Once in a						Training Once in a						Training Once in a					
		Month	Six Month	Year	Never	M.S.	Rank	Month	Six Month	Year	Never	M.S.	Rank	Month	Six Month	Year	Never	M.S.	Rank
[IV]	**AGRICULTURE ENGINEERING**																		
	(a) Farm Implements																		
1.	Repairing and maintaining of pump set, oil engine, electric motor, tractor etc.	–	–	10	5	0.67	5.5	–	–	–	15	0.00	8.5	–	–	9	6	0.60	7.0
2.	Pump and irrigation equipments	–	–	10	5	0.67	5.5	–	–	–	15	0.00	8.5	–	12	–	3	1.60	4.0
3.	General repairs of sprayer and dusters	–	–	15	–	1.00	2.5	–	–	–	15	0.00	8.5	–	–	10	5	0.67	6.0
4.	Operation of land leveling equipments	–	–	9	6	0.67	7.0	–	–	–	15	0.00	8.5	–	–	12	3	0.80	5.0
5.	Farm power and mechanization	–	–	8	7	0.53	8.0	–	–	–	15	0.00	8.5	–	10	5	–	1.67	2.5
6.	Sprinkler, irrigation engineering	–	–	15	–	1.00	2.5	–	–	–	15	0.00	8.5	–	–	–	15	0.00	11.0
7.	Processing equipments of agriculture production	–	–	–	15	0.00	11.5	–	–	–	15	0.00	8.5	–	10	5	–	1.67	2.5
8.	Farm carpentry	–	–	–	15	0.00	11.5	–	–	–	15	0.00	8.5	–	–	–	15	0.00	11.0
	(b) Home Equipment																		
9.	Preparation and working of smokeless Chula	–	15	–	–	2.0	1.0	–	–	15	–	1.00	1.0	10	5	–	–	2.33	1.0
10.	Working with solar heater	–	–	–	15	0.00	11.5	–	–	–	15	0.00	8.5	–	–	–	15	0.00	11.0
11.	Working with Gobar gas plant	–	–	12	3	0.80	4.0	–	–	–	15	0.00	8.5	–	–	–	15	0.00	11.0

Contd...

Table 51.1–Contd...

Sl.No.	Activity/Item	Number of Farmers Under Each Response Category																	
		KVK, Udaipur						KVK, Jodhpur						KVK, Banswara					
		Training Once in a						Training Once in a						Training Once in a					
		Month	Six Month	Year	Never	M.S.	Rank	Month	Six Month	Year	Never	M.S.	Rank	Month	Six Month	Year	Never	M.S.	Rank
	(c) Cottage industries of																		
12.	Agriculture	–	–	–	15	0.00	11.5	–	–	–	15	0.00	8.5	–	–	–	15	0.00	11.0
13.	Animal Husbandry	–	–	–	15	0.00	11.5	–	–	–	15	0.00	8.5	–	–	–	15	0.00	11.0
14.	Forest	–	–	–	15	0.00	11.5	–	–	–	15	0.00	8.5	–	–	–	15	0.00	11.0
	Average (IV)	**0**	**1**	**6**	**8**	**0.52**		**0**	**0**	**1**	**15**	**0.07**		**1**	**3**	**3**	**9**	**0.67**	
[V]	**HOME SCIENCE**																		
	(a) Vocational Training																		
1.	Knitting and embroidery	–	–	15	–	1.00	10.0	–	10	–	5	1.33	8.0	–	10	5	–	1.67	9.0
2.	Tailoring	15	–	–	–	3.00	1.5	–	–	10	5	0.67	11.5	–	15	–	–	2.00	7.5
3.	Working with charkha	–	–	–	15	0.00	13.5	–	–	–	15	0.00	14.0	–	–	–	15	0.00	12.5
4.	Bag making from jute	–	–	–	15	0.00	13.5	–	–	–	15	0.00	14.0	–	–	–	15	0.00	12.5
5.	Flower making from paper	–	–	–	15	0.00	13.5	–	–	10	5	0.67	11.5	–	–	–	15	0.00	12.5
	(b) Home Science Project																		
6.	Care to be taken at the time of pregnancy	12	3	–	–	2.80	4.0	12	–	3	–	2.60	3.5	15	–	–	–	3.00	3.5
7.	Balance diet	12	3	–	–	2.80	4.0	12	–	3	–	2.60	3.5	15	–	–	–	3.00	3.5
8.	First aid	–	–	6	9	0.40	11.0	–	3	12	–	1.20	9.0	–	–	–	15	0.00	12.5
9.	Child and family welfare	12	3	–	–	2.80	4.0	–	10	5	–	1.67	6.5	15	–	–	–	3.00	3.5
10.	Better cooking with local product	10	5	–	–	2.67	6.0	15	–	–	–	3.00	1.5	15	–	–	–	3.00	3.5
11.	Cleanness of home	15	–	–	–	3.00	1.5	15	–	–	–	3.00	1.5	15	–	–	–	3.00	3.5
12.	Community health and cleanness	8	7	–	–	2.53	7.0	–	1	14	–	1.07	10.0	15	–	–	–	3.00	3.5

Contd...

Table 51.1–Contd...

Sl.No.	Activity/Item	Number of Farmers Under Each Response Category																	
		KVK, Udaipur						KVK, Jodhpur						KVK, Banswara					
		Training Once in a						Training Once in a						Training Once in a					
		Month	Six Month	Year	Never	M.S.	Rank	Month	Six Month	Year	Never	M.S.	Rank	Month	Six Month	Year	Never	M.S.	Rank
13.	Vegetable processing by dehydration	–	10	5	–	1.67	8.0	2	12	1	–	2.07	5.0	–	15	–	–	2.00	7.5
14.	Home decoration	–	5	10	–	1.33	9.0	–	10	5	–	1.67	6.5	–	–	–	15	0.00	12.5
15.	Vegetable processing by canning	–	–	–	15	0.00	13.5	–	–	–	15	0.00	14.0	–	–	–	15	0.00	12.5
	Average (V)	**6**	**2**	**2**	**5**	**1.60**		**4**	**3**	**4**	**4**	**1.44**		**6**	**3**	**0**	**6**	**1.58**	

Agricultural Engineering Activities

Training were conducted more frequently in 'Preparation and working of smokeless Chula' (2.0) and it was ranked first. 'General repairs of sprayers and dusters' (1.00) were considered equally important and combined ranked second. The third ranked was awarded to working with Gobar gas plant.

However, training was not conducted at all those activities whose mean score were zero and as such these were ranked at the bottom place.

Home Science Activities

A careful study of the table bring to the fore that 'Tailoring' (3.00) as well as 'cleanness of home' (3.00) were ranked top place. It signifies that trainings were conducted more frequently on these two activities as compared to the remaining other activities. 'Care to be taken at the time of pregnancy' (2.80), 'Balance diet' (2.80), 'Child and family welfare' (2.80) were combined ranked at second place followed by 'Better cooking with local products' (2.67), 'Community health and cleanness' (2.53) and 'Vegetable processing by dehydration' and so on.

However, trainings were not conducted at all on 'Working with charkha,' 'Bag making from jute,' 'Flower making from paper' and 'Vegetable processing by canning' activities as per the opinion of all the respondents and as such these were ranked at the bottom.

KVK, Jodhpur

Agronomical Activities

A close look to the mean score in table explains that the trainings were conducted more frequently in 'Seed treatment'(1.87) which was ranked first followed by 'Handling of plant protection equipment'(1.73), 'Method of using insecticides and pesticides' (1.67) and so on.

However, the trainings were not conducted at all on 'Graded bunding', 'Bench terracing', 'Soil survey and mapping,' 'Green manuring' and 'Control of katra' activities and as such these were ranked at the last place.

Horticultural Activities

The training were conducted more frequently in 'Nursery raising and transplanting'(2.93) which was given top priority followed by 'Fruit production technology of ber' (2.60), 'Kitchen gardening'(2.12) and so on.

However, the training were not conducted at all on 'Fruit production technology of lemon', 'Seed production technology', 'Flower cultivation' and 'Bee keeping' activities and these were ranked at the last place.

Animal Husbandry and Allied Field Activities

The training was conducted more frequently in 'Preparation of feed mixture' (3.00), 'Preparation of nutrition recipes for balance nutrition' (3.00) as well as increasing 'Milk production' (3.00). Since these three activities were having highest same mean scores so these were ranked combined at the top place. 'Cross breeding' (1.80) as well as 'Sheep and goat rearing' (1.80) were ranked combined at the second place. 'First aid of animals' (1.67), 'Animal vaccination' (1.67) 'Animal hygiene' (1.67) and 'Hygiene milking production' were ranked combined at the third place.

However, the trainings were not conducted at all on those activities which having zero mean scores and ranked at the bottom.

Agricultural Engineering Activities

Trainings were conducted only on 'Preparation and working of smokeless Chula' (1.00). It was interesting to note that the trainings were not held on remaining activities.

Home Science Activities

The trainings were conducted more frequently in 'Better cooking with local products' (3.00) as well as 'Cleanness of home' (3.00). Since both the activities were ranked combined at the top. 'Care to be taken at the time of pregnancy'((2.60) and 'Balance diet'(2.60) were considered as equally important and as such these were ranked at second place followed by 'Vegetable processing by dehydration' (2.07), 'Child and family welfare' (1.67) and so on.

However, trainings were not conducted at all on 'Working with charkha,' 'Bag making from jute' and 'Vegetable processing by canning.' Thus, these were ranked at the bottom.

KVK, Banswara

Agronomical Activities

Mean scores in Table 51.1 revealed that in KVK, Banaswara trainings were conducted frequently in 'Methods of using insecticides and pesticides' (2.67) hence, it was ranked first followed by 'Improved practices of irrigated area' (2.53), 'Handling of plant protection equipment' (2.40) and so on.

However, trainings were not conducted at all on 'Soil testing,' 'Soil reclamation,' 'Land conservation,' 'Land leveling,' 'Development of shelter belts,' 'Bench terracing,' 'Green manuring,' 'Control of whitegrub' and 'Control of katra' activities and as such these were ranked at the bottom.

Horticultural Activities

Trainings were conducted frequently in 'Vegetable cultivation' (3.00) as well as 'kitchen gardening' (3.00).Since both the activities were ranked combined at the top. 'Nursery raising and transplanting' (2.67) was ranked at the second place. 'Pruning and training' (1.67), 'Propagation by cutting, budding and grafting' *etc.* (1.67), 'Fruit production technology of lemon' (1.67) were having same mean score and as such these were ranked combined at the third place.

However, trainings were not conducted at all on 'Fruit and vegetable preservation', 'Food preservation', 'Energy plantation' and 'Bee keeping' activities as per the opinion of respondent and as such these were ranked at the bottom.

Animal Husbandry and Allied Field Activities

All the respondents reported that trainings were conducted monthly on 'Cross breeding' (3.00) of animals. This activity was ranked first, 'Preparation of feed mixture' (2.00), 'Fodder crop and varieties recommended for rabi and kharif' (2.00), 'Urea feeding to animals' (2.00), 'Silage making' (2.00), 'First aid of animals' (2.00), 'Animal vaccination' (2.00), 'Animal hygiene' (2.00), 'Increasing milk production' (2.00) were having equal mean score and as such these were ranked combined at the second place.

Further, it may also be noted that many activities were having zero mean score which have the indicative of the fact that trainings were never conducted on these aspects.

Agricultural Engineering Activities

'Preparation and working on smokeless Chula' was having the highest mean score (2.33). It means trainings were conducted frequently on this aspect. 'Farm power and machinery' (1.67) as well as 'Processing equipment of agricultural production' (1.67) were ranked combined at second place followed by 'Pump and irrigation equipments' (0.80), 'General repairs of sprayers and dusters' (0.67) and 'Repairing and maintenance of pumpset and oil engine' (0.60).

About remaining aspects the respondents viewed that the trainings were never conducted.

Home Science Activities

The trainings were conducted frequently on 'Care to be taken at the time of pregnancy' (3.00), 'Balance diet' (3.00), 'Child and family welfare' (3.00), 'Better cooking with local products' (3.00), 'Cleanness of home' (3.00) and 'Community health and cleanness' (3.00). Since all these activities were ranked combined at the top place. 'Tailoring' (2.00) as well as 'Vegetable processing by dehydration' (2.00) were considered equal in getting training by the respondents and as such these were ranked combined at second place, whereas, 'Knitting and embroidery' (1.67) was ranked at third place.

However, trainings were never conducted on 'Working with charkha,' 'Bag making from jute,' 'Flower making from paper; 'First-aid,' 'Home decoration' and 'Vegetable processing by canning' as per the opinion of all the respondents and as such these were ranked at last place.

Overall Subject-wise Frequency of Trainings Organised by KVK as Perceived by Farmers

Data in Table 51.2 reflects the overall subject-wise frequency of training organized by KVKs as perceived by farmers.

KVK, Udaipur

Looking to the pooled mean score, it was revealed that the 'Agronomy' (1.70) was given top priority amongst five subjects. The next in the order was 'Home science' (1.60) followed by 'Horticulture' (1.40), 'Animal husbandry and allied fields' (1.37) and 'Agricultural engineering' (0.52).

The interference can be drawn that the farmers got trainings more frequently on agronomical activities as compared to rest of the four major activities.

KVK, Jodhpur

An examination of data in Table 51.2 explains that training about the 'Home science' (1.44) was given top priority and was ranked first. The second place was given to 'Horticulture' activities (1.42). The third and fourth ranks were given in 'Animal husbandry and allied fields' (1.14) and 'Agronomical' activities (0.70) and 'Agricultural engineering activities was placed at the bottom.

It may be deduced from above analysis that the trainings were conducted frequently in 'Home science' activities while, the trainings were conducted less frequently in 'Agricultural engineering' activities.

KVK, Banswara

A close look to the pooled mean score in Table 51.2. It was revealed that amongst five major subjects the trainings were conducted more frequently in 'Home Science' activities (1.58) which was given top priority followed by 'Horticulture' (1.51), 'Agronomy' (1.33), 'Animal husbandry and allied fields' (1.01) and 'Agricultural engineering' activities (0.67).

From the above findings, it may be concluded that the trainings were conducted more frequently on 'Home science' activities while, the trainings were conducted less frequently in 'Agricultural Engineering' activities.

Table 51.2: Analysis of Overall Subject Wise Frequency of Trainings Organized by KVKs as Perceived by Farmers

Sl.No.	Activity/Item	Number of Farmers Under Each Response Category																				
		KVK, Udaipur							KVK, Jodhpur							KVK, Banswara						
		Training Once in a							Training Once in a							Training Once in a						
		15 Days	Month	Six Month	Year	Never	M.S.	Rank	15 Days	Month	Six Month	Year	Never	M.S.	Rank	15 Day	Month	Six Month	Year	Never	M.S.	Rank
1.	Agronomy	8	3	8	2	3	1.70	I	0	0	3	5	7	0.70	IV	2	2	7	1	6	1.33	II
2.	Horticulture	5	1	7	4	3	1.40	III	5	3	2	7	3	1.42	II	2	2	2	6	5	1.15	III
3.	Animal Husbandry and allied fields	4	2	7	2	5	1.37	IV	6	2	4	4	6	1.14	III	1	1	6	2	7	1.01	IV
4.	Agriculture Engineering	0	0	1	6	8	0.52	V	0	0	0	1	15	0.07	V	0	1	3	3	9	0.67	V
5.	Home science	7	6	2	2	5	1.60	II	4	4	3	4	4	1.44	I	3	6	3	0	6	1.58	I

References

Dash, A.K. and Mishra, M., 2004. Krishi Vigyan Kendras: The light house for rural people. *Orissa Review*, pp. 52–56.

Dwivedi, S.N., 1985. Possibilities of increased income through training of farm women, farm youth and school dropout at KVKs held at Conoor, Tamil Nadu, Oct. 6–8, ICAR, New Delhi, pp. 12–13.

Joshi, G.G., 1979. An evaluation of the institutional training programme conducted by the farmers training centre, Vijapur in Gujarat State. *M.Sc. (Agri.) Thesis*, Gujarat Agricultural University, Anand.

Mishra, A.S. and Bhatt, P.N., 1988. Evaluation of officer training in soil and water conservation. *Indian J. Ext. Edu.*, 24(3 and 4): 71–74.

Patel, A.R., 1991. Evaluation of an institutional training programme conducted by KVK Deesa in Banaskantha District of Gujarat State. *M.S. (Agri.) Thesis*, Gujarat Agricultural University, Sardarkrushinagar.

Author Index

Subject Index

N

O

About the Book

The present book *"Recent Advances in Seed Spices"* is an attempt to compile information on status of seed spices, utilization of genetic resources, production technologies for enhancement of productivity and quality, protection technologies for the management of various diseases and insects, technologies for farm mechanization in seed spices, post harvest technologies for minimization of post harvest losses, interventions of biotechnological tools for spices improvement' in terms of quality, transfer of technologies from research farm to farmers field, and use of information technology for spices trade.

Hope that this book will be very useful to the students, spices research workers, biotechnologists, traders and exporters.

About the Authors

Dr. Y. Ravindrababu is presently working as Research Scientist (Spices) at Centre for Research on Seed Spices, S.D. Agricultural University, Jagudan. He has 22 research publications to his credit in national and international journals and three technical bulletins. He guided 11 M.Sc. (Ag.) and 1 Ph.D. students in the discipline of Plant Breeding and Genetics. He is CPI to the NAIP project on "Value Chain in Major Seed Spices for Domestic and Export Promotion". He was associated with identification of seven verities in various crops *viz.*, Sorghum, Jatropha, Karingada, Amaranth, Chilli and Fennel.

Dr. R.K. Jaiman is presently working as Assistant Research Scientist at CRSS, SDAU, Jagudan. He has 12 research publications in various national and international journals. He was associated with development of disease management technologies in cumin, fennel and coriander. He is Co-PI to the project" Referral Laboratory for Seed Spices". He is recognized P.G. teacher.

Dr. Kirti D. Patel is presently working as Research Scientist (Pl. pathology) at CRSS, SDAU, Jagudan. So far he has 20 research papers in various national and international journals. He was instrumental in development of blight management technology in cumin crop. He was associated with development of control measures for various diseases in seed spices. He is recognized P.G. teacher and so far guided 5 M.Sc. (Ag.) and 1 Ph.D. students in the discipline of Plant Pathology.

Prof. Naresh R. Patel is presently working as Assistant Research Scientist at CRSS, SDAU, Jagudan. He has published 15 research articles in different national and international journals. He was associated with development of wilt resistant variety GC-4 and Hari Om awardee. He was also associated with number of recommendations on management of various diseases in seed spices.

Dr. S.B.S. Tikka is presently working as Director of Research and Dean, P.G. Studies, S.D. Agricultural University, Sardarkrushinagar. He has more than 200 papers to his credit in national and international journals and published numbers of technical bulletins. He guided 20 M.Sc. (Ag.) and 15 Ph.D. students. He also contributed number of chapters in various books and published six books. He has vast experience in agro forestry and pulses crops and developed number of high yielding verities in pigeonpea, black gram, green gram and moth bean. He was also associated with production and protection technologies in pulse crops.